Charles E. Rickart

Natural Function Algebras

Springer-Verlag
New York Heidelberg Berlin

Dr. Charles E. Rickart
Department of Mathematics
Yale University
New Haven, Connecticut 06520
USA

AMS Classifications: 32E25, 32F05, 46J10, 46J15

Library of Congress Cataloging in Publication Data

Rickart, Charles Earl, 1913–
Natural function algebras.

(Universitext)
Bibliography: p.
Includes indexes.
1. Function algebras. 2. Banach algebras.
I. Title.
QA326.R53 512′.55 79-20915

Printed in the United States of America.

9 8 7 6 5 4 3 2 1

ISBN 0-387-90449-2 Springer-Verlag New York Heidelberg Berlin
ISBN 3-540-90449-2 Springer-Verlag Berlin Heidelberg New York

To Ann

PREFACE

The term "function algebra" usually refers to a uniformly closed algebra of complex valued continuous functions on a compact Hausdorff space. Such Banach algebras, which are also called "uniform algebras", have been much studied during the past 15 or 20 years. Since the most important examples of uniform algebras consist of, or are built up from, analytic functions, it is not surprising that most of the work has been dominated by questions of analyticity in one form or another. In fact, the study of these special algebras and their generalizations accounts for the bulk of the research on function algebras. We are concerned here, however, with another facet of the subject based on the observation that very general algebras of continuous functions tend to exhibit certain properties that are strongly reminiscent of analyticity. Although there exist a variety of well-known properties of this kind that could be mentioned, in many ways the most striking is a local maximum modulus principle proved in 1960 by Hugo Rossi [R11]. This result, one of the deepest and most elegant in the theory of function algebras, is an essential tool in the theory as we have developed it here. It holds for an arbitrary *Banach* algebra of functions defined on the spectrum (maximal ideal space) of the algebra. These are the algebras, along with appropriate generalizations to algebras defined on noncompact spaces, that we call "natural function algebras".

At first it was generally believed that the analytic type properties observed in uniform algebras would turn out to be a consequence of a genuine analytic structure somehow imposed by the given algebra on its spectrum in such a way that the functions involved would become analytic in the usual sense. For certain important classes of uniform algebras, the conjectured analytic structure does indeed exist. However, the general conjecture was demolished in 1963 by G. Stolzenberg [S8] who constructed a nontrivial uniform algebra whose spectrum does not admit any relevant analytic structure whatsoever. (See also examples by J. Garnett [G4].) Therefore the analytic phenomena exhibited by general uniform algebras appear to be independent of analyticity in the usual sense. This is the starting point for our investigations, and the main purpose of this monograph is to provide a systematic account of some of the algebraically induced analytic phenomena associated with "natural" function algebras. Although much of the work on uniform algebras might be included in such an account, most of it has been well-reported elsewhere [G1, S10], so will not be discussed here except as needed. In fact, our function algebras will usually be defined on Hausdorff spaces that need not be even locally compact. The bulk of the published results for these general function algebras considered from the above point of view have appeared during the last 10 or 15 years in a series of papers primarily by the author [R2-R10]. (See also [B9-B13], [K3, K4], [M3, M4].) The present account, strongly biased by the author's special interests and prejudices, includes numerous improvements on old results along with a number of previously unpublished results.

Our approach to the study of analytic phenomena in general function algebras may be described briefly as follows. Note first that the classical holomorphy theory, based on n-dimensional complex space $\mathbb{C}^n$, is ultimately determined by the algebra $\mathcal{P}$ of all polynomials in $\mathbb{C}^n$. In the abstract situation the space $\mathbb{C}^n$ is replaced by a more-or-less arbitrary Hausdorff space Σ and the algebra $\mathcal{P}$ by a given "structure algebra" $\mathfrak{a}$ of continuous complex-valued functions on Σ. Then $\mathfrak{a}$ determines an "$\mathfrak{a}$-holomorphy" theory based on Σ roughly analogous to the way $\mathcal{P}$ determines the classical theory. Therefore, from this point of view, the fundamental object is the space-algebra "pair" $[\Sigma, \mathfrak{a}]$. However, in order to obtain interesting results, one must impose some rather general but crucial conditions on $[\Sigma, \mathfrak{a}]$ distinguishing an important category of pairs that we call "natural systems". In the first place, $\mathfrak{a}$ is assumed to determine the topology of Σ in the sense that the given topology is equivalent to the weakest under which elements of $\mathfrak{a}$ are continuous. In this case $[\Sigma, \mathfrak{a}]$ is called a "system". Secondly, $[\Sigma, \mathfrak{a}]$ (or the algebra $\mathfrak{a}$) is assumed to be "natural" in the sense that every homomorphism of $\mathfrak{a}$ onto C, that is continuous relative to the compact-open topology in $\mathfrak{a}$, is a point evaluation in the space Σ.

Included, of course, is the pair $[\mathbb{C}^n, \rho]$ as well as the pair associated with a commutative Banach algebra *via* the Gelfand representation of the algebra on its spectrum. The naturality condition, which has more effect than might be expected on the surface, is the one that ensures occurance of the analytic phenomena in which we are interested. In particular, it enables us to generalize and apply the Rossi Local Maximum Modulus Principle, which plays a key role throughout our discussion. The proof of the latter, involving the solution of a Cousin I problem, represents the main *technical* dependence of the subject on the theory of several complex variables. There is, of course, even greater *general* dependence on the classical theory for both ideas and motivation. Even in the most general case, we are able to establish a variety of nontrivial results, many of which are full or partial generalizations of results in Several Complex Variables. Therefore, although the initial motivation for this study came from the theory of uniform algebras, the end-result is a kind of abstract complex function theory. However, even in the case of results that parallel closely familiar topics, proofs tend to be rather different, with greater dependence on function algebra methods.

The $\mathfrak{a}$-holomorphy theory, as we have developed it, might also be considered as another approach to "Infinite Dimensional Holomorphy". The latter subject, which already has an extensive literature (see [C1],[H2],[N3]), involves the study of functions on infinite dimensional linear topological spaces and takes off from the fact that the classical theory is based on the linear space structure of $\mathbb{C}^n$. The theory is accordingly more traditional in character and tends to become involved in a variety of technical problems concerning linear topological spaces. It consequently does not have a great deal in common with the algebra approach developed here, even for functions defined in a vector space. Aside from a substantial overlap in the important special case of dual vector spaces, the precise connection between the two approaches even for functions in vector spaces has not been worked out and appears to involve some rather difficult questions concerning linear topological spaces and their generalized polynomials.

The material is presented in seventy four sections grouped into chapters. The first ten chapters are devoted to functions in general Hausdorff spaces and the last five to functions in linear vector spaces, along with applications to the general case. The material in Chapters XI and XII constitutes an introduction to holomorphy theory in dual pairs of vector spaces. In many respects, dual pairs provide an ideal setting for infinite dimensional holomorphy and we hope to extend and develop this subject more completely at another time.

There is no doubt material in the literature that we have overlooked and which should have been included here. There are also certain obvious topics, such as the question of analytic structure in the space Σ associated with certain natural systems $[\Sigma, \mathfrak{A}]$, that obviously deserve more attention. Analytic structure is an important subject in the theory of uniform algebras and it is plausible that many of the results there could be generalized. Although some progress along these lines has been made by Brooks [B13] and Kramm [K3,K4], much remains to be done in the general setting. Another topic that needs to be explored in the general setting concerns cohomology in natural systems. The compact case is covered by results of Arens [A2] and Royden [R12]. The cohomology results that we have obtained to date for the general case are rather incomplete so have not been included. Another potentially important program of study is suggested by the fact that the category of *all* natural systems is obviously too inclusive to exhibit some of the more interesting structure found in the classical situation. It would therefore be interesting to distinguish a subcategory that might exclude some of the pathology admited by the general case. Although a number of ideas are suggested by the familiar examples, it is still not clear just how an appropriate subcategory should be defined.

Much of the material included here has been presented in one form or another in graduate courses offered by the author at Yale University. We recall in particular lectures given during the academic year 1966-67 which were faithfully attended by a number of graduate students and faculty. Included were F. F. Bonsall, who was visiting Yale at the time, plus M. E. Shauck, S. Sidney and E. L. Stout, who were on the staff. Also included were our former Ph.D. students R. G. Blumenthal and Brian Cole. We are much indebted to these mathematicians for numerous helpful comments and criticisms. A special debt is owed to Stu Sidney whose talent for coming up with crucial examples more or less on demand has done much to add some concreteness to a necessarily rather abstract subject. We also wish to express appreciation to the National Science Foundation for its generous support of the research contained herein. Finally, we take this opportunity to thank Cathy Belton, for typing the first draft of the manuscript, and Donna Belli, for the excellent though very difficult job of preparing the final copy.

New Haven, Connecticut
June 28, 1979

CONTENTS

CHAPTER I
THE CATEGORY OF PAIRS

§1. PAIRS AND SYSTEMS

Let Σ be a Hausdorff space and denote by $C(\Sigma)$ the algebra of all complex-valued continuous functions on Σ. If X is an arbitrary subset of Σ and f is a function defined on X let $|f|_X = \sup\{|f(x)| : x \in X\}$. If $X = \Sigma$ write $|f|$ instead of $|f|_\Sigma$. When X is compact $(X \subset\subset \Sigma)$ the function $f \to |f|_X$ is a semi-norm on the algebra $C(\Sigma)$. The family of all such semi-norms associated with compact subsets of Σ defines a Hausdorff topology on $C(\Sigma)$ with respect to which $C(\Sigma)$ is a topological algebra. This topology is equivalent to the compact-open topology, the topology of uniform convergence on compact subsets of Σ. A neighborhood basis at a point $f \in C(\Sigma)$ is given by sets of the form

$$N_f(K,\varepsilon) = \{f' \in C(\Sigma) : |f-f'|_K < \varepsilon\}$$

where $\varepsilon > 0$ and $K \subset\subset \Sigma$. If Σ is compact then we obviously obtain the usual norm topology on $C(\Sigma)$. Note that if Σ is not locally compact then a function on Σ may be a uniform limit of continuous functions on compact subsets of Σ without being itself continuous, so $C(\Sigma)$ need not be closed with respect to the compact-open topology in the space of all functions on Σ. In all that follows, the topology assumed for functions defined on Σ will always be the c-o (compact-open) topology unless the contrary is explicitly indicated.

The objects with which we are primarily concerned are *pairs* $[\Sigma,\mathfrak{A}]$ consisting of a Hausdorff space Σ along with a subalgebra $\mathfrak{A}$ of $C(\Sigma)$. For convenience, $\mathfrak{A}$ is always assumed to contain the constant functions. Given a pair $[\Sigma,\mathfrak{A}]$ we have, in addition to the given topology on Σ, the $\mathfrak{A}$-*topology* which is defined to be the coarsest (weakest) topology with respect to which each function in $\mathfrak{A}$ is continuous. This is a uniform topology and, since $\mathfrak{A} \subseteq C(\Sigma)$, it is obviously coarser than the given topology on Σ. Note that the $\mathfrak{G}$-topology will be Hausdorff iff $\mathfrak{G}$ separates the points of Σ; i.e. iff $a(\sigma_1) = a(\sigma_2)$, for every $a \in \mathfrak{G}$, implies $\sigma_1 = \sigma_2$. A neighborhood basis of a point $\sigma \in \Sigma$ for the $\mathfrak{G}$-topology is given by sets of the form

$$N_\sigma(a_1,\ldots,a_n;\varepsilon) = \{\sigma' \in \Sigma: |a_i(\sigma)-a_i(\sigma')| < \varepsilon, i=1,\ldots,n\}$$

where $\varepsilon > 0$ and $\{a_1,\ldots,a_n\}$ is an arbitrary finite subset of $\mathfrak{G}$. It is sufficient here to consider neighborhoods determined by linearly independent elements $a_1,\ldots,a_n$. If $\mathfrak{G}_0$ is any subset of $\mathfrak{G}$ which generates $\mathfrak{G}$ (i.e. the smallest subalgebra of $\mathfrak{G}$ that contains $\mathfrak{G}_0$ is $\mathfrak{G}$ itself) then it is not difficult to prove that the $\mathfrak{G}_0$-topology is equivalent to the $\mathfrak{G}$-topology on Σ. Finally we note that the $\mathfrak{G}$-topology has the elementary but important property that an arbitrary basic neighborhood $N_\sigma(a_1,\ldots,a_n;\varepsilon)$ of a point σ contains one whose closure is contained in N_σ, e.g. $N_\sigma(a_1,\ldots,a_n;\varepsilon/2)$.

1.1 DEFINITION. *A pair* $[\Sigma,\mathfrak{G}]$ *is called a system if the given topology in* Σ *is equivalent to the* $\mathfrak{G}$*-topology. The collection of all system pairs is denoted by* S.

Systems play a central role in our theory. However, since it will be necessary on occasion to consider pairs that are not systems, we avoid the restriction where possible and reasonable. Observe that if Σ is compact then $[\Sigma,\mathfrak{G}]$ will be a system iff $\mathfrak{G}$ separates points. If Σ is not compact then this is no longer true, as is shown by taking $\Sigma = \mathbb{C}$ and $\mathfrak{G}$ the algebra of all continuous functions on $\mathbb{C}$ whose limits at infinity exist equal to their values at the origin. In this case $\mathfrak{G}$ separates points but every $\mathfrak{G}$-neighborhood of the origin contains points outside every compact set.

§2. MORPHISMS AND EXTENSIONS OF PAIRS

Consider any two pairs $[\Sigma_1,\mathfrak{G}_1]$, $[\Sigma_2,\mathfrak{G}_2]$ and a continuous map $\rho : \Sigma_1 \to \Sigma_2$. If $\mathfrak{G}_2 \circ \rho \subseteq \mathfrak{G}_1$ then we call ρ a *pair morphism* and write

$$\rho : [\Sigma_1,\mathfrak{G}_1] \to [\Sigma_2,\mathfrak{G}_2] .$$

We thus obtain the *category* P of all pairs. Observe that the function map $\mathfrak{G}_2 \to \mathfrak{G}_2 \circ \rho$ dual to ρ is an algebra homomorphism of $\mathfrak{G}_2$ into $\mathfrak{G}_1$ which is automatically continuous with respect to the compact-open topologies in $\mathfrak{G}_1$ and $\mathfrak{G}_2$. If $[\Sigma_2,\mathfrak{G}_2]$ is a system then the condition $\mathfrak{G}_2 \circ \rho \subseteq \mathfrak{G}_1$ already implies continuity of $\rho : \Sigma_1 \to \Sigma_2$. It is obvious that the system pairs constitute a full subcategory S of P.

If a morphism $\rho : [\Sigma_1,\mathfrak{G}_1] \to [\Sigma_2,\mathfrak{G}_2]$ satisfies the equality $\mathfrak{G}_2 \circ \rho = \mathfrak{G}_1$ then $[\Sigma_2,\mathfrak{G}_2]$, or $([\Sigma_2,\mathfrak{G}_2],\rho)$, is called an *extension* of $[\Sigma_1,\mathfrak{G}_1]$. In this case each element of $\mathfrak{G}_2$ is an "extension" *via* ρ of an element of $\mathfrak{G}_1$. If $a_2 \circ \rho = 0$ implies $a_2 = 0$ then the extension is said to be *faithful*. If $[\Sigma_2,\mathfrak{G}_2]$ is a system then the extension of $[\Sigma_1,\mathfrak{G}_1]$ is called a *system extension*. We shall indicate an extension by writing

$$\rho : [\Sigma_1,\mathfrak{G}_1] \Rightarrow [\Sigma_2,\mathfrak{G}_2] .$$

If $\mathfrak{G}_1$ separates points then the map ρ in an extension is automatically injective. Moreover if $[\Sigma_1,\mathfrak{G}_1]$ and $[\Sigma_2,\mathfrak{G}_2]$ are both systems then ρ is a homeomorphism. If $\rho : \Sigma_1 \to \Sigma_2$ is a surjective homeomorphism then the extension is called an *isomorphism* and the pairs $[\Sigma_1,\mathfrak{G}_1]$, $[\Sigma_2,\mathfrak{G}_2]$ are said to be *isomorphic under* ρ. Note that in this case the dual of ρ maps $\mathfrak{G}_2$ isomorphically and bicontinuously onto $\mathfrak{G}_1$. If $[\Sigma,\mathfrak{G}] \in P$ and Σ_0 is a subspace of Σ then the pair $[\Sigma_0, \mathfrak{G}|\Sigma_0]$ will be denoted simply by $[\Sigma_0,\mathfrak{G}]$ and said to be *contained* in $[\Sigma,\mathfrak{G}]$. Under the identity map $\iota : \Sigma_0 \to \Sigma$ it is obvious that $[\Sigma,\mathfrak{G}]$ is an extension of $[\Sigma_0,\mathfrak{G}]$. More generally, one extension $\rho_1 : [\Sigma,\mathfrak{G}] \Rightarrow [\Sigma_1,\mathfrak{G}_1]$ is said to be *contained in* a second extension $\rho_2 : [\Sigma,\mathfrak{G}] \Rightarrow [\Sigma_2,\mathfrak{G}_2]$ if there exists $\rho : [\Sigma_1,\mathfrak{G}_1] \Rightarrow [\Sigma_2,\mathfrak{G}_2]$ such that $\rho_2 = \rho \circ \rho_1$; i.e. the following diagram commutes

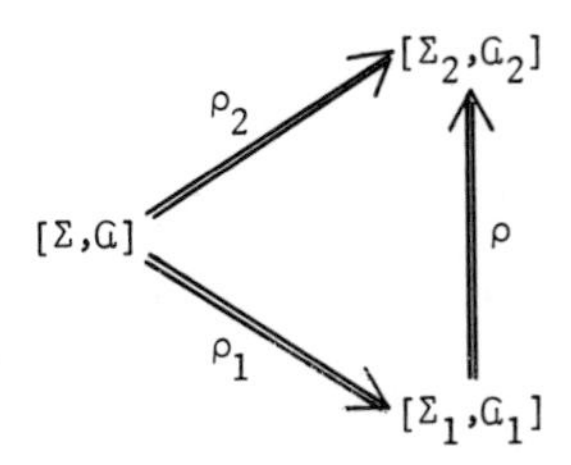

If in addition $\rho : [\Sigma_1,\mathfrak{G}_1] \Rightarrow [\Sigma_2,\mathfrak{G}_2]$ is an isomorphism then the two extensions are said to be *equivalent*.

Let $[\Sigma,\mathfrak{G}]$ be an arbitrary pair, so $\mathfrak{G}$ need not separate points of Σ, and define $\sigma_1 \sim \sigma_2$ iff $a(\sigma_1) = a(\sigma_2)$ for each $a \in \mathfrak{G}$. Then "$\sim$" is an equivalence relation in Σ and the equivalence class $[\sigma]$ containing a point $\sigma \in \Sigma$ is a closed set. Denote by $\Sigma|\mathfrak{G}$ the collection of all these equivalence classes. Then the map $\varkappa : \Sigma \to \Sigma|\mathfrak{G}$, $\sigma \mapsto [\sigma]$, is surjective. If $\Sigma|\mathfrak{G}$ is given the usual equivalence class topology (i.e. a set $E \subseteq \Sigma|\mathfrak{G}$ is open iff $\varkappa^{-1}(E)$ is open in Σ) then $\Sigma|\mathfrak{G}$ is a Hausdorff space and $\varkappa$ is a continuous map. For each $a \in \mathfrak{G}$ and $\sigma \in \Sigma$ define $\tilde{a}([\sigma]) = a(\sigma)$. Then $\tilde{a}$ is a well-defined continuous function on $\Sigma|\mathfrak{G}$ and $\tilde{\mathfrak{G}}$ is a point-separating algebra. Thus we obtain an extension $\varkappa : [\Sigma,\mathfrak{G}] \Rightarrow [\Sigma|\mathfrak{G},\tilde{\mathfrak{G}}]$ called the *separating extension* of $[\Sigma,\mathfrak{G}]$.

§3. NATURAL SYSTEMS

Consider next for a given pair $[\Sigma,\mathfrak{G}]$ a homomorphism $\varphi : a \mapsto \hat{a}(\varphi)$ of the algebra $\mathfrak{G}$ onto the complex field $\mathbb{C}$. The homomorphism is said to be *dominated by* a compact set $K \subset\subset \Sigma$ if $|\hat{a}(\varphi)| \le |a|_K$ for every $a \in \mathfrak{G}$. This notion provides an important criterion for continuity of homomorphisms. Continuity is understood of course to be with respect to the compact-open topology in $\mathfrak{G}$.

3.1 PROPOSITION. *A necessary and sufficient condition for a homomorphism $\varphi : \mathfrak{G} \to \mathbb{C}$ to be continuous is that it be dominated by some compact set $K_\varphi \subset\subset \Sigma$.*

Proof. Assume first that the condition is satisfied. Let $a \in \mathfrak{G}$ and $\varepsilon > 0$ be arbitrary and consider the "compact-open" neighborhood $N_a(K_\varphi,\varepsilon) = \{a' \in \mathfrak{G} : |a-a'|_{K_\varphi} < \varepsilon\}$ of the point a. Then $a' \in N_a(K_\varphi,\varepsilon)$ implies $|\hat{a}(\varphi)-\hat{a}'(\varphi)| = |\widehat{a-a'}(\varphi)| \le |a-a'|_{K_\varphi} < \varepsilon$, so φ is continuous at a, proving the sufficiency.

Now assume that φ is continuous. Then in particular there exists a compact set $K \subset\subset \Sigma$ and $\delta > 0$ such that $a \in N_0(K,\delta)$ implies $|\hat{a}(\varphi)| < 1$. For arbitrary $a \in \mathfrak{G}$ and $\varepsilon > 0$ set $b = (|a|_K + \varepsilon)^{-1}\delta a$. Then $b \in N_0(K,\delta)$, so $|\hat{b}(\varphi)| < 1$. Hence $|\hat{a}(\varphi)| < \delta^{-1}(|a|_K + \varepsilon)$. Since this inequality holds for all $\varepsilon > 0$ it follows that

$$|\hat{a}(\varphi)| \le \delta^{-1}|a|_K, \quad a \in \mathfrak{G} .$$

Observe that for all n

$$|\hat{a}(\varphi)|^n = |\widehat{a^n}(\varphi)| \le \delta^{-1}|a^n|_K = \delta^{-1}|a|_K^{\,n}$$

so $|\hat{a}(\varphi)| \le \delta^{-1/n}|a|_K$. Letting $n \to \infty$ we obtain $|\hat{a}(\varphi)| \le |a|_K$, $a \in \mathfrak{a}$. ♦

An important class of continuous homomorphisms of $\mathfrak{a}$ onto $\mathbb{C}$ are the *point evaluations*. These are homomorphism φ of the form

$$\hat{a}(\varphi) = a(\sigma_\varphi), \quad a \in \mathfrak{a}$$

where σ_φ is a point of Σ. Further properties of homomorphisms will be obtained in §12. At this point we introduce the class of those pairs which are the main objects of interest in our investigations.

3.2 DEFINITION. *A pair* $[\Sigma,\mathfrak{a}]$ *is said to be* ***natural*** *if every continuous homomorphism of* $\mathfrak{a}$ *onto* $\mathbb{C}$ *is a point evaluation. If every homomorphism of* $\mathfrak{a}$ *onto* $\mathbb{C}$ *(continuous or not) is a point evaluation then* $[\Sigma,\mathfrak{a}]$ *is said to be* ***strictly natural.***

When appropriate we shall also apply the terms "natural" and "strictly natural" to the algebra $\mathfrak{a}$ as well as the pair $[\Sigma,\mathfrak{a}]$. A strictly natural algebra $\mathfrak{a}$ is called a "generic subalgebra of $C(\Sigma)$" by F. Quigley [Q1]. Note that since point evaluations are always continuous, *every* homomorphism of a strictly natural algebra $\mathfrak{a}$ onto $\mathbb{C}$ is automatically continuous. Naturality is actually a much stronger condition than might appear on the surface. It has the character of an "analytic" type restriction since it guarantees the occurance of the "analytic phenomena" which we are interested in studying. The collection of all natural pairs will be denoted by N. It is obvious that N is a full subcategory of P and that natural systems constitute a full subcategory of N. Observe that naturality is a "global" property of a pair $[\Sigma,\mathfrak{a}]$. We also have the corresponding "local" property.

3.3 DEFINITION. *The pair* $[\Sigma,\mathfrak{a}]$ *is said to be* ***locally natural*** *if there exists for each point of* Σ *arbitrarily small neighborhoods* U *such that* $[U,\mathfrak{a}]$ *is natural.*

Some of the most important examples of natural pairs (actually natural systems) are described briefly at the end of the section.

3.4 PROPOSITION. *The pair* $[\Sigma,\mathfrak{a}]$ *will be natural iff* $[\Sigma,\mathfrak{a}']$ *is natural, where* $\mathfrak{a}'$ *is any algebra that contains* $\mathfrak{a}$ *and is contained in* $\bar{\mathfrak{a}}$, *the c-o closure of* $\mathfrak{a}$ *in* $C(\Sigma)$.

Proof. Let φ be a continuous homomorphism of $\mathfrak{G}'$ onto $\mathbb{C}$. Then the restriction of φ to $\mathfrak{G}$ is also continuous. Therefore if $[\Sigma,\mathfrak{G}]$ is natural $\sigma_\varphi \in \Sigma$ exists such that $\hat{a}(\varphi) = a(\sigma_\varphi)$ for $a \in \mathfrak{G}$. Now choose a compact set $K_\varphi \subset\subset \Sigma$ such that $|\hat{f}(\varphi)| \leq |f|_K$ for all $f \in \mathfrak{G}'$. We may obviously choose K_φ to contain the point σ_φ. For arbitrary $f \in \mathfrak{G}'$ and $\varepsilon > 0$ choose $a \in \mathfrak{G}$ such that $|f-a|_{K_\varphi} < \varepsilon$. Then

$$|\hat{f}(\varphi)-f(\sigma_\varphi)| \leq |\hat{f}(\varphi)-\hat{a}(\varphi)| + |a(\sigma_\varphi)-f(\sigma_\varphi)|$$
$$\leq |f-a|_{K_\varphi} + |a-f|_{K_\varphi} < 2\varepsilon \ .$$

Since ε is arbitrary it follows that $\hat{f}(\varphi) = f(\sigma_\varphi)$, so $[\Sigma,\mathfrak{G}']$ is natural.

Now assume that $[\Sigma,\mathfrak{G}']$ is natural and let φ be a continuous homomorphism of $\mathfrak{G}$ onto $\mathbb{C}$. Since $\mathfrak{G}$ is dense in $\mathfrak{G}'$ with respect to the compact-open topology the homomorphism φ obviously extends uniquely to $\mathfrak{G}'$. Moreover if K_φ dominates φ in $\mathfrak{G}$ then it will also dominate φ in $\mathfrak{G}'$. In other words the extension of φ is also continuous on $\mathfrak{G}'$. Since $[\Sigma,\mathfrak{G}']$ is natural φ is a point evaluation on $\mathfrak{G}'$ and hence on $\mathfrak{G}$, so $[\Sigma,\mathfrak{G}]$ is natural. ♦

If Σ is compact and $\mathfrak{G}$ separates points then the above proposition asserts that $[\Sigma,\mathfrak{G}]$ will be natural iff Σ is equal to the spectrum $\Phi_{\bar{\mathfrak{G}}}$ of the uniform algebra $\bar{\mathfrak{G}}$. Moreover $[\Phi_{\bar{\mathfrak{G}}},\hat{\bar{\mathfrak{G}}}]$ is even strictly natural. On the other hand if $\mathfrak{G}$ is not closed in $C(\Sigma)$ then $[\Sigma,\mathfrak{G}]$ will not in general be strictly natural even though it is natural. This is shown by the system $[D,\mathfrak{P}]$, where D is the closed unit disc in $\mathbb{C}$ and $\mathfrak{P}$ consists of all polynomials on D. In this case $[D,\overline{\mathfrak{P}|D}]$ is natural (see Example 5.1.), so $[D,\mathfrak{P}]$ is natural. However every point $\zeta \in \mathbb{C} \setminus D$ defines a homomorphism, $P|D \to P(\zeta)$ of $\mathfrak{P}|D$ onto $\mathbb{C}$ which is not a point evaluation at any point of D and in particular is not continuous.

3.5 PROPOSITION. *If* $\rho : [\Sigma,\mathfrak{G}] \Rightarrow [\Omega,\mathfrak{B}]$ *is an extension of* $[\Sigma,\mathfrak{G}]$ *such that* $\rho(\Sigma) = \Omega$ *then* $[\Omega,\mathfrak{B}]$ *natural implies* $[\Sigma,\mathfrak{G}]$ *natural.*

Proof. Let $\varphi : \mathfrak{G} \to \mathbb{C}$ be a continuous homomorphism dominated by $K_\varphi \subset\subset \Sigma$. Since $\mathfrak{B}\circ\rho \subseteq \mathfrak{G}$ the composition of ρ and φ

$$b \mapsto b\circ\rho \mapsto \widehat{b\circ\rho}(\varphi), \quad b \in \mathfrak{B}$$

defines a homomorphism of $\mathfrak{B}$ onto $\mathbb{C}$. Furthermore

$$|\widehat{b\circ\rho}(\varphi)| \le |b\circ\rho|_{K_\varphi} = |b|_{\rho(K_\varphi)}, \quad b \in \mathfrak{B}$$

and $\rho(K_\rho)$ is compact in Ω since ρ is continuous. Hence if $[\Omega,\mathfrak{B}]$ is natural then there exists $\omega_\varphi \in \Omega$ such that $\widehat{b\circ\rho}(\varphi) = b(\omega_\varphi)$, $b \in \mathfrak{B}$. Since $\rho(\Sigma) = \Sigma$ there exists $\sigma_\varphi \in \Sigma$ such that $\rho(\sigma_\varphi) = \omega_\varphi$. Also since $\mathfrak{A} = \mathfrak{B}\circ\rho$ there exists for each $a \in \mathfrak{A}$ an element $b \in \mathfrak{B}$ such that $a = b\circ\rho$. Thus

$$\hat{a}(\varphi) = \widehat{b\circ\rho}(\varphi) = b(\omega_\varphi) = (b\circ\rho)(\sigma_\varphi) = a(\sigma_\varphi)\ .$$

Therefore $[\Sigma,\mathfrak{A}]$ is natural. ♦

3.6 COROLLARY. *If the system extension associated with a pair* $[\Sigma,\mathfrak{A}]$ *is natural then* $[\Sigma,\mathfrak{A}]$ *is natural.*

The condition $\rho(\Sigma) = \Omega$ in the above proposition implies that the extension is faithful, so the dual of ρ, *via* $b \mapsto b\circ\rho$, is an isomorphism of the algebra $\mathfrak{B}$ with $\mathfrak{A}$. The map ρ of course need not be one-to-one, but even if it is, one still cannot deduce naturality of $[\Sigma,\mathfrak{B}]$ from that of $[\Sigma,\mathfrak{A}]$. The point here is that ρ is only continuous and need not be a homeomorphism. Therefore Ω may contain compact subsets that are not images of compact subsets of Σ, so $\mathfrak{B}$ may admit continuous homomorphisms that are not induced by continuous homomorphisms of $\mathfrak{A}$.

§4. PRODUCTS OF PAIRS

Next we give a general construction for a product of pairs that preserves naturality. Consider an arbitrary family $\{[\Sigma_\lambda,\mathfrak{A}_\lambda] : \lambda \in \Lambda\}$ of pairs and form the cartesian product $\Sigma^\Lambda = \prod\Sigma_\lambda$ of the spaces Σ_λ. Under the usual product space topology Σ^Λ is also a Hausdorff space. We denote an arbitrary element of Σ^Λ by $\check{\sigma} = \{\sigma_\lambda\}$. Now for each $\lambda \in \Lambda$ consider the projection

$$p_\lambda : \Sigma^\Lambda \to \Sigma_\lambda,\ \check{\sigma} \mapsto \sigma_\lambda$$

of the space Σ^Λ onto Σ_λ. The dual $f \to f\circ p_\lambda$ of p_λ maps functions on Σ_λ to functions on Σ^Λ and in particular maps $\mathfrak{A}_\lambda$ isomorphically onto a subalgebra $\mathfrak{A}_\lambda\circ p_\lambda$ of $C(\Sigma^\Lambda)$. Denote by ${}^\Lambda\mathfrak{A}$ the subalgebra of $C(\Sigma^\Lambda)$ spanned by the algebras $\mathfrak{A}_\lambda\circ p_\lambda$,

$\lambda \in \Lambda$. We call $[\Sigma^\Lambda, {}^\Lambda\mathfrak{G}]$ the *product* of the family of pairs $[\Sigma_\lambda, \mathfrak{G}_\lambda]$, $\lambda \in \Lambda$. If each of the pairs $[\Sigma_\lambda, \mathfrak{G}_\lambda]$ is a system then it is easy to see that $[\Sigma^\Lambda, {}^\Lambda\mathfrak{G}]$ is also a system.

4.1 PROPOSITION. *The product pair* $[\Sigma^\Lambda, {}^\Lambda\mathfrak{G}]$ *will be (strictly) natural iff each of the pairs* $[\Sigma_\lambda\ \mathfrak{G}_\lambda]$, $\lambda \in \Lambda$, *is (strictly) natural.*

Proof. It will be sufficient to make the proof for naturality. Observe first that if K is a compact subset of Σ^Λ then it is contained in a compact set of the form $\prod K_\lambda$, where K_λ is a compact subset of Σ_λ for each λ. In fact, let $K_\lambda = p_\lambda(K)$. Then K_λ is compact and $K \subseteq \prod K_\lambda$. Now assume that each of the pairs $[\Sigma_\lambda, \mathfrak{G}_\lambda]$ is natural and let φ be a continuous homomorphism of ${}^\Lambda\mathfrak{G}$ onto $\mathbb{C}$ dominated by a compact set K. By the preceding observation one may assume that $K = \prod K_\lambda$, where $K_\lambda \subset\subset \Sigma_\lambda$. For each $\lambda \in \Lambda$ define $\varphi_\lambda : \mathfrak{G}_\lambda \to \mathbb{C}$ by setting

$$\hat{a}(\varphi_\lambda) = \widehat{a \circ p_\lambda}(\varphi), \quad a \in \mathfrak{G}_\lambda \ .$$

Then φ_λ is a homomorphism of $\mathfrak{G}_\lambda$ onto $\mathbb{C}$. Moreover for each $a \in \mathfrak{G}_\lambda$

$$|\hat{a}(\varphi_\lambda)| = |\widehat{a \circ p_\lambda}(\varphi)| \leq |a \circ p_\lambda|_K = |a|_{p_\lambda(K)} = |a|_{K_\lambda} \ .$$

Therefore φ_λ is continuous, so there exists $\delta_\lambda \in \Sigma_\lambda$ such that $\hat{a}(\varphi_\lambda) = a(\delta_\lambda)$, $a \in \mathfrak{G}_\lambda$. Since ${}^\Lambda\mathfrak{G}$ is generated by the subalgebras $\mathfrak{G}_\lambda \circ p_\lambda$ it follows that $\hat{f}(\varphi) = f(\check{\delta})$ for all $f \in {}^\Lambda\mathfrak{G}$, where $\check{\delta} = \{\delta_\lambda\}$. Thus $[\Sigma^\Lambda, {}^\Lambda\mathfrak{G}]$ is natural.

Now assume that $[\Sigma^\Lambda, {}^\Lambda\mathfrak{G}]$ is natural. Let μ be an arbitrary element of Λ and φ_μ a continuous homomorphism of $\mathfrak{G}_\mu$ onto $\mathbb{C}$ dominated by a compact set $K_\mu \subset\subset \Sigma_\mu$. For each $\lambda \neq \mu$ choose an arbitrary point $\delta_\lambda \in \Sigma_\lambda$. Now each element $f \in {}^\Lambda\mathfrak{G}$ may be written in the form

$$f = \sum_\nu (a_\nu \circ p_\mu) \prod_{\lambda \neq \mu} (a_{\nu\lambda} \circ p_\lambda)$$

where a_ν and $a_{\nu\lambda}$ belong to $\mathfrak{G}_\lambda$ for all $\nu, \lambda \in \Lambda$ and only a finite number are different from zero. Define

$$\hat{f}(\varphi) = \sum_\nu \hat{a}_{\nu\mu}(\varphi_\mu) \prod_{\lambda \neq \mu} a_{\nu\lambda}(\delta_\lambda) \ .$$

We must prove first that $\hat{f}(\varphi)$ is well-defined. This is equivalent to showing that $f = 0$ implies $\hat{f}(\varphi) = 0$. If $f = 0$ then in particular

$$\sum_\nu a_{\nu\mu}(\sigma_\mu)\prod_{\lambda\neq\mu} a_{\nu\lambda}(\delta_\lambda) = 0$$

for all $\sigma_\mu \in \Sigma_\mu$. Therefore the function $\sum_\nu(\prod_{\lambda\neq\mu} a_{\nu\lambda}(\delta_\lambda))a_{\nu\mu}$ is equal to zero as an element of $\mathfrak{A}_\mu$. Hence

$$\sum_\nu(\prod_{\lambda\neq\mu} a_{\nu\lambda}(\delta_\lambda))\hat{a}_{\nu\mu}(\varphi_\mu) = 0$$

so $\hat{f}(\varphi) = 0$. It follows that $\varphi : f \mapsto \hat{f}(\varphi)$ is a homomorphism of ${}^\Lambda\mathfrak{A}$ onto $\mathbb{C}$. Moreover, if we define

$$K = \{\check{\sigma} \in \Sigma^\Lambda : \sigma_\mu \in K_\mu,\ \sigma_\lambda = \delta_\lambda \text{ for } \lambda \neq \mu\}$$

then K is a compact subset of Σ^Λ and $p_\mu(K) = K_\mu$ while $p_\lambda(K) = \delta_\lambda$ for $\lambda \neq \mu$. Therefore

$$f(\check{\sigma}) = \sum_\nu a_{\nu\mu}(\sigma_\mu)\prod_{\lambda\neq\mu} a_{\nu\lambda}(\delta_\lambda)$$

for $\check{\sigma} \in K$, so

$$|f|_K = |\sum_\nu a_{\nu\mu}\prod_{\lambda\neq\mu} a_{\nu\lambda}(\delta_\lambda)|_{K_\mu} .$$

We thus have

$$\begin{aligned} |\hat{f}(\varphi)| &= |\sum_\nu \hat{a}_{\nu\mu}(\varphi_\mu)\prod_{\lambda\neq\mu} a_{\nu\lambda}(\delta_\lambda)| \\ &= |\widehat{\sum_\nu a_{\nu\mu}\prod_{\lambda\neq\mu} a_{\nu\lambda}(\delta_\lambda)}(\varphi_\mu)| \\ &\leq |\sum_\nu a_{\nu\mu}\prod_{\lambda\neq\mu} a_{\nu\lambda}(\delta_\lambda)|_{K_\mu} = |f|_K . \end{aligned}$$

In other words $\varphi : {}^\Lambda\mathfrak{A} \to \mathbb{C}$ is continuous, so there exists $\check{\sigma}^\varphi \in \Sigma^\Lambda$ such that $\hat{f}(\varphi) = f(\check{\sigma}^\varphi)$ for all $f \in {}^\Lambda\mathfrak{A}$. In particular for $a \in \mathfrak{A}_\mu$

$$\hat{a}(\varphi_\mu) = \widehat{a\circ\rho_\mu}(\varphi) = a\circ\rho(\check{\sigma}^\varphi) = a(\sigma_\mu^\varphi)$$

so $[\Sigma_\mu,\mathfrak{A}_\mu]$ is natural. ♦

We prove next that naturality is also preserved under a projective limit of pairs with respect to a family of morphisms. More precisely, let $\{[\Sigma_\lambda,\mathfrak{A}_\lambda] : \lambda \in \Lambda\}$ be an arbitrary family of pairs where the index set Λ is now assumed to be a directed set under a given partial order "$\leq$". Also assume given for arbitrary $\lambda \leq \lambda'$ a pair morphism

$$\rho_{\lambda\lambda'} : [\Sigma_{\lambda'},\mathfrak{A}_{\lambda'}] \to [\Sigma_\lambda,\mathfrak{A}_\lambda] .$$

Then $\{\rho_{\lambda\lambda'},\Sigma_{\lambda'}\}$ is a projective (or inverse) limit system with limit $\Sigma_\infty = \varprojlim \rho_{\lambda\lambda'}(\Sigma_{\lambda'})$, where

$$\Sigma_\infty = \{\check{\sigma} \in \Sigma^\Lambda : \sigma_\lambda = \rho_{\lambda\lambda'}(\sigma_{\lambda'}), \quad \lambda \leq \lambda'\}$$

and Σ_∞ is given the topology induced from the product space Σ^Λ. Without further restriction the set Σ_∞ may be empty. Therefore in order to avoid trivialities we shall assume outright that Σ_∞ is not empty. Note that in this case ${}^\Lambda\mathfrak{a}|\Sigma_\infty$ is actually equal to the union of the algebras $(\mathfrak{a}_\lambda \circ p_\lambda)|\Sigma_\infty$ and may be identified with the direct limit $\varinjlim \mathfrak{a}_\lambda \circ \rho_{\lambda\lambda'}$. The desired naturality result may now be stated.

4.2 PROPOSITION. *If for each* $\lambda \in \Lambda$ *the pair* $[\Sigma_\lambda, \mathfrak{a}_\lambda]$ *is (strictly) natural and* $\mathfrak{a}_\lambda$ *separates the points of* Σ_λ *then the pair* $[\Sigma_\infty, {}^\Lambda\mathfrak{a}]$ *is also (strictly) natural.*

Proof. Again it will be sufficient to make the proof for the natural case. Therefore let $\psi : {}^\Lambda\mathfrak{a}|\Sigma_\infty \to \mathbb{C}$ be a continuous homomorphism dominated by a compact set $K \subseteq \Sigma_\infty$. Then $f \mapsto \widehat{f|\Sigma_\infty}(\psi)$, $f \in {}^\Lambda\mathfrak{a}$, defines a continuous homomorphism of ${}^\Lambda\mathfrak{a}$ also dominated by K. Hence by Proposition 4.1 there exists $\check{\delta} \in \Sigma^\Lambda$ such that

$$\widehat{f|\Sigma_\infty}(\psi) = f(\check{\delta}), \quad f \in {}^\Lambda\mathfrak{a} .$$

It remains to prove that $\check{\delta} \in \Sigma_\infty$.

Observe that for $\lambda \leq \lambda'$ and $\check{\sigma} \in \Sigma_\infty$

$$a_\lambda(\rho_{\lambda\lambda'}(\sigma_{\lambda'})) = a_\lambda(\sigma_\lambda), \quad a_\lambda \in \mathfrak{a}_\lambda \cdot$$

Therefore since $K \subseteq \Sigma_\infty$ it follows that

$$|a_\lambda \circ \rho_{\lambda\lambda'} \circ p_{\lambda'} - a_\lambda \circ p_\lambda|_K = 0$$

and hence

$$(a_\lambda \circ \rho_{\lambda\lambda'} \circ p_{\lambda'})(\check{\delta}) = (a_\lambda \circ p_\lambda)(\check{\delta}) \cdot$$

In other words

$$a_\lambda(\rho_{\lambda\lambda'}(\delta_{\lambda'})) = a_\lambda(\delta_\lambda), \quad a_\lambda \in \mathfrak{a}_\lambda \cdot$$

Since $\mathfrak{a}_\lambda$ separates points it follows that $\delta_\lambda = \rho_{\lambda\lambda'}(\delta_{\lambda'})$, so $\check{\delta} \in \Sigma_\infty$. ♦

An important special instance of the above construction of a product of pairs occurs when there are only two pairs $[\Sigma_1, \mathfrak{a}_1]$ and $[\Sigma_2, \mathfrak{a}_2]$ involved. In this case the product of the two algebras will be recognized as the ordinary Kronecker product

$\mathfrak{A}_1 \otimes \mathfrak{A}_2$ consisting of all finite sums of terms each of which is the product of an element of $\mathfrak{A}_1$ with an element of $\mathfrak{A}_2$ regarded in the obvious way as a function on $\Sigma_1 \times \Sigma_2$.

§5. EXAMPLES AND REMARKS

We close this chapter with some of the main examples of natural pairs. Each is in fact a natural system.

5.1 EXAMPLE. *The system* $[\Phi_{\mathfrak{B}}, \hat{\mathfrak{B}}]$ *associated with a commutative Banach algebra* $\mathfrak{B}$: Let $\mathfrak{B}$ denote a complex commutative Banach algebra with a unit element 1. Recall [R1, p. 75] that every homomorphism $\varphi : b \mapsto \hat{b}(\varphi)$ of $\mathfrak{B}$ onto $\mathbb{C}$ is automatically continuous with respect to the given norm in $\mathfrak{B}$. In fact $|\hat{b}(\varphi)| \leq \|b\|$ for all $b \in \mathfrak{B}$. Denote by $\Phi_{\mathfrak{B}}$ the collection of *all* homomorphism of $\mathfrak{B}$ onto $\mathbb{C}$. Then $b \mapsto \hat{b}$ defines a homomorphism of the algebra $\mathfrak{B}$ onto a subalgebra $\hat{\mathfrak{B}}$ of the algebra of all complex-valued functions on $\Phi_{\mathfrak{B}}$. If $\Phi_{\mathfrak{B}}$ is given the $\hat{\mathfrak{B}}$-topology then it becomes a compact Hausdorff space, commonly called the *spectrum* of $\mathfrak{B}$. Moreover $|\hat{b}|_{\Phi_{\mathfrak{B}}} \leq \|b\|$ for all $b \in \mathfrak{B}$. Since every homomorphism of $\hat{\mathfrak{B}}$ obviously induces a homomorphism of $\mathfrak{B}$ it follows that $[\Phi_{\mathfrak{B}}, \hat{\mathfrak{B}}]$ is a strictly natural system.

If $[X, \mathcal{C}]$ is a system in which X is a compact Hausdorff space then the closure $\bar{\mathcal{C}}$ of the algebra $\mathcal{C}$ in $C(X)$ is a Banach algebra under the sup norm $|c|_X$. An arbitrary homomorphism of $\bar{\mathcal{C}}$ onto $\mathbb{C}$ restricts to a *continuous* homomorphism of $\mathcal{C}$ onto $\mathbb{C}$. A homomorphism of $\mathcal{C}$ obviously need not be continuous, but if it is continuous then it extends uniquely to a homomorphism of $\bar{\mathcal{C}}$. Therefore the restriction map for homomorphisms of $\bar{\mathcal{C}}$ takes $\Phi_{\bar{\mathcal{C}}}$ one-to-one onto the collection $\Phi_{\mathcal{C}}$ of all continuous homomorphisms of $\mathcal{C}$ onto $\mathbb{C}$. Moreover this correspondence is a homeomorphism if $\Phi_{\mathcal{C}}$ is given the $\hat{\mathcal{C}}$-topology. To each point $x \in X$ we associate the point evaluation $\varphi_x : c \mapsto c(x)$ for $\mathcal{C}$ and thus obtain a homeomorphic embedding $\rho : x \mapsto \varphi_x$ of X in $\Phi_{\mathcal{C}}$. It is obvious that $\hat{\mathcal{C}} \circ \rho = \mathcal{C}$. Therefore $\rho : [X, \mathcal{C}] \Rightarrow [\Phi_{\mathcal{C}}, \hat{\mathcal{C}}]$ is a faithful extension of $[X, \mathcal{C}]$. Since $|c|_{\Phi_{\mathcal{C}}} = |c|_X$ for each $c \in \mathcal{C}$ it follows that $\Phi_{\hat{\mathcal{C}}} = \Phi_{\mathcal{C}}$. In this sense $\Phi_{\mathcal{C}}$ is the "natural" domain of existence for the functions in $\mathcal{C}$ and justifies our calling $[\Phi_{\mathcal{C}}, \hat{\mathcal{C}}]$ natural. If $\mathcal{C} = C(X)$ then $\Phi_{\mathcal{C}} = X$, so $[X, C(X)]$ is natural. Also if $X = D$, the closed unit disc in C, and $\mathcal{A}(D)$ is the *disc algebra* consisting of all functions continuous on D and holomorphic on the open unit disc then $\mathcal{A}(D) = \overline{\mathcal{P}|D}$, where $\mathcal{P}$ denotes the algebra of polynomials, and $[D, \mathcal{A}(D)]$ is natural. (See R1, p. 303 .) We also have the *boundary value algebra* $\mathcal{A}_T$ obtained by restricting elements of $\mathcal{A}(D)$ to the unit circle T. In this case $[T, \mathcal{A}_T]$ is a system but is not natural since $\Phi_{\mathcal{A}_T} = D$, $\hat{\mathcal{A}}_T = \mathcal{A}(D)$.

5.2 EXAMPLE. *The system* $[\mathbb{C}^\Lambda, \mathcal{P}]$. Let Λ be an arbitrary index set and denote by $\mathbb{C}^\Lambda$ the cartesian product of "Λ" complex planes $\mathbb{C}_\lambda$. Elements of $\mathbb{C}^\Lambda$ are denoted by $\check{\zeta} = \{\zeta_\lambda : \lambda \in \Lambda\}$. Next let $\mathcal{P}$ denote the algebra of all polynomials in $\mathbb{C}^\Lambda$. Thus each element of $\mathcal{P}$ is an ordinary complex polynomial in a finite number of the variables ζ_λ. Under the usual product space topology on $\mathbb{C}^\Lambda$ each element of $\mathcal{P}$, regarded as a function on $\mathbb{C}^\Lambda$, is obviously continuous. Furthermore the topology on $\mathbb{C}^\Lambda$ is clearly equivalent to the $\mathcal{P}$-topology, so $[\mathbb{C}^\Lambda, \mathcal{P}]$ is a system. Note that $\mathbb{C}^\Lambda$ is never compact and is not even locally compact when Λ is infinite.

The system $[\mathbb{C}^\Lambda, \mathcal{P}]$ is obviously the product of the pairs $[\mathbb{C}_\lambda, \mathcal{P}_\lambda]$, where $\mathcal{P}_\lambda$ denotes the algebra of all polynomials in one variable in the complex plane $\mathbb{C}_\lambda$. Moreover $[\mathbb{C}_\lambda, \mathcal{P}_\lambda]$ is a strictly natural system. In fact, if z_λ denotes the coordinate function $z_\lambda(\zeta) = \zeta$, $\zeta \in \mathbb{C}_\lambda$, then $\mathcal{P}_\lambda$ is generated by z_λ. Let $\varphi : \mathcal{P}_\lambda \to \mathbb{C}$ be an arbitrary homomorphism of $\mathcal{P}_\lambda$ onto $\mathbb{C}$. Then $\hat{P}(\varphi) = P(\hat{z}_\lambda(\varphi))$ for each $P \in \mathcal{P}_\lambda$, so φ is a point evaluation and it follows that $[\mathbb{C}_\lambda, \mathcal{P}_\lambda]$ is strictly natural. Therefore by Proposition 4.1 the system $[\mathbb{C}^\Lambda, \mathcal{P}]$ is also strictly natural. Observe that the coordinate polynomials

$$Z_\lambda(\check{\zeta}) = \zeta_\lambda, \ \check{\zeta} \in \mathbb{C}^\Lambda, \ \lambda \in \Lambda$$

constitute a system of generators for the algebra $\mathcal{P}$; *i.e.* $\mathcal{P}$ is the smallest subalgebra (with 1) that contains $\{Z_\lambda : \lambda \in \Lambda\}$. This fact may be used to prove directly that $[\mathbb{C}^\Lambda, \mathcal{P}]$ is strictly natural.

The example $[\mathbb{C}^n, \mathcal{P}]$ for n a positive integer motivates much of the theory developed below. On the other hand the general case $[\mathbb{C}^\Lambda, \mathcal{P}]$ turns out to play an essential role in the theory. In fact, involvement with $[\mathbb{C}^\Lambda, \mathcal{P}]$ seems to be more-or-less inevitable even in the study of pairs $[\Sigma, \mathcal{G}]$ with compact Σ. The necessity of including $[\mathbb{C}^\Lambda, \mathcal{P}]$ among the admissible pairs $[\Sigma, \mathcal{G}]$ thus precludes any blanket compactness restrictions on the spaces Σ.

5.3 EXAMPLE. *The system* $[S, \mathcal{O}_S]$, S *a Stein space.* Let S be a Stein space and denote by $\mathcal{O}_S$ the algebra of all global holomorphic functions on S. Then $[S, \mathcal{O}_S]$ is a natural system. This is a nontrivial fact which may be deduced from well-known results from several complex variables. [See G7, Corollary 7, p. 213].

If G is a domain of holomorphy in $\mathbb{C}^n$ then $[G, \mathcal{O}_G]$ turns out to be *strictly* natural. This is a special case of a result due to E. Michael [M5, proof of Theorem 12.7] which is in turn contained in a more abstract result also due to Michael [M5, Proposition 12.5]. The latter concerns a pair $[T, A]$, where T is a completely regular space and asserts (in our terminology) that $[T, A]$ will be strictly natural if A satisfies the following three conditions:

(i) If $f \in A$ and f is nowhere zero in T then $f^{-1} \in A$.

(ii) If $f_1,\dots,f_m$ are elements of A whose zero sets have empty intersection then there exist elements $g_1,\dots,g_m \in A$ such that $f_1g_1 + \dots + g_mg_m = 1$.

(iii) There exist elements $z_1,\dots,z_n \in A$ such that for each $\check{\zeta} \in \mathbb{C}^n$ the set $\{t \in T : \{z_i(t)\} = \check{\zeta}\}$ is compact.

It was pointed out to us by S. Sidney that strict naturality of the system $[\mathbb{C}^n, \mathcal{O}_n]$, where $\mathcal{O}_n$ consists of all entire functions in $\mathbb{C}^n$, may be proved by the following elementary direct argument: Let $\varphi : h \mapsto \hat{h}(\varphi)$ be an arbitrary homomorphism of $\mathcal{O}_n$ onto $\mathbb{C}$ and set $\zeta_k^\varphi = \hat{Z}_k(\varphi)$ $(k = 1,\dots,n)$, so $\check{\zeta}^\varphi \in \mathbb{C}^n$ and $\hat{P}(\varphi) = P(\check{\zeta}^\varphi)$ for each $P \in \mathcal{P}$. An arbitrary $h \in \mathcal{O}_n$ may be written in the form

$$h(\check{\zeta}) = h(\check{\zeta}^\varphi) + \sum_{k=1}^n (\zeta_k - \zeta_k^\varphi) g_k(\check{\zeta}), \quad \check{\zeta} \in \mathbb{C}^n$$

where $g_k \in \mathcal{O}_n$ for each k. Thus

$$h = h(\check{\zeta}^\varphi) + \sum_{k=1}^n (\hat{Z}_k - \zeta_k^\varphi)\hat{g}_k(\varphi).$$

Since

$$\hat{h}(\varphi) = h(\check{\zeta}^\varphi) + \sum_{k=1}^n (\hat{Z}_k(\varphi) - \zeta_k^\varphi)\hat{g}_k(\varphi)$$

we have

$$\hat{h}(\varphi) = h(\check{\zeta}^\varphi), \quad h \in \mathcal{O}_n$$

showing that $[\mathbb{C}^n, \mathcal{O}_n]$ is strictly natural.

CHAPTER II
CONVEXITY AND NATURALITY

§6. $\mathfrak{G}$-CONVEX HULLS. HULL-KERNEL TOPOLOGY

The importance of polynomial convexity, or $\mathcal{P}$-convexity, in $\mathbb{C}^n$ is well-known for the theory of analytic functions of several complex variables. The analogous concept is also of central importance in the study of arbitrary pairs. We begin with a precise definition of the general concept and some of its properties.

Consider an arbitrary pair $[\Sigma,\mathfrak{G}]$ and let K be a compact subset of Σ. Then the set

$$\hat{K} = \{\sigma \in \Sigma : |a(\sigma)| \leq |a|_K, \ a \in \mathfrak{G}\}$$

is called the $\mathfrak{G}$-*convex hull* of the set K. If L is a compact subset of $\hat{K}$ then $|a|_L \leq |a|_{\hat{K}} = |a|_K$, $a \in \mathfrak{G}$, so $\hat{L} \subseteq \hat{K}$. Therefore $\hat{K}$ is $\mathfrak{G}$-convex in the sense of the following definition which is a direct generalization of the familiar notion of polynomial convexity in $\mathbb{C}^n$.

6.1 DEFINITION. *A set* $\Omega \subseteq \Sigma$ *is said to be* $\mathfrak{G}$-*convex if for every compact set* $K \subset\subset \Omega$ *it is true that* $\hat{K} \subset \Omega$.

It is obvious that the empty set and the entire space are $\mathfrak{G}$-convex. In particular every subset $X \subseteq \Sigma$ is contained in at least one $\mathfrak{G}$-convex set, *viz* Σ. Moreover the intersection of an arbitrary family of $\mathfrak{G}$-convex sets is clearly also $\mathfrak{G}$-convex. Hence the intersection of all $\mathfrak{G}$-convex sets that contain a given set X is $\mathfrak{G}$-convex and is clearly a uniquely determined smallest $\mathfrak{G}$-convex set that contains X. It is called the $\mathfrak{G}$-*convex hull* of X in Σ and is denoted by $\hat{X}$. If X is compact then this $\mathfrak{G}$-convex hull of X is obviously equal to the $\mathfrak{G}$-convex hull defined above for compact sets. It is also obvious that a set will be $\mathfrak{G}$-convex iff it coincides with its $\mathfrak{G}$-convex hull. Observe that these convexity notions do not depend on the fact that $\mathfrak{A}$ is an algebra and are meaningful for an arbitrary family $\mathfrak{F}$ of function on Σ. When $\mathfrak{F} \neq \mathfrak{A}$ we shall denote the $\mathfrak{F}$-convex hull of a set X by $\hat{X}^{\mathfrak{F}}$ or some other special symbol. We note also that the $\mathfrak{A}$-convex hull $\hat{X}$ contains the union of all the $\mathfrak{A}$-convex hulls of compact subsets of X. This union may however be a *proper* subset of $\hat{X}$. (See Examples 11.2 and 11.3 below.) The precise relationship between X and $\hat{X}$ will be discussed later in this section.

An important special class of $\mathfrak{A}$-convex subsets of Σ consists of the "polyhedra" which may be defined as follows: Let $\mathcal{E}$ be an arbitrary subset of $\mathfrak{A}$ and let $r : \mathcal{E} \to [0, \infty]$ be any non-negative extended real-valued function defined on $\mathcal{E}$. Then the set

$$P(\mathcal{E}, r) = \{\sigma \in \Sigma : |a(\sigma)| \leq r(a),\ a \in \mathcal{E}\}$$

is called a *closed* $\mathfrak{A}$-*polyhedron* in Σ. If $r(a) > 0$ for all $a \in \mathcal{E}$ then the set

$$P^0(\mathcal{E}, r) = \{\sigma \in \Sigma : |a(\sigma)| < r(a),\ a \in \mathcal{E}\}$$

is called an *open* $\mathfrak{A}$-polyhedron in Σ. Note that $P(\mathcal{E}, r)$ is always a closed set but need not be equal to the closure of $P^0(\mathcal{E}, r)$, and if $\mathcal{E}$ is infinite $P^0(\mathcal{E}, r)$ need not be an open set. If $r(a) = 0$ for all $a \in \mathcal{E}$ then $P(\mathcal{E}, r)$ is called a *hull set* and denoted by $h(\mathcal{E})$. If K is any compact subset of $P^0(\mathcal{E}, r)$ then $|a|_K < r(a)$ for each $a \in \mathcal{E}$. Therefore $\hat{K} \subseteq P^0(\mathcal{E}, r)$, so $P^0(\mathcal{E}, r)$ is $\mathfrak{A}$-convex. Similarly $P(\mathcal{E}, r)$ is also $\mathfrak{A}$-convex. Observe that every basic $\mathfrak{A}$-neighborhood of a point of $\mathcal{E}$ is an open $\mathfrak{A}$-polyhedron and is therefore $\mathfrak{A}$-convex. Thus we have the important fact that a system $[\Sigma, \mathfrak{A}]$ is *locally* $\mathfrak{A}$-*convex* in the sense that each point of Σ admits arbitrarily small $\mathfrak{A}$-convex neighborhoods.

If E is an arbitrary subset of Σ then the set

$$k(E) = \{a \in \mathfrak{A} : a|E = 0\}$$

is called the *kernel* of the set E. It is an ideal in $\mathfrak{A}$ equal to the intersection of those maximal ideals in $\mathfrak{A}$ which are kernels of those homomorphisms of $\mathfrak{A}$ onto $\mathbb{C}$ defined by point evaluations in E. The hull sets in Σ are precisely those sets E for which $E = h(k(E))$, and the mapping $E \to h(k(E))$, defined on subsets of Σ, is a closure operation in Σ that determines a topology in E called the *hull-kernel topology associated with* $\mathfrak{A}$. Since hull sets are automatically closed in the given topology of E the hull-kernel topology is in general coarser than the former. Observe that a hull-set is always $\mathfrak{G}$-convex. Therefore since $E \subseteq h(k(E))$, also $\hat{E} \subseteq h(k(E))$. In particular if $a \in \mathfrak{G}$ and $a|E = 0$ then $a|\hat{E} = 0$.

§7. $\mathfrak{G}$-CONVEXITY IN A NATURAL PAIR $[\Sigma, \mathfrak{G}]$

The next theorem is of fundamental importance in the theory of natural pairs.

7.1 THEOREM. *Let* $[\Sigma, \mathfrak{G}]$ *be an arbitrary pair and* Ω *a subset of* Σ.

(i) If $[\Sigma, \mathfrak{G}]$ *is natural (strictly natural) and* Ω *is* $\mathfrak{G}$*-convex (hull-kernel closed) then* $[\Omega, \mathfrak{G}]$ *is natural (strictly natural).*

(ii) If $[\Omega, \mathfrak{G}]$ *is natural (strictly natural) and* $\mathfrak{G}$ *separates points then* Ω *is* $\mathfrak{G}$*-convex (hull-kernel closed).*

Proof. It will be sufficient to make the proof for the case in which $[\Sigma, \mathfrak{G}]$ is natural since the argument for the strictly natural case is essentially the same.

Assume first that Ω is $\mathfrak{G}$-convex and let $\varphi : \mathfrak{G}|\Omega \to \mathbb{C}$ be a continuous homomorphism of $\mathfrak{G}|\Omega$ onto $\mathbb{C}$ dominated by a compact set $K_\varphi \subset\subset \Omega$. Then

$$a \mapsto a|\Omega \mapsto \widehat{a|\Omega}(\varphi),\ a \in \mathfrak{G}$$

defines a continuous homomorphism of $\mathfrak{G}$ onto $\mathbb{C}$. Therefore there exists $\sigma_\varphi \in \Sigma$ such that $\widehat{a|\Omega}(\varphi) = a(\sigma_\varphi)$, $a \in \mathfrak{G}$. Moreover since $|a(\sigma_\varphi)| \leq |a|_{K_\varphi}$, $a \in \mathfrak{G}$, it follows that $\sigma_\varphi \in \hat{K}_\varphi \subseteq \Omega$. Therefore $[\Omega, \mathfrak{G}]$ is natural, proving (i).

Now assume that $[\Omega, \mathfrak{G}]$ is natural and that $\mathfrak{G}$ separates the points of Σ. Let K be a compact subset of Ω. If $\delta \in \hat{K}$ then $|a(\delta)| \leq |a|\Omega|_K$, $a \in \mathfrak{G}$, so $a|\Omega = 0$ implies $a(\delta) = 0$. Therefore $a|\Omega \mapsto a(\delta)$, $a \in \mathfrak{G}$, is a well-defined continuous homomorphism of $\mathfrak{G}|\Omega$ onto $\mathbb{C}$. Hence there exists $\omega \in \Omega$ such that $a(\omega) = a(\delta)$ for all $a \in \mathfrak{G}$. Since $\mathfrak{G}$ separates points $\delta = \omega \in \Omega$, so $\delta \in \Omega$ and it follows that Ω is $\mathfrak{G}$-convex, proving (ii). ◆

Since $\mathfrak{G}$-neighborhoods are $\mathfrak{G}$-convex we have the following corollary to the above theorem.

7.2 COROLLARY. *A natural system is automatically locally natural (Definition 3.3). A given system* $[\Sigma, \mathfrak{A}]$ *will be locally natural iff each point of* Σ *admits a neighborhood* U *such that* $[\Sigma, \mathfrak{A}]$ *is natural.*

The next result is also very important because it enables us to apply methods from the theory of uniform algebras to the study of arbitrary natural systems.

7.3 THEOREM. *If* $[\Sigma, \mathfrak{A}]$ *is a natural system, then the* $\mathfrak{A}$*-convex hull of every compact set is compact.*

Proof. Let K be an arbitrary compact subset of Σ and consider the system $[\hat{K}, \mathfrak{A}]$. Since $|a|_{\hat{K}} = |a|_K < \infty$ the algebra $\mathfrak{A}|\hat{K}$ consists of bounded functions. Denote by $\mathfrak{A}_{\hat{K}}$ the closure of $\mathfrak{A}|\hat{K}$ in $C(\hat{K})$ with respect to uniform convergence on $\hat{K}$. Then $\mathfrak{A}_{\hat{K}}$ is a Banach algebra under the supnorm. Since $\hat{K}$ is $\mathfrak{A}$-convex in Σ and $[\Sigma, \mathfrak{A}]$ is natural it follows by the preceding theorem that $[\hat{K}, \mathfrak{A}]$ is also natural. Moreover since $\mathfrak{A}|\hat{K} \subseteq \mathfrak{A}_{\hat{K}} \subseteq \overline{\mathfrak{A}|\hat{K}}$ it follows by Proposition 4.2 that the system $[\hat{K}, \mathfrak{A}_{\hat{K}}]$ is natural. This means that $\hat{K}$ is the spectrum of $\mathfrak{A}_{\hat{K}}$ in the $\mathfrak{A}_{\hat{K}}$-topology and is hence compact in this topology. But the $\mathfrak{A}_{\hat{K}}$-topology on $\hat{K}$ is clearly equivalent to the $\mathfrak{A}$-topology on $\hat{K}$ which is the topology induced on $\hat{K}$ by Σ since $[\Sigma, \mathfrak{A}]$ is a system. Therefore $\hat{K}$ is a compact subset of Σ. ♦

7.4 COROLLARY. *If* $[\Sigma, \mathfrak{A}]$ *is a natural system and* K *is any compact subset of* Σ *then* $[\hat{K}, \mathfrak{A}_{\hat{K}}]$ *is a strictly natural system.*

Theorem 7.3 suggests the following definition.

7.5 DEFINITION. *A pair* $[\Sigma, \mathfrak{A}]$ *is said to be* convex *if the* $\mathfrak{A}$*-convex hull of every compact subset of* Σ *is compact.*

Although every natural system is convex not every convex system is natural. This is already shown by the boundary value algebra mentioned in Example 5.1. The question of when convexity implies naturality will be discussed in its proper context later (Chapter VIII).

§8. CLOSURE OPERATIONS

It will be useful to examine more closely the relationship between a subset X of Σ and the $\mathfrak{G}$-convex hull $\hat{X}$ of X in Σ for the case of an arbitrary pair $[\Sigma, \mathfrak{G}]$. First denote by $X^{\#}$ the union of all those sets $\hat{K}$ where K ranges over compact subsets of X. Then obviously $X \subseteq X^{\#} \subseteq \hat{X}$. Although $X^{\#}$ may be a proper subset of $\hat{X}$ (see Examples 11.2, 11.3) it is nevertheless possible to obtain the $\mathfrak{G}$-convex hull of X through iteration of the operation "#". Since this iteration process is useful and will be needed in other situations, it is worthwhile to describe it in general terms.

Let S be an arbitrary set of points and let $E \mapsto E^*$ denote a mapping which associates with each subset E of S another subset E^*. Then "*" is called a *closure operation in* S if it satisfies the following properties:

(i) $\emptyset^* = \emptyset$, where $\emptyset$ is the empty set.

(ii) $E \subseteq E^*$ for every $E \subseteq S$.

(iii) $E_1 \subseteq E_2$ implies $E_1^* \subseteq E_2^*$.

A closure operation is said to be *proper* if it satisfies the additional property

(iv) $(E^*)^* = E^*$, for every $E \subseteq S$.

Observe that "#" is a closure operation in Σ.

A set E is said to be **-closed* if $E^* = E$. It is obvious that both $\emptyset$ and S are *-closed. Also the intersection of an arbitrary collection of *-closed sets is *-closed. Therefore each set $E \subseteq S$ is contained in a unique smallest *-closed set, *viz* the intersection of all *-closed sets that contain it. The smallest *-closed set that contains E is called the **-closure* of E and denoted by $E^{(*)}$. Note that "(*)" is a proper closure operation. In the case of "#" a set is #-closed iff it is $\mathfrak{G}$-convex, so $X^{(\#)} = \hat{X}$ for arbitrary X.

The following proposition, which in one form or another is well-known, enables us to use induction arguments to deduce certain properties of the *-closure of a set from properties of the closure operation "*". We include a proof for the sake of completeness. (Cf. [H1, p. 191].)

8.1 PROPOSITION. *For arbitrary* $E \subseteq S$ *there exists an ordinal* μ *such that to each ordinal* $\nu \le \mu$ *there corresponds a set* $E_\nu \subseteq S$ *with the following properties:*

(i) $E_0 = E$, $E_\mu = E^{(*)}$, *and* $E_\alpha \subsetneq E_\beta$ *for* $0 \le \alpha < \beta \le \mu$.

(ii) If $\nu \le \mu$ *then* $E_\nu = (\bigcup_{\alpha<\nu} E_\alpha)^*$.

Proof. First define $E_0 = E$ and then by transfinite induction define E_ν for *every* ordinal ν so that

$$E_\nu = (\bigcup_{\alpha<\nu} E_\alpha)^*.$$

Then $\alpha < \beta$ obviously implies $E_\alpha \subseteq E_\alpha^* \subseteq E_\beta$. Furthermore the number of distinct sets E_ν cannot exceed the cardinality of the set S. Hence there exists a first ordinal μ such that $E_\mu = E_\rho$ for some $\rho > \mu$. Since $\mu < \mu + 1 \le \rho$, we have

$$E_\mu \subseteq E_\mu^* \subseteq E_{\mu+1} \subseteq E_\rho = E_\mu$$

Therefore $E_\mu^* = E_\mu$, so E_μ is *-closed. Moreover μ is the first ordinal for which E_μ is *-closed. It only remains to prove that $E_\mu = E^{(*)}$. Observe that by definition $E_0 \subseteq E^{(*)} \subseteq E_\mu$. Now if ν is an ordinal such that $E_\alpha \subseteq E^{(*)}$ for all $\alpha < \nu$ then

$$E_\nu = (\bigcup_{\alpha<\nu} E_\alpha)^* \subseteq (E^{(*)})^* = E^{(*)}.$$

Therefore by induction $E_\nu \subseteq E^{(*)}$ for all ν. In particular $E_\mu \subseteq E^{(*)}$ and hence $E_\mu = E^{(*)}$. ◆

The family $\{E_\nu : \nu \le \mu\}$ of sets associated with $E^{(*)}$ by the above proposition is obviously uniquely determined. We call it the **-resolution* of $E^{(*)}$.

An application of the above proposition to the closure operation "#" in Σ gives the desired iterative "construction" of the $\mathfrak{a}$-convex hull. We now use the result to prove a theorem which extends another well-known property of convex hulls in the compact case.

8.2 THEOREM. *Let* $[\Sigma, \mathfrak{a}]$ *be an arbitrary pair and* X *any subset of* Σ. *Then* $|a|_{\hat{X}} = |a|_X$ *for every* $a \in \mathfrak{a}$.

Proof. Observe first that since $|a|_{\hat{K}} = |a|_K$ for every compact set K it follows immediately that $|a|_{X\#} = |a|_X$ for arbitrary $X \subseteq \Sigma$. Now let $\{X_\nu\}$ be the

#-resolution of $\hat{X}$ and note that $|a|_{X_0} = |a|_X$. Therefore assume that $|a|_{X_\alpha} = |a|_X$ for each $\alpha < \nu$. Then

$$|a|_{X_\nu} = |a|_{\bigcup\limits_{\alpha<\nu} X_\alpha} = \sup_{\alpha<\nu} |a|_{X_\alpha} = |a|_X.$$

Hence by induction $|a|_{X_\nu} = |a|_X$ for all ν, so $|a|_{\hat{X}} = |a|_{X_\mu} = |a|_X$. ◆

§9. CONVEXITY AND EXTENSIONS

The next theorem gives the behavior of $\mathfrak{a}$-convex hulls under extensions.

9.1 THEOREM. *Let* $\rho : [\Sigma, \mathfrak{a}] \Rightarrow [\Sigma', \mathfrak{a}']$ *be an arbitrary extension of* $[\Sigma, \mathfrak{a}]$ *and let* X *be an arbitrary subset of* Σ. *Then* $\rho(\hat{X}) \subseteq \widehat{\rho(X)}$. *If either* X *is compact or* $\rho : \Sigma \to \rho(\Sigma)$ *is a homeomorphism then* $\rho(\hat{X}) = \widehat{\rho(X)} \cap \rho(\Sigma)$.

Proof. Consider first an arbitrary compact set $K \subset\subset \Sigma$ and let $\sigma \in \hat{K}$. Since $\mathfrak{a}' \circ \rho = \mathfrak{a}$ we have for each $a' \in \mathfrak{a}'$

$$|a'(\rho(\sigma))| = |(a' \circ \rho)(\sigma)| \leq |a' \circ \rho|_K = |a'|_{\rho(K)}$$

so $\rho(\sigma) \in \widehat{\rho(K)}$ and hence $\rho(\hat{K}) \subseteq \widehat{\rho(K)}$. Moreover if $\sigma' \in \widehat{\rho(K)} \cap \rho(\Sigma)$ and $a \in \mathfrak{a}$ then there exist $\sigma \in \Sigma$ and $a' \in \mathfrak{a}'$ such that $\rho(\sigma) = \sigma'$ and $a' \circ \rho = a$. Thus

$$|a(\sigma)| = |(a' \circ \rho)(\sigma)| = |a'(\rho(\sigma))| = |a'(\sigma')|$$
$$\leq |a'|_{\rho(K)} = |a' \circ \rho|_K = |a|_K.$$

Therefore $\sigma \in \hat{K}$ and hence $\sigma' \in \rho(\hat{K})$. In other words $\widehat{\rho(K)} \cap \rho(\Sigma) \subseteq \rho(\hat{K})$, so we obtain $\rho(\hat{K}) = \widehat{\rho(K)} \cap \rho(\Sigma)$. This establishes the assertion in the theorem concerning compact sets.

Next let X be an arbitrary subset of Σ. It follows immediately from the result just proved for compact sets that $\rho(X^\#) \subseteq \rho(X)^\#$. Furthermore if $\rho : \Sigma \to \rho(\Sigma)$ is a homeomorphism then every compact subset of $\rho(X)$ is of the form $\rho(K)$, where $K \subset\subset X$, which implies that $\rho(X^\#) = \rho(X)^\# \cap \rho(\Sigma)$ in this case. Now consider the #-resolutions $\{X_\nu\}$ and $\{\rho(X)_\nu\}$ of the convex hulls $\hat{X}$ and $\widehat{\rho(X)}$. It is trivial that $\rho(X_0) = \rho(X)_0$, so we assume for all $\alpha < \nu$ that $\rho(X_\alpha) \subseteq \rho(X)_\alpha$, with $\rho(X_\alpha) = \rho(X)_\alpha \cap \rho(\Sigma)$ in case $\rho : \Sigma \to \rho(\Sigma)$ is a homeomorphism. Then

$$\rho(\bigcup_{\alpha<\nu} X_\alpha) \subseteq \bigcup_{\alpha<\nu} \rho(X)_\alpha$$

with

$$\rho(\bigcup_{\alpha<\nu} X_\nu) = (\bigcup_{\alpha<\nu} \rho(X)_\alpha) \cap \rho(\Sigma)$$

in the homeomorphism case. An application of the results for "#" obtained above thus gives

$$\rho(X_\nu) \subseteq \rho(X)_\nu, \ \rho(X_\nu) = \rho(X)_\nu \cap \rho(\Sigma)$$

respectively. Therefore by induction $\rho(\hat{X}) \subseteq \widehat{\rho(X)}$, with $\rho(\hat{X}) = \widehat{\rho(X)} \cap \rho(\Sigma)$ if $\rho : \Sigma \to \rho(\Sigma)$ is a homeomorphism. ♦

Before stating the next theorem, in which the induction technique is used to prove a result on extension of pair morphisms, we introduce a definition which will also be needed in the subsequent discussion.

9.2 DEFINITION. *An extension* $\rho : [\Sigma, \mathcal{G}] \Rightarrow [\Omega, \mathcal{B}]$ *is said to be minimal if* $\widehat{\rho(\Sigma)} = \Omega$.

9.3 THEOREM. *Let* $[\Sigma, \mathcal{G}]$, $[\Omega, \mathcal{B}]$ *and* $[\Gamma, \mathcal{C}]$ *be systems, where* $[\Gamma, \mathcal{C}]$ *is natural. Assume given a minimal faithful extension* $\rho : [\Sigma, \mathcal{G}] \Rightarrow [\Omega, \mathcal{B}]$. *Then every morphism* $\mu : [\Sigma, \mathcal{G}] \to [\Gamma, \mathcal{C}]$ *extends to* $[\Omega, \mathcal{B}]$ *via* ρ; *i.e. there exists a morphism* $\bar{\mu} : [\Omega, \mathcal{B}] \to [\Gamma, \mathcal{C}]$ *such that* $\mu = \bar{\mu}\circ\rho$. *If* $[\Gamma, \mathcal{C}]$ *is strictly natural then the condition that* ρ *be minimal may be dropped.*

$$\begin{array}{ccc} & [\Gamma, \mathcal{C}] & \\ \mu \nearrow & & \nwarrow \bar{\mu} \\ [\Sigma, \mathcal{G}] & \xrightarrow[\rho]{} & [\Omega, \mathcal{B}] \end{array}$$

Proof. Let c be an arbitrary element of $\mathcal{C}$. Then $c\circ\mu \in \mathcal{G}$ and, since $\mathcal{B}\circ\rho = \mathcal{G}$, there exists $b_c \in \mathcal{B}$ such that $b_c\circ\rho = c\circ\mu$. Since ρ is faithful (*i.e.*, $b\circ\rho = 0$ implies $b = 0$) the element b_c is uniquely determined by c. Now denote by $\{\Omega_\nu\}$ the #-resolution of $\Omega = \widehat{\rho(\Sigma)}$ with $\Omega_0 = \rho(\Sigma)$. We shall prove by induction the existence for each ordinal ν a mapping $\mu_\nu : \Omega_\nu \to \Gamma$ such that $c\circ\mu_\nu = b_c|\Omega_\nu$, $c \in \mathcal{C}$. Note first that since ρ is faithful $\rho : \Sigma \to \rho(\Sigma)$ is a homeomorphism. Thus we may define $\mu_0 = \mu\circ\rho^{-1}$ observing that

$$c\circ\mu_0 = c\circ\mu\circ\rho^{-1} = b_c|\Omega_0$$

so μ_ν exists for $\nu = 0$. We accordingly assume that μ_ν exists for all $\alpha < \nu$. If $\alpha < \beta < \nu$ then $\Omega_\alpha \subseteq \Omega_\beta$ and

$$(c \circ \mu_\beta)(\omega) = b_c(\omega) = (c \circ \mu_\alpha)(\omega), \quad \omega \in \Omega_\alpha$$

so $\mu_\beta | \Omega_\alpha = \mu_\alpha$. Therefore if we set

$$\mu_{(\nu)}(\omega) = \mu_\alpha(\omega), \quad \omega \in \Omega_\alpha, \ \alpha < \nu$$

then $\mu_{(\nu)}$ is a well-defined map of $\bigcup_{\alpha<\nu} \Omega_\nu$ into Γ and

$$c \circ \mu_{(\nu)} = b_c \mid \bigcup_{\alpha<\nu} \Omega_\alpha.$$

Now let $\omega \in \Omega_\nu$. Then there exists a compact set $K \subseteq \bigcup_{\alpha<\nu} \Omega_\alpha$ such that $\omega \in \hat{K}$. Define $\varphi_\omega : \mathcal{C} \to \mathbb{C}$, $c \mapsto b_c(\omega)$. Then φ_ω is a homomorphism of $\mathcal{C}$ onto $\mathbb{C}$. Moreover $|b|_{\hat{K}} = |b|_K$ for each $b \in \mathcal{B}$, so

$$\begin{aligned} |\hat{c}(\varphi_\omega)| &= |b_c(\omega)| \le |b_c|_K \\ &= |c \circ \mu_{(\nu)}|_K = |c|_{\mu_{(\nu)}}(K), \quad c \in \mathcal{C}. \end{aligned}$$

Since $\mu_{(\nu)}$ is continuous $\mu_{(\nu)}(K)$ is compact in Γ and it follows that $\varphi_\omega : \mathcal{C} \to \mathbb{C}$ is continuous. Hence there exists $\gamma_\omega \in \Gamma$ such that $b_c(\omega) = c(\gamma_\omega)$ for each $c \in \mathcal{C}$. Define $\mu_\nu : \Omega_\nu \to \Gamma$, $\omega \mapsto \gamma_\omega$. Then $c \circ \mu_\nu = b_c | \Omega_\nu$, so μ_ν exists for all ν by induction. Since $\Omega_\nu = \Omega$ for sufficiently large ν there exists a map $\bar{\mu} : \Omega \to \Gamma$ such that $c \circ \bar{\mu} = b_c$, so $\mathcal{C} \circ \bar{\mu} \subseteq \mathcal{B}$. Moreover

$$c \circ \bar{\mu} \circ \rho = b_c \circ \rho = c \circ \mu, \quad c \in \mathcal{C}$$

which implies $\bar{\mu} \circ \rho = \mu$ and completes the proof of the first part of the theorem.

Now assume that $[\Gamma, \mathcal{C}]$ is strictly natural and for each $\omega \in \Omega$ define $\varphi_\omega : \mathcal{C} \to \mathbb{C}$, $c \mapsto b_c(\omega)$. Then $\varphi_\omega : \mathcal{C} \to \mathbb{C}$ is a homomorphism, so there exists $\gamma_\omega \in \Gamma$ such that $b_c(\omega) = c(\gamma_\omega)$, $c \in \mathcal{C}$. Define $\bar{\mu} : \Omega \to \Gamma$, $\omega \mapsto \gamma_\omega$. Then $\mathcal{C} \circ \bar{\mu} \subseteq \mathcal{B}$ and $\bar{\mu} \circ \rho = \mu$, completing the proof of the theorem. ♦

§10. NATURAL EXTENSIONS

We turn next to the construction of "natural extensions" of an arbitrary pair $[\Sigma, \mathcal{A}]$ in accordance with the following definition.

10.1 DEFINITION. *An extension* $\mu : [\Sigma, \mathcal{A}] \Rightarrow [\Omega, \mathcal{B}]$ *is said to be natural if the pair* $[\Omega, \mathcal{B}]$ *is natural.*

If $\mu : [\Sigma, \mathfrak{A}] \Rightarrow [\Omega, \mathfrak{B}]$ is an arbitrary natural extension of $[\Sigma, \mathfrak{A}]$ then by Theorem 7.1 the extension $\mu : [\Sigma, \mathfrak{A}] \Rightarrow [\widehat{\mu(\Sigma)}, \mathfrak{A}]$ is a *minimal* natural extension. Observe that, by Theorem 9.1 the property of being a minimal natural extension is preserved under extension equivalence (§2). We shall prove below that any two minimal natural *system* extensions of a pair $[\Sigma, \mathfrak{A}]$ are equivalent. First however we describe a well-known and very useful method of constructing extensions that will also provide the existence of natural, and hence minimal natural, extensions of an arbitrary pair $[\Sigma, \mathfrak{A}]$.

Let $\check{z} = \{z_\lambda : \lambda \in \Lambda\}$ be a system of generators for the algebra $\mathfrak{A}$; *i.e.* $\{z_\lambda\} \subseteq \mathfrak{A}$ and the smallest subalgebra of $\mathfrak{A}$ that contains $\{z_\lambda\}$ is equal to $\mathfrak{A}$. Thus each element of $\mathfrak{A}$ is equal to a polynomial in a finite number of the generators z_λ. In particular we could have $\{z_\lambda\} = \mathfrak{A}$.

10.2 PROPOSITION. *The map*

$$\check{z} : \Sigma \to \mathbb{C}^\Lambda, \quad \sigma \mapsto \check{z}(\sigma) = \{z_\lambda(\sigma)\}$$

defines a natural system extension

$$\check{z} : [\Sigma, \mathfrak{A}] \Rightarrow [\mathbb{C}^\Lambda, \mathcal{P}]$$

of the pair $[\Sigma, \mathfrak{A}]$ *and in particular defines a minimal natural system extension* $\check{z} : [\Sigma, \mathfrak{A}] \Rightarrow [\widehat{\check{z}(\Sigma)}, \mathcal{P}]$.

Proof. It is obvious that the map $\check{z} : \Sigma \to \mathbb{C}^\Lambda$ is continuous. Consider an arbitrary polynomial $P \in \mathcal{P}$

$$P(\check{\zeta}) = P(\zeta_{\lambda_1}, \ldots, \zeta_{\lambda_n}), \quad \check{\zeta} \in \mathbb{C}^\Lambda$$

and set

$$P(\check{z}) = P(z_{\lambda_1}, \ldots, z_{\lambda_n}).$$

Then $P(\check{z}) \in \mathfrak{A}$. Since $\check{z}$ generates $\mathfrak{A}$ it follows that $\mathcal{P} \circ \check{z} = \mathfrak{A}$. Therefore since $[\mathbb{C}^\Lambda, \mathcal{P}]$ is a natural system the proposition follows. ◆

Observe that $\check{z}$ will be an injection iff $\mathfrak{A}$ separates points and will be a homeomorphism iff $[\Sigma, \mathfrak{A}]$ is a system. The dual map $P \mapsto P(\check{x}) = P \circ \check{z}$ is a homomorphism of $\mathcal{P}$ onto $\mathfrak{A}$ with kernel

$$k = \{P \in \mathcal{P} : P|\check{z}(\Sigma) = 0\}.$$

Thus k is the ideal of all polynomial relations among the generators $\check{z}$.

10.3 THEOREM. *Let* $[\Sigma, \mathfrak{G}]$ *be an arbitrary pair. Then any two minimal natural system extensions of* $[\Sigma, \mathfrak{G}]$ *are equivalent.*

Proof. It will be sufficient to prove that an arbitrary minimal natural system extension, say $\mu : [\Sigma, \mathfrak{G}] \Rightarrow [\Omega, \mathfrak{B}]$, is equivalent to the special one constructed above. Since $\widehat{\mu(\Sigma)} = \Omega$ it follows that $b|\mu(\Sigma) = 0$ implies $b = 0$ by Theorem 8.2. Therefore the map $b \mapsto b\circ\mu$ of $\mathfrak{B}$ onto $\mathfrak{G}$ is an isomorphism. Hence if u_λ is chosen in $\mathfrak{B}$ for each $\lambda \in \Lambda$ so that $u_\lambda\circ\mu = z_\lambda$ then $\{u_\lambda : \lambda \in \Lambda\}$ is a system of generators for $\mathfrak{B}$. Therefore the map

$$\check{u} : \Omega \to \mathbb{C}^\Lambda, \quad \omega \mapsto \check{u}(\omega) = \{u_\lambda(\omega)\}$$

is a homeomorphism of Ω into $\mathbb{C}^\Lambda$ and $\check{z} = \check{u}\circ\mu$. Since $\mathfrak{P}\circ u = \mathfrak{B}$ we have an isomorphism

$$\check{u} : [\Omega, \mathfrak{B}] \Rightarrow [\check{u}(\Omega), \mathfrak{P}].$$

Therefore $[\check{u}(\Omega), \mathfrak{P}]$ is natural and by Theorem 7.1 (ii) the set $\check{u}(\Omega)$ is $\mathfrak{P}$-convex in $\mathbb{C}^\Lambda$. Now by Theorem 9.1

$$\widehat{\check{z}(\Sigma)} = \widehat{(\check{u}\circ\mu)(\Sigma)} = \widehat{\check{u}(\mu(\Sigma))}$$
$$= \check{u}(\widehat{\mu(\Sigma)}) = \check{u}(\Omega).$$

In other words $\check{u}$ maps Ω homeomorphically onto $\check{z}(\Sigma)$. This, with the fact that $\check{z} = \check{u}\circ\mu$, means that the extensions $\mu : [\Sigma, \mathfrak{G}] \Rightarrow [\Omega, \mathfrak{B}]$ and $\check{z} : [\Sigma, \mathfrak{G}] \Rightarrow [\widehat{\check{z}(\Sigma)}, \mathfrak{P}]$ are equivalent. ♦

Since all minimal natural system extensions of a given pair are equivalent, we shall denote a representative one for a given pair $[\Sigma, \mathfrak{G}]$ by the special notation

$$\phi : [\Sigma, \mathfrak{G}] \Rightarrow [\Phi_{[\Sigma, \mathfrak{G}]}, \hat{\mathfrak{G}}]$$

and, by analogy with the case of uniform algebras, call the space $\Phi_{[\Sigma, \mathfrak{G}]}$ the *spectrum of* $[\Sigma, \mathfrak{G}]$ or *of* $\mathfrak{G}$.

The preceeding discussion of natural extensions could be paralleled by a similar discussion of *strictly* natural extensions. In the latter case $\mathfrak{G}$-convex hulls are replaced by hull-kernel closures. For example, an arbitrary strictly natural extension $\mu : [\Sigma, \mathfrak{G}] \Rightarrow [\Omega, \mathfrak{B}]$ contains the *minimal* strictly natural extension

$$\mu : [\Sigma, \mathfrak{G}] \Rightarrow [hk(\mu(\Sigma)), \mathfrak{B}].$$

Also, as in the natural case, any two minimal strictly natural system extensions are equivalent. Note that for the proof of this we must use the fact that the system $[\mathbb{C}^\Lambda, \mathcal{P}]$ is actually strictly natural.

If Σ is compact recall that the spectrum of $\mathcal{G}$ is the space $\Phi_{\mathcal{G}}$ of all continuous homomorphisms $\varphi : a \mapsto \hat{a}(\varphi)$ of $\mathcal{G}$ onto $\mathbb{C}$, where $\Phi_{\mathcal{G}}$ is given the $\hat{\mathcal{G}}$-topology. If $\phi_0 : \Sigma \to \Phi_{\mathcal{G}}$ is the map that takes points of Σ to the corresponding point evaluations then $\phi_0 : [\Sigma, \mathcal{G}] \Rightarrow [\Phi_{\mathcal{G}}, \hat{\mathcal{G}}]$ is a minimal natural system extension of $[\Sigma, \mathcal{G}]$. Therefore when Σ is compact $[\Phi_{\mathcal{G}}, \hat{\mathcal{G}}] = [\Phi_{[\Sigma, \mathcal{G}]}, \hat{\mathcal{G}}]$.

For pairs $[\Sigma, \mathcal{G}]$ with non-compact Σ we shall continue to denote by $\Phi_{\mathcal{G}}$ the family of all continuous homomorphisms of $\mathcal{G}$ onto $\mathbb{C}$ and by $\phi_0 : \Sigma \to \Phi_{\mathcal{G}}$ the point evaluation map. If $\Phi_{\mathcal{G}}$ is given the $\hat{\mathcal{G}}$-topology, then ϕ_0 is continuous and will be a homeomorphism iff $[\Sigma, \mathcal{G}]$ is a system. As before, ϕ_0 defines a minimal extension $\phi_0 : [\Sigma, \mathcal{G}] \Rightarrow [\Phi_{\mathcal{G}}, \hat{\mathcal{G}}]$, but if Σ is not compact the extension need not be natural. This is shown by Example 11.3. On the other hand, as might be expected, this extension is contained in every natural system extension. Thus we have the following proposition.

10.4 PROPOSITION. *Every natural system extension* $\mu : [\Sigma, \mathcal{G}] \Rightarrow [\Omega, \mathcal{B}]$ *contains the extension* $\phi_0 : [\Sigma, \mathcal{G}] \Rightarrow [\Phi_{\mathcal{G}}, \hat{\mathcal{G}}]$.

Proof. By definition of "contained in" (§2) we must produce an extension $\rho : [\Phi_{\mathcal{G}}, \hat{\mathcal{G}}] \Rightarrow [\Omega, \mathcal{B}]$ such that $\mu = \rho \circ \phi_0$. Let $\varphi \in \Phi_{\mathcal{G}}$ and K_φ be a compact subset of Σ that majorizes φ. Then $\mu(K_\varphi)$ is a compact subset of Ω that majorizes the homomorphism $\varphi \circ \mu$ of $\mathcal{B}$ onto $\mathbb{C}$. Hence there exists $\omega_\varphi \in \Omega$ such that

$$\hat{b}(\varphi \circ \mu) = \widehat{b \circ \mu}(\varphi) = b(\omega_\varphi), \ b \in \mathcal{B}.$$

The map $\rho : \varphi \mapsto \omega_\varphi$ is clearly a homeomorphism of $\Phi_{\mathcal{G}}$ onto Ω since $[\Omega, \mathcal{B}]$ is a system. Moreover for each $b \in \mathcal{B}$

$$\begin{aligned} b((\rho \circ \phi_0)(\sigma)) &= (b \circ \rho)(\phi_0(\sigma)) = b(\rho(\phi_0(\sigma))) \\ &= \widehat{b \circ \mu}(\phi_0(\sigma)) = (b \circ \mu)(\sigma) = b(\mu(\sigma)) \end{aligned}$$

so $\mu = \rho \circ \phi_0$ and $\mathcal{B} \circ \rho = \widehat{\mathcal{B} \circ \mu} = \hat{\mathcal{G}}$. Therefore $\rho : [\Phi_{\mathcal{G}}, \hat{\mathcal{G}}] \Rightarrow [\Omega, \mathcal{B}]$. ♦

10.5 PROPOSITION. *Let* $[\Sigma, \mathcal{G}]$ *be a natural system. Then for an arbitrary set* $X \subseteq \Sigma$ *the two extensions,*

$$\iota : [X, \mathfrak{a}] \Rightarrow [X^{\#}, \mathfrak{a}], \qquad \phi_0 : [X, \mathfrak{a}] \Rightarrow [\Phi_{\mathfrak{a}|X}, \widehat{\mathfrak{a}|X}]$$

where $\iota : X \hookrightarrow X^{\#}$, *are equivalent.*

Proof. Let $\delta \in X^{\#}$. Then there exists a compact set $K \subseteq X$ such that $\delta \in \hat{K}$. Therefore the point evaluation at δ defines an element $\varphi_\delta \in \Phi_{\mathfrak{a}|X}$ such that

$$\widehat{a|X}(\varphi_\delta) = a(\delta), \ a \in \mathfrak{a}.$$

Similarly if $\varphi \in \Phi_{\mathfrak{a}|X}$ and K is a compact subset of X that majorizes φ then

$$a \mapsto a|X \mapsto \widehat{a|X}(\varphi), \ a \in \mathfrak{a}$$

defines a continuous homomorphism of $\mathfrak{a}$ onto C. Hence there exists $\delta_\varphi \in \Sigma$ such that $\widehat{a|X}(\varphi) = a(\delta_\varphi)$, $a \in \mathfrak{a}$. Since $|a(\delta_\varphi)| \le |a|_K$ for all $a \in \mathfrak{a}$ it follows that $\delta_\varphi \in \hat{K}$, so $\delta_\varphi \in X^{\#}$. Moreover

$$\widehat{a|X}(\varphi_{\delta_\varphi}) = a(\delta_\varphi) = \widehat{a|X}(\varphi), \ a \in \mathfrak{a}$$

so $\varphi_{\delta_\varphi} = \varphi$. This means that $\rho : \delta \mapsto \varphi_\delta$ defines a one-to-one map of $X^{\#}$ onto $\Phi_{\mathfrak{a}|X}$. Since $[\Sigma, \mathfrak{a}]$ and $[\Phi_{\mathfrak{a}|X}, \widehat{\mathfrak{a}|X}]$ are systems, it follows that $\rho : X^{\#} \to \Phi_{\mathfrak{a}|X}$ is a surjective homeomorphism. Furthermore

$$(\widehat{a|X}\circ\rho)(\delta) = \widehat{a|X}(\varphi_\delta) = a(\delta) = (a|X^{\#})(\delta)$$

for arbitrary $\delta \in X^{\#}$ and $a \in \mathfrak{a}$, so $\widehat{\mathfrak{a}|X}\circ\rho = \mathfrak{a}|X^{\#}$ and hence $\rho : [X^{\#}, \mathfrak{a}] \Rightarrow [\Phi_{\mathfrak{a}|X}, \widehat{\mathfrak{a}|X}]$ is an isomorphism. Finally since $\phi_0(\sigma) = \varphi_\sigma$ for $\sigma \in X$ and $\rho(\sigma) = \varphi_\sigma$ for $\sigma \in X^{\#}$ it follows that $\phi_0 = \rho\circ\iota$. Therefore ρ defines an equivalence of the two extensions. ♦

We have already observed that if Σ is compact then $\phi_0 : [\Sigma, \mathfrak{a}] \Rightarrow [\Phi_{\mathfrak{a}}, \hat{\mathfrak{a}}]$ is a minimal natural system extension of $[\Sigma, \mathfrak{a}]$. More generally if Σ is σ-compact and $\mathfrak{a}$ is closed in $C(\Sigma)$ with respect to the compact-open topology then $[\Phi_{\mathfrak{a}}, \hat{\mathfrak{a}}]$ is also natural, so again $\phi_0 : [\Sigma, \mathfrak{a}] \Rightarrow [\Phi_{\mathfrak{a}}, \hat{\mathfrak{a}}]$ is a minimal natural system extension and $\Phi_{\mathfrak{a}} = \Phi_{[\Sigma, \mathfrak{a}]}$. In this case the compact-open topology on $\mathfrak{a}$ is complete metric, so by a closed graph argument the map $a \mapsto \hat{a}$ from $\mathfrak{a}$ to $\hat{\mathfrak{a}}$ is continuous and the naturality of $[\Phi_{\mathfrak{a}}, \hat{\mathfrak{a}}]$ follows. This fact was pointed out to us by F. Bonsall. (cf. also Michael [M5, Proposition 8.1].) On the other hand it is not true that $[\Phi_{\mathfrak{a}}, \hat{\mathfrak{a}}]$ is always natural. This is shown in Example 11.3 below.

§11. EXAMPLES

In Theorem 7.3 we required that $[\Sigma, \mathfrak{G}]$ be a system. The following example shows that it is not sufficient in general to require only that $\mathfrak{G}$ separate points.

11.1 EXAMPLE. Let D be the closed unit disc and T the unit circle in $\mathbb{C}$. Let $\mathfrak{C}$ be the algebra of *all* continuous functions on D such that $\mathfrak{C}|T = \mathcal{A}_T$, the boundary value algebra (see the end of the discussion of Example 5.1). It is not difficult to show that $\Phi_{\mathfrak{C}}$, the spectrum of $\mathfrak{C}$, is equal to the sphere S obtained by joining the disc D to a copy D' of itself along the unit circle T, and that the image $\hat{\mathfrak{C}}$ of $\mathfrak{C}$ in $C(S)$ consists of all continuous functions on S whose restrictions to D' belong to the disc algebra $\mathcal{A}(D')$ on D'. Next let $\hat{\mathfrak{C}}^0$ denote the subalgebra of $\hat{\mathfrak{C}}$ consisting of those functions that take the same value at the origins 0 and $0'$ of the two discs D and D'. The spectrum of $\hat{\mathfrak{C}}^0$ is then equal to S with the two points 0 and $0'$ identified. Finally denote by S^0 the sphere S with the point $0'$ deleted. Observe that S^0 is homeomorphic with the spectrum of $\hat{\mathfrak{C}}^0$ with respect to the $\hat{\mathfrak{C}}^0$-topology on S^0 but not with respect to the given "euclidean" topology on S^0. In fact S^0 is not compact in the latter topology but is compact in the $\hat{\mathfrak{C}}^0$-topology. Since *every* homomorphism of $\hat{\mathfrak{C}}^0$ onto $\mathbb{C}$ is a point evaluation in S^0 it follows that $[S^0, \hat{\mathfrak{C}}^0]$ is strictly natural independently of the topology in S^0. Moreover $\hat{\mathfrak{C}}^0$ separates the points of S^0. On the other hand T is a compact subset of S^0 with respect to either topology and its $\hat{\mathfrak{C}}^0$-convex hull in S^0 is equal to $D' \setminus \{0'\}$ plus the point 0. Thus $\hat{T}$, though compact when S^0 is given the $\hat{\mathfrak{C}}^0$-topology, is not compact in the euclidean topology.

This example may be used to illustrate the fact that the restriction in Theorem 10.3 to system extensions cannot be dropped in general. Consider the deleted sphere S^0 of the example and for convenience denote by S^0_e and S^0_c the set S^0 under the euclidean topology and the $\hat{\mathfrak{C}}^0$-topology respectively. Then $[S^0_c, \hat{\mathfrak{C}}^0]$ is a strictly natural system. The pair $[S^0_e, \hat{\mathfrak{C}}^0]$ is obviously also strictly natural but is not a system since the two topologies in S^0 are not equivalent. Now consider the unit disc $D \subset S^0$ and observe that the identity embedding $\iota : D \hookrightarrow S^0$ is a homeomorphism with respect to both topologies. Therefore both

$$\iota : [D, \hat{C}^0] \Rightarrow [S_c^0, \hat{C}^0] \quad \text{and} \quad \iota : [D, \hat{C}^0] \Rightarrow [S_e^0, \hat{C}^0]$$

are natural extensions of $[D, \hat{C}^0]$. Moreover since the $\hat{C}^0$-convex hull of D in S^0 is equal to S^0 both extensions are minimal. However they are not equivalent since S_c^0 and S_e^0 are not homeomorphic.

The next two examples show that the closure operation "#" (§8) need not be proper; i.e. $(X^\#)^\#$ may be larger than $X^\#$. They were suggested by an example constructed for the same purpose by S. Sidney.

11.2 EXAMPLE. This example is constructed within the system $[\mathbb{C}^2, \mathcal{P}]$ and "∧" denotes the $\mathcal{P}$-convex hull. Consider first the following sets in $\mathbb{C}^2$: The two unit discs

$$D_0 = \{(\zeta_1, 0) : |\zeta_1| \le 1\} \quad \text{and} \quad D_1 = \{(1, \zeta_2) : |\zeta_2| \le 1\}$$

the solid horn-shaped region

$$H = \{(\zeta_1, \zeta_2) : \zeta_1 = e^{2\pi it} \ (0 \le t \le 1), \ |\zeta_2| \le t\}$$

the set H minus the deleted disc $D_1' = \{(1, \zeta_2) : 0 < |\zeta_2| \le 1\}$

$$H_0 = \{(\zeta_1, \zeta_2) : \zeta_1 = e^{2\pi it} \ (0 \le t < 1), \ |\zeta_2| \le t\}$$

and the surface of the set H_0

$$\Gamma = \{(\zeta_1, \zeta_2) : \zeta_1 = e^{2\pi it} \ (0 \le t < 1), \ |\zeta_2| = t\}.$$

Note that D_1 is the large end of the horn H and the point $(1, 0)$ is its vertex. Also H intersects the disc D_0 in the unit circle $T_0 = \{(\zeta_1, 0) : |\zeta_1| = 1\}$. The set Γ is equal to the curved surface of H minus the circle $T_1 = \{(1, \zeta_2) : |\zeta_2| = 1\}$. It is not very difficult to prove that $\Gamma^\# = H_0$. On the other hand H_0 contains the compact set T_0 which is not contained in Γ. Therefore $D_0 = \hat{T}_0 \subseteq H_0^\#$, so $\Gamma^\# \subsetneq (\Gamma^\#)^\#$. (*Question*: Is the $\mathcal{P}$-convex hull of Γ in $\mathbb{C}^2$ equal to $H_0 \cup D_0$? It can be shown that $\hat{\Gamma} \cap (H \cup D_0) = H_0 \cup D_0$;

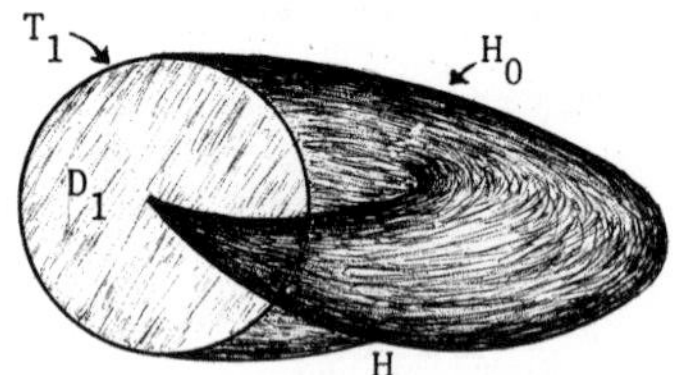

i.e. $\hat{\Gamma}$ cannot contain any points of H outside of $H_0 \cup D_0$. Also is the set $H \cup D_0$ $\wp$-convex?)

We may consider in place of $[\mathbb{C}^2, \wp]$ a uniform algebra on the compact set H. Let $\mathcal{C}$ denote the algebra of all functions $f(\zeta_1, \zeta_2)$ defined and continuous on H with the following properties: (1) As a function of ζ_2, f is holomorphic on each of the discs

$$D_t = \{(e^{2\pi it}, \zeta_2) : |\zeta_1| < t\}, \ 0 < t \leq 1.$$

(2) f reduces to an element of the boundary value algebra $\mathcal{A}_{T_0}$ (see Example 5.1) on the circle T_0. Then $\mathcal{C}$ is obviously a uniform algebra on H and its spectrum $\Phi_{\mathcal{C}}$ can be shown to be equal to $H \cup D_0$. Hence $[H \cup D_0, \hat{\mathcal{C}}]$ is a natural system. Taking $\hat{\mathcal{C}}$-convex hulls of sets within $H \cup D_0$ we obtain as before that $\Gamma^{\#} = H_0$ while $H_0^{\#} = H_0 \cup D_0$. Moreover one can prove here that $(\Gamma^{\#})^{\#} = \hat{\Gamma} = H_0 \cup D_0$.

11.3 EXAMPLE. This is an example of a natural system and a set to which the operation "#" may be applied an arbitrary finite number of times without reaching the #-closure (convex hull).

Consider the following sequence of closed discs in $\mathbb{C}$:

$$D_n = \{\zeta : |\zeta - \delta_n| \leq \frac{1}{2^n}\}, \ n = 1,2,...$$

where for each n, $\delta_n = 2 - (3/2^n)$. Note that successive discs D_n, D_{n+1} are tangent at the point

$$\tau_n = 2 - (1/2^{n-1}),$$

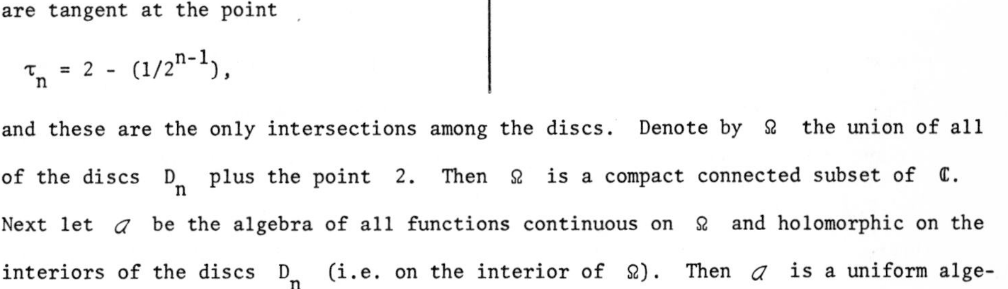

and these are the only intersections among the discs. Denote by Ω the union of all of the discs D_n plus the point 2. Then Ω is a compact connected subset of $\mathbb{C}$. Next let $\mathcal{A}$ be the algebra of all functions continuous on Ω and holomorphic on the interiors of the discs D_n (i.e. on the interior of Ω). Then $\mathcal{A}$ is a uniform algebra on Ω and $[\Omega, \mathcal{A}]$ is natural.

Consider next the sequence of pairs of points

$$\alpha_n = \delta_n + \frac{1}{2^{n+1}} i, \ \beta_n = \delta_{n+1} + \frac{1}{2^{n+1}} i$$

for $n = 1, 2, \ldots$ and denote by $\mathcal{A}_0$ the set of all $f \in \mathcal{A}$ such that $f(\alpha_n) = f(\beta_n)$ for each $n \geq 1$. Then $\mathcal{A}_0$ is a closed subalgebra of $\mathcal{A}$ that "identifies" α_n with β_n for each n. It can be shown that the spectrum $\Phi_{\mathcal{A}_0}$ is equal to the continuous image Ω_0 of the set Ω obtained by identifying α_n with β_n for each n, so the system $[\Omega_0, \mathcal{A}_0]$ is natural.

Now let T_n denote the circle that bounds the disc D_n, and let T_n' be the circle T_n minus the point β_{n-1}, $n \geq 2$. Denote by Γ the union of the circle T_1 with all of the deleted circles T_n' for $n \geq 2$. Then Γ is a connected subset of Ω that maps homeomorphically into Ω_0 since it does not contain any of the points β_n. We accordingly regard Γ as a subset of Ω_0 and consider the closure operation "#" for the system $[\Omega_0, \mathcal{A}_0]$. Since Γ contains T_1 it is immediate that $\Gamma^\#$ will contain the image of D_1 in Ω_0. On the other hand Γ does not contain any other complete circles, so $\Gamma^\#$ does not contain the images of any of the discs D_n for $n > 1$. However $\Gamma^\#$ contains the image of α_1 and hence the image of β_1, so $\Gamma^\#$ contains the circle T_2 but only deleted circles T_n' for $n > 2$. Therefore

$$\Gamma \subsetneq \Gamma^\# \subsetneq (\Gamma^\#)^\#.$$

Repeating these arguments we see that

$$\Gamma \subsetneq \Gamma^\# \subsetneq (\Gamma^\#)^\# \subsetneq \cdots \subsetneq \Gamma^{\#n} \subsetneq \Gamma^{(\#)} = \Omega_0.$$

where $\Gamma^{\#n}$ denotes the result of applying "#" n-times to the set Γ.

We have seen that an arbitrary system $[\Sigma, \mathfrak{A}]$ admits a minimal natural system extension $\emptyset : [\Sigma, \mathfrak{A}] \Rightarrow [\Phi_{[\Sigma, \mathfrak{A}]}, \hat{\mathfrak{A}}]$ which is uniquely determined up to extension equivalence. Moreover by Proposition 10.4 this minimal extension contains the usual extension $\emptyset_0 : [\Sigma, \mathfrak{A}] \Rightarrow [\Phi_{\mathfrak{A}}, \hat{\mathfrak{A}}]$ in which $\Phi_{\mathfrak{A}}$ is the space of all continuous homomorphisms of $\mathfrak{A}$ onto $\mathbb{C}$. The above example illustrates the fact that $[\Phi_{\mathfrak{A}}, \hat{\mathfrak{A}}]$ is not always natural. Consider the system $[\Gamma, \mathcal{A}_0]$. Then by Proposition 10.5, $[\Phi_{\mathcal{A}_0}|\Gamma, \hat{\mathcal{A}}_0]$ is isomorphic with $[\Gamma^\#, \mathcal{A}_0]$. However $\Gamma^\#$ is not $\mathcal{A}_0$-convex in Ω_0, so by Theorem 7.1 the system $[\Gamma^\#, \mathcal{A}_0]$, and hence $[\Phi_{\mathcal{A}_0}|\Gamma, \hat{\mathcal{A}}_0]$, is not natural. Furthermore the system $[\Gamma^{\#n}, \mathcal{A}_0]$ also fails to be natural for every n. Of course $[\hat{\Gamma}, \mathcal{A}_0]$ is natural and is isomorphic with $[\Phi_{[\Gamma, \mathcal{A}_0]}, \hat{\mathcal{A}}_0]$.

CHAPTER III
THE ŠILOV BOUNDARY AND LOCAL MAXIMUM PRINCIPLE

§12. INDEPENDENT POINTS

Let $[\Sigma, \mathfrak{G}]$ be an arbitrary pair and consider a homomorphism $\varphi : a \to \hat{a}(\varphi)$ of $\mathfrak{G}$ onto $\mathbb{C}$. By Proposition 3.1, φ will be continuous iff there exists a compact set $K \subset\subset \Sigma$ that dominates φ; *i.e.*

$$|\hat{a}(\varphi)| \le |a|_K, \; a \in \mathfrak{G} .$$

In general the dominating compact set K will not be uniquely determined. For example any larger compact set will also serve. Denote by $\mathcal{K}_\varphi$ the collection of all compact subsets of Σ that dominate φ. A set $K_0 \in \mathcal{K}_\varphi$ is called a *support* for φ if it is minimal; *i.e.* no compact set properly contained in K_0 dominates φ. It is easy to prove by a routine application of Zorn's lemma that each element of $\mathcal{K}_\varphi$ contains a support for φ. Note that φ may be supported by many distinct compact sets. For example in $[\mathbb{C}, \mathcal{P}]$ a point evaluation is supported by every circle that contains the point in its interior.

12.1 PROPOSITION. *A support for* φ *is either a perfect set or reduces to a single point* σ_φ. *In the latter case* φ *is the point evaluation at* σ_φ.

Proof. Observe first that if φ is supported by the point σ_φ then $a(\sigma_\varphi) = 0$ implies $\hat{a}(\varphi) = 0$, so

$$0 = \widehat{a-a(\sigma_\varphi)}(\varphi) = \hat{a}(\varphi) - a(\sigma_\varphi)$$

and hence $\hat{a}(\varphi) = a(\sigma_\varphi)$ for all $a \in \mathfrak{G}$. Therefore let K be a support for φ that contains more than one point and suppose that δ were an isolated point of K. Then the set $K' = K \setminus \{\delta\}$ is a non-empty proper compact subset of K so cannot support φ. Hence there exists $u \in \mathfrak{G}$ such that $|u|_{K'} < |\hat{u}(\varphi)| = 1$. Similarly there exists $v \in \mathfrak{G}$ such that $0 = |v(\delta)| < |\hat{v}(\varphi)| = 1$. Now choose a positive integer k such that $|u^k|_{K'} < (|v|_{K'})^{-1}$ and set $w = u^k v$. Then $|\hat{w}(\varphi)| = 1$ and $w(\delta) = 0$, so

$$1 = |w(\varphi)| \leq |w|_K = |w|_{K'} \leq |u^k|_{K'} |v|_{K'} < 1$$

a contradiction. Therefore the support K must be perfect. ◊

We specialize now to the point evaluations and denote as usual the homomorphism associated with a point $\sigma \in \Sigma$ by $\varphi_\sigma : \mathfrak{G} \to \mathbb{C}$. Note that the map $\sigma \mapsto \varphi_\sigma$ will be one-to-one iff $\mathfrak{G}$ separates points. When $\varphi = \varphi_\sigma$ the collection of dominating sets $\mathcal{K}_\varphi$ will be denoted by $\mathcal{K}_\sigma$. We will also say that a set in $\mathcal{K}_\sigma$ *dominates the point* σ and that K *supports* σ if it supports φ_σ. The collection $\mathcal{K}_\sigma$ obviously consists of all $K \subset\subset \Sigma$ such that $\sigma \in \hat{K}$. Each φ_σ is clearly supported by the singleton $\{\sigma\}$; *i.e.* points are, so-to-speak, "self-supporting". If $a(\sigma) = a(\sigma')$ for all $a \in \mathfrak{G}$ then σ is also supported by $\{\sigma'\}$. On the other hand if $\mathfrak{G}$ separates points then $\{\sigma\}$ is the only singleton that supports σ. These remarks suggest the following definition.

12.2 DEFINITION. *A point $\delta \in \Sigma$ is called an independent point for $[\Sigma, \mathfrak{G}]$ if every support for δ reduces to a singleton. The set of independent points will be denoted by $\partial_0[\Sigma, \mathfrak{G}]$. If G is a subset of Σ then δ is said to be independent relative to G if every support for δ which is contained in G reduces to a singleton.*

Observe that if $\mathfrak{G}$ separates points and $\delta \notin G$ then δ will be independent relative to G iff it is not dominated by any compact set contained in G.

Bochner and Martin [B6, p. 84] introduced a notion of *frontier property* for a point δ on the boundary of a domain G in $\mathbb{C}^n$. It asserts that for each compact $K \subset\subset G$ and each neighborhood U of δ there exists a point $\xi \in U \cap G$ and a function h holomorphic in G such that $|h(\xi)| > |h|_K$. Also F. Quigley [Q1, p. 85] in considering a pair $[\Sigma, \mathfrak{G}]$ for which Σ is locally compact and $\mathfrak{G}$ separates points introduced an extension of the notion of frontier property to points in the closure $\bar{G}$ of a relatively compact set G in Σ. He defines a point $\delta \in \bar{G}$ to be a *frontier point of* G if for each compact $K \subset\subset G$ and each neighborhood U of δ there exists $\sigma \in U \cap G$ and $a \in \mathfrak{G}$ such that $|a(\sigma)| > |a|_K$. It is obvious that every point of $\bar{G}$ that is independent relative to G is also a frontier point of G.

However a frontier point of G need not be independent relative to G. We note that Quigley also investigated the relationship between the set of all frontier points of G and the Šilov boundary of $\bar{G}$ relative to $\mathcal{G}$.

12.3 PROPOSITION. *A point* $\delta \in \Sigma$ *will be independent for* $[\Sigma, \mathcal{G}]$ *iff for an arbitrary set* $X \subseteq \Sigma$, $\delta \in \hat{X}$ *implies* $\varphi_\delta = \varphi_{\delta'}$ *for some* $\delta' \in X$.

Proof. If K is a support for δ and $\varphi_\delta = \varphi_{\delta'}$ for some $\delta' \in K$ then $\{\delta'\}$ supports δ so K must reduce to $\{\delta'\}$. Therefore the condition implies independence. Now assume that δ is independent and let X be a subset of Σ such that $\delta \in X$. Consider the #-resolution $\{X_\nu\}$ of $\hat{X}$ given by Proposition 8.1 and denote by ν_0 the first ordinal such that $\varphi_\delta = \varphi_{\delta'}$ for some $\delta' \in X_{\nu_0}$. If $\nu_0 = 0$, then $\delta' \in X$ and there is nothing to prove. If $\nu_0 > 0$ then there exists a compact set $K \subset\subset \bigcup_{\alpha<\nu_0} X_\alpha$ such that $\delta \in \hat{K}$. Therefore K contains a support for δ which must be a singleton, say $\{\delta'\}$, since δ is independent. Thus $\varphi_\delta = \varphi_\delta'$ and $\delta' \in X_\alpha$ for some $\alpha < \nu_0$. But this contradicts the definition of ν_0 which means that $\nu_0 = 0$. ♦

12.4 PROPOSITION. *Let* $\rho : [\Sigma, \mathcal{G}] \Rightarrow [\Sigma', \mathcal{G}']$ *be an arbitrary extension of* $[\Sigma, \mathcal{G}]$. *Then a point* $\delta \in \Sigma$ *such that* $\rho(\delta)$ *is independent for* $[\rho(\Sigma), \mathcal{G}]$ *must be independent for* $[\Sigma, \mathcal{G}]$.

Proof. Let K be an arbitrary support for δ. Then $\rho(K)$ is a compact subset of $\rho(\Sigma)$. Since $\mathcal{G}' \circ \rho = \mathcal{G}$, $\rho(K)$ obviously dominates $\rho(\delta)$ so must contain a support for $\rho(\delta)$. But $\rho(\delta)$ is independent for $[\rho(\Sigma), \mathcal{G}']$, so there must exist $\delta' \in K$ such that $\{\rho(\delta')\}$ supports $\rho(\delta)$. This clearly implies that $\{\delta'\}$ supports δ. ♦

§13. THE SILOV BOUNDARY OF A PAIR

It is possible that a pair $[\Sigma, \mathcal{G}]$ may not have any independent points whatsoever. This is true for example in the case of $[\mathbb{C}^n, \mathcal{P}]$. However the situation is quite different when Σ is compact, as we shall now see. We assume for the moment that $\mathfrak{a}$ separates points. This, together with compactness for Σ, implies that $[\Sigma, \mathfrak{a}]$ is a system. Recall that a closed set $B \subseteq \Sigma$ is called an $\mathfrak{a}$-*boundary* for Σ if $|a|_B = |a|_\Sigma$ for each $a \in \mathfrak{a}$, and the *Šilov boundary* $\partial_{\mathfrak{a}}\Sigma$ *of* Σ *relative to* $\mathfrak{a}$ is a uniquely determined minimal $\mathfrak{a}$-boundary [A3; B7, p. 113; R1, p. 133]. Observe that since $\widehat{\partial_{\mathfrak{a}}\Sigma} = \Sigma$ the Šilov boundary contains $\partial_0[\Sigma, \mathfrak{a}]$, the set of independent points for $[\Sigma, \mathfrak{a}]$. A point $\delta \in \Sigma$ is called a *strong boundary point* of Σ relative to $\mathfrak{a}$ if for each neighborhood U of δ there exists $u \in \mathfrak{a}$ such that $|u|_\Sigma = |u(\delta)| > |u|_{\Sigma\setminus U}$. It is obvious that every strong boundary point belongs to the Šilov boundary. Moreover if $\mathfrak{a}$ is closed in $C(\Sigma)$ then strong boundary points are dense in $\partial_{\mathfrak{a}}\Sigma$. In fact, the maximum set $\{\sigma \in \Sigma : |a(\sigma)| = |a|_\Sigma\}$ for each $a \in \mathfrak{a}$ contains a strong boundary point. Therefore each element of $\mathfrak{a}$ actually assumes its maximum absolute value on the set of strong boundary points [R1, p. 141]. It is readily verified that if $\bar{\mathfrak{a}}$ is the closure of $\mathfrak{a}$ in $C(\Sigma)$ then $\partial_{\bar{\mathfrak{a}}}\Sigma = \partial_{\mathfrak{a}}\Sigma$. Also, every strong boundary point δ of Σ relative to $\bar{\mathfrak{a}}$ is an independent point for $[\Sigma, \mathfrak{a}]$. In fact, let K be an arbitrary compact subset of Σ with $\delta \notin K$. Then $\Sigma\setminus K$ is a neighborhood of δ so there exists $f \in \bar{\mathfrak{a}}$ such that $|f(\delta)| > |f|_{\Sigma\setminus U} = |f|_K$. Therefore there exists $a \in \mathfrak{a}$ such that $|a(\delta)| > |a|_K$. Hence $\delta \notin \hat{K}$ so δ is an independent point for $[\Sigma, \mathfrak{a}]$. On the other hand, whether or not every independent point for $[\Sigma, \mathfrak{a}]$ must be a strong boundary point for $[\Sigma, \bar{\mathfrak{a}}]$ appears to be a rather difficult open question. (See remarks in §21.) In any case, when Σ is compact $\overline{\partial_0[\Sigma, \mathfrak{a}]} = \partial_{\mathfrak{a}}\Sigma$. Now an application of Proposition 12.4 to the separating extension of $[\Sigma, \mathfrak{a}]$ (§2) shows that if Σ is compact then there always exists a relative abundance of independent points for $[\Sigma, \mathfrak{a}]$ whether or not $\mathfrak{a}$ separates points. These remarks suggest the following general definition.

13.1 DEFINITION. *The closure of the set* $\partial_0[\Sigma, \mathfrak{a}]$ *of independent points for an arbitrary pair* $[\Sigma, \mathfrak{a}]$ *is called the Šilov boundary of* $[\Sigma, \mathfrak{a}]$ *and denoted by* $\partial[\Sigma, \mathfrak{a}]$.

We note in passing that definitions of "boundaries" for non-compact spaces that generalize the Šilov boundary have also been given by various other authors [B9, B11, B12, G6, K2, Q2, R5].

The next two propositions contain generalizations of familiar results for the compact case.

13.2 PROPOSITION. *Let* $\bar{\mathfrak{a}}$ *denote the closure of* $\mathfrak{a}$ *in* $C(\Sigma)$ *with respect to the compact-open topology. Then* $\partial_0[\Sigma, \bar{\mathfrak{a}}] = \partial_0[\Sigma, \mathfrak{a}]$ *and in particular* $\partial[\Sigma, \bar{\mathfrak{a}}] = \partial[\Sigma, \mathfrak{a}]$.

Proof. It is obvious from the definition that $\partial_0[\Sigma, \mathfrak{a}] \subseteq \partial_0[\Sigma, \bar{\mathfrak{a}}]$. Therefore let δ be an arbitrary point of $\partial_0[\Sigma, \bar{\mathfrak{a}}]$ and let K be an arbitrary support for δ relative to $\mathfrak{a}$. Then $|a(\delta)| \le |a|_K$ for all $a \in \mathfrak{a}$. Since elements of $\bar{\mathfrak{a}}$ are uniform limits of elements of $\mathfrak{a}$ on compact sets and the set $K \cup \{\delta\}$ is compact it follows that $|f(\delta)| \le |f|_K$ for all $f \in \bar{\mathfrak{a}}$. Therefore K contains a support for δ relative to $\bar{\mathfrak{a}}$. Since δ is independent for $[\Sigma, \bar{\mathfrak{a}}]$ it follows that there exists $\delta' \in K$ such that $|f(\delta)| \le |f(\delta')|$ for all $f \in \bar{\mathfrak{a}}$. In particular $|a(\delta)| \le |a(\delta')|$ for all $a \in \mathfrak{a}$. In other words δ is supported by $\{\delta'\} \subseteq K$ and therefore $K = \{\delta'\}$. Hence δ is an independent point for $[\Sigma, \mathfrak{a}]$, so $\partial_0[\Sigma, \bar{\mathfrak{a}}] \subseteq \partial_0[\Sigma, \mathfrak{a}]$. ◊

13.3 PROPOSITION. *Assume that* $\mathfrak{a}$ *separates points and let* X *be an arbitrary subset of* Σ. *Then* $\partial_0[\hat{X}, \mathfrak{a}] \subseteq X$. *In particular if* X *is closed then* $\partial[\hat{X}, \mathfrak{a}] \subseteq X$.

Proof. Let $\delta \in \hat{X}\setminus X$. Then by the minimality of the $\mathfrak{a}$-convex hull the set $\hat{X}\setminus\{\delta\}$ is not $\mathfrak{a}$-convex. Therefore there exists $K \subset\subset \hat{X}\setminus\{\delta\}$ such that $\delta \in \hat{K}$. Since $\mathfrak{a}$ separates points and $\delta \notin K$ it follows that δ cannot be independent for $[\hat{X}, \mathfrak{a}]$. In other words $\partial_0[X, \mathfrak{a}] \subseteq X$. ◊

13.4 PROPOSITION. *Let* $\rho : [\Sigma, \mathfrak{a}] \Rightarrow [\Sigma', \mathfrak{a}']$ *be any extension of* $[\Sigma, \mathfrak{a}]$ *such that* $\mathfrak{a}'$ *separates the point of* $\widehat{\rho(\Sigma)}$. *Then*

$$\partial_0[\widehat{\rho(\Sigma)}, \mathfrak{a}'] \subseteq \rho(\partial_0[\Sigma, \mathfrak{a}]).$$

Proof. Let δ' be an arbitrary independent point for $[\widehat{\rho(\Sigma)}, \mathfrak{a}']$. Then by Proposition 13.3 $\delta' \in \rho(\Sigma)$, so there exists $\delta \in \Sigma$ such that $\rho(\delta) = \delta'$. Let K be an arbitrary compact set in Σ that supports the point δ; i.e. $\delta \in \hat{K}$. Then by Theorem 9.1 $\rho(\hat{K}) \subseteq \widehat{\rho(K)}$. Therefore $\delta' \in \widehat{\rho(K)}$. Since $\mathfrak{a}'$ separates the points of $\widehat{\rho(\Sigma)}$ and δ' is independent it follows that $\delta' \in \rho(K)$. Hence there exists $\delta_0 \in K$

such that $\rho(\delta_0) = \delta'$. Also if $a \in \mathfrak{a}$ then there exists $a' \in \mathfrak{a}'$ such that $a = a' \circ \rho$, so

$$a(\delta) = (a' \circ \rho)(\delta) = a'(\delta') = (a' \circ \rho)(\delta_0) = a(\delta_0).$$

Therefore $\{\delta_0\}$ supports δ. But since K is a support for δ it follows that $K = \{\delta_0\}$; i.e. every support for δ must reduce to a singleton, proving that $\delta \in \partial_0[\Sigma, \mathfrak{a}]$ and hence that $\delta' \in \rho(\partial_0[\Sigma, \mathfrak{a}])$.◊

13.5 DEFINITION. *Let* P *be a closed subset of* Σ, U *an open neighborhood of* P, *and* u *a continuous function defined on* U *such that* $u(\sigma) = 1$ *for* $\sigma \in P$ *while* $|u(\sigma)| < 1$ *for* $\sigma \in U \backslash P$. *Then* P *is called a* ***peak set in*** U *with* ***peaking function*** u. *If* u *is the restriction to* U *of an element of* $\mathfrak{a}$ *then* P *is called a* ***local peak set for*** $[\Sigma, \mathfrak{a}]$. *If also* $U = \Sigma$ *then* P *is called a* ***global peak set for*** $[\Sigma, \mathfrak{a}]$. *When* P *is a singleton the term "peak point" is used in place of "peak set".*

If $a \in \mathfrak{a}$ and U is open set in Σ that contains a point σ_0 with

$$|a(\sigma_0)| = |a|_U > 0$$

then the set $P = \{\sigma \in U : a(\sigma) = a(\sigma_0)\}$ is a local peak set for $[\Sigma, \mathfrak{a}]$ provided it is closed. The required neighborhood of P is the open set U and the peaking function may be taken equal to $2^{-1}[a(\sigma_0)^{-1}a + 1]$. The next result is an extension to the general situation of a property of strong boundary points in the compact case [R1, pp. 139, 141].

13.6 PROPOSITION. *Let* S *be a subset of* Σ *which is an arbitrary intersection of (global) peak sets. Then* $\partial_0[S, \mathfrak{a}] = S \cap \partial_0[\Sigma, \mathfrak{a}]$, *so* $\partial[S, \mathfrak{a}] \subseteq S \cap \partial[\Sigma, \mathfrak{a}]$.

Proof. If δ is independent for $[\Sigma, \mathfrak{a}]$ then it is *a fortiori* independent for $[S, \mathfrak{a}]$. Therefore assume that δ is independent for $[S, \mathfrak{a}]$ and let K be an arbitrary support for δ in Σ relative to $\mathfrak{a}$. Consider the compact (possibly empty) set $K' = K \cap S$. Since a support for δ relative to $\mathfrak{a}|S$ is automatically a support relative to $\mathfrak{a}$ it follows that either $\delta \notin \hat{K}'$ or K must be a singleton. If $\delta \notin \hat{K}'$ then there exists $u \in \mathfrak{a}$ such that $|u(\delta)| > 1 > |u|_{K'}$, where we take $|u|_{K'} = 0$ if $K' = \emptyset$. Now define $K'' = \{\sigma \in K : |u(\sigma)| \geq 1\}$. Then $K'' \cap S = \emptyset$. Since

S is an intersection of peak sets and K'' is compact there exists a peak set $P \subseteq \Sigma$ such that $K'' \cap P = \emptyset$. Let $v \in \mathfrak{A}$ be a peaking function for P. Then $|v|_{K''} < 1$. Hence there exists k such that $|v^k|_{K''} < (|u|_K + 1)^{-1}$. Define $w = uv^k$. Then $|w(\delta)| > 1$ while $|w(\sigma)| \leq |u|_K |v|_{K''}^k < 1$, $\sigma \in K''$, and $|w(\sigma)| = |u(\sigma)||v(\sigma)|^k < 1$, $\sigma \in K \setminus K''$. Therefore $|w(\delta)| > 1 > |w|_K$, contradicting the fact that K supports δ. ◆

The following example shows that one cannot in general replace the inclusion by equality in the assertion of Proposition 13.6 concerning the Šilov boundaries.

13.7 EXAMPLE. Let Σ equal the solid cylinder $D \times [0, 1]$, where D is the closed unit disc in $\mathbb{C}$, and take $\mathfrak{A}$ to be the algebra of all functions $f(\zeta, t)$ continuous on the cylinder and, for $t = 0$, holomorphic as a function of $\zeta \in D^0$, the open unit disc. It is not difficult to prove that $\mathfrak{A}$ is a uniform algebra on Σ and $[\Sigma, \mathfrak{A}]$ is natural. Moreover $\partial[\Sigma, \mathfrak{A}] = \Sigma$. Let $S = D \times \{0\}$. Then S is clearly a peak set in $\mathfrak{A}$ (peaking function $f(\zeta, t) = 1 - t$) and by the maximum modulus principle $\partial[S, \mathfrak{A}] = T$, the unit circle in $\mathbb{C}$. Therefore in this case

$$\partial[S, \mathfrak{A}] \subsetneq S \cap \partial[\Sigma, \mathfrak{A}].$$

13.8 PROPOSITION. *If* $\partial[\Sigma, \mathfrak{A}] = \emptyset$ *then an element of* $\mathfrak{A}$ *will have a limit at infinity iff it is constant.*

Proof. Let a be an element of $\mathfrak{A}$ which has a limit at infinity. Then there is no loss in assuming that $\lim_{\sigma \to \infty} a(\sigma) = 0$. Now suppose that $a(\sigma_0) \neq 0$ for some $\sigma_0 \in \Sigma$. Then there exists a compact set $K_0 \subset\subset \Sigma$ such that $|a(\sigma)| < |a(\sigma_0)|$, $\sigma \in \Sigma \setminus K_0$, and we may assume that $\sigma_0 \in K_0$. It follows that $|a(\sigma_0)| \leq |a|_{K_0} = |a|_\Sigma < \infty$. This implies that K_0 contains a compact peak set and will therefore contain an independent point for $[\Sigma, \mathfrak{A}]$ by Proposition 13.6. This contradicts the assumption that $\partial[\Sigma, \mathfrak{A}] = \emptyset$, so we conclude that $a = 0$. ◆

§14. A LOCAL MAXIMUM PRINCIPLE FOR NATURAL SYSTEMS

This section is devoted to the discussion of a local maximum principle for natural systems along with some of its consequence. The general principle rests on a local maximum modulus principle for the compact case due to Hugo Rossi [R11] who obtained the result as a consequence of a more general theorem concerning local peak sets. In the present setting the latter theorem may be stated as follows:

14.1 THE ROSSI LOCAL PEAK SET THEOREM. *Let* $[\Omega, \mathfrak{B}]$ *be a natural system in which* Ω *is compact and* $\mathfrak{B}$ *is a Banach algebra under some given norm. Then every local peak set for* $[\Omega, \mathfrak{B}]$ *is a global peak set.*

The proof of this theorem depends on non-trivial results from the theory of analytic functions of several complex variables and will not be reproduced here. (See A1, G7, H4, R11, S9, W1.) It should be noted that the naturality condition is essential for the proof of the theorem. As a consequence of the local peak set theorem we have the following *local maximum modulus principle.*

14.2 THEOREM. *Let* $[\Sigma, \mathfrak{a}]$ *be a natural system in which* Σ *is compact and let* U *be an open subset of* $\Sigma \backslash \partial_{\mathfrak{a}}\Sigma$. *Then* $|a|_{\bar{U}} = |a|_{bdU}$ *for every* $a \in \mathfrak{a}$, *where* bdU *denotes the topological boundary of* U *in* Σ.

Proof. Since Σ is compact the inequality $|a|_{\bar{U}} > |a|_{bdU}$ implies the existence of a point $\sigma_0 \in U$ such that $|a(\sigma_0)| = |a|_U$. Also the set

$$P = \{\sigma \in U : a(\sigma) = a(\sigma_0)\}$$

is closed and is accordingly a local peak set for $[\Sigma, \mathfrak{a}]$ with neighborhood U and peaking function $2^{-1}[a(\sigma_0)^{-1}a + 1]$. Since $\bar{\mathfrak{a}}$ is a Banach algebra under the supnorm and $[\Sigma, \bar{\mathfrak{a}}]$ is also natural the local peak set theorem applies. Therefore P must be a global peak set for $[\Sigma, \bar{\mathfrak{a}}]$. But this is impossible since $\partial_{\bar{\mathfrak{a}}}\Sigma = \partial_{\mathfrak{a}}\Sigma$ and $P \cap \partial_{\mathfrak{a}}\Sigma = \emptyset$. Therefore $|a|_{\bar{U}} = |a|_{bdU}$ for every $a \in \mathfrak{a}$. ◊

14.3 COROLLARY. *Let* U *be an arbitrary non-vacuous open subset of* $\Sigma \backslash \partial_{\mathfrak{a}}\Sigma$. *Then* $bdU \neq \emptyset$; *i.e.* U *cannot be open and closed in* Σ. *In particular* $\Sigma \backslash \partial_{\mathfrak{a}}\Sigma$ *contains no isolated points.*

We obtain next a generalization of the local peak point theorem to the non-compact case. Further generalizations are given in §20 for $\mathfrak{a}$-holomorphic functions. (Theorems 20.1, 20.5, 20.7.) First we need a definition.

14.4 DEFINITION. *Let* $[\Sigma, \mathfrak{G}]$ *be an arbitrary pair. Then a point* $\delta \in \Sigma$ *is called a* ***locally independent point for*** $[\Sigma, \mathfrak{G}]$ *if there exists a neighborhood* U *of* δ *such that* δ *is an independent point for* $[U, \mathfrak{G}]$. *If* $U = \Sigma$ *(i.e.* δ *is simply an independent point for* $[\Sigma, \mathfrak{G}]$*) then* δ *is called a* ***globally independent point for*** $[\Sigma, \mathfrak{G}]$.

14.5 THEOREM. *If* $[\Sigma, \mathfrak{G}]$ *is a natural system then every locally independent point for* $[\Sigma, \mathfrak{G}]$ *is globally independent.*

Proof. Suppose that δ were a locally but not globally independent point for $[\Sigma, \mathfrak{G}]$. Then there exists compact $K \subseteq \Sigma$ such that $\delta \in \hat{K}\backslash K$. Since δ is locally independent there exists an open neighborhood U of δ such that δ is independent for $[U, \mathfrak{G}]$. Since $[\Sigma, \mathfrak{G}]$ is a system there exists a neighborhood V of δ such that $\bar{V} \subset U$ and $V \cap \hat{K} = \emptyset$. In particular $\overline{V \cap \hat{K}} \subset U \cap \hat{K}$. By Theorem 7.3 the set $\hat{K}$ is compact. Hence if $\Gamma = bd_{\hat{K}}(V \cap \hat{K})$ then Γ is a compact subset of U which does not contain δ. Therefore by the local independence there exists $u \in \mathfrak{G}$ such that $|u(\delta)| > |u|_\Gamma$. Now consider the natural system $[\hat{K}, \mathfrak{G}]$. Then $\partial_{\mathfrak{G}}\hat{K} \subseteq K$ and $V \cap \hat{K}$ is an open subset of $\hat{K}\backslash\partial_{\mathfrak{G}}\hat{K}$. Applying Theorem 14.2 to $[\hat{K}, \mathfrak{G}]$ we obtain $|u|_{V \cap \hat{K}} = |u|_\Gamma$. But this contradicts the inequality $|u(\delta)| > |u|_\Gamma$ and proves that local independence implies global independence. ◊

If P is compact local peak set for $[\Sigma, \mathfrak{G}]$ then by Proposition 13.6 and Theorem 14.5 the set P must contain an independent point for $[\Sigma, \mathfrak{G}]$. We also have as a consequence the following result: For $a \in \mathfrak{G}$ and $\alpha > 0$ set

$$M(\alpha) = \{\sigma \in \Sigma : |a(\sigma)| \geq \alpha\}.$$

Then any nonempty compact subset M_0 of $M(\alpha)$ such that $M(\alpha)\backslash M_0$ is closed contains an independent point for $[\Sigma, \mathfrak{G}]$ so intersects $\partial[\Sigma, \mathfrak{G}]$. These are of course well-known results for the compact case. (cf. [M4].)

The next theorem may be regarded as a general version of the local maximum modulus principle. If G is an open set in Σ then we denote by the symbol "$\sigma \underset{G}{\to} \infty$" the usual notion of a "limit at infinity" in the space G for elements of $\mathfrak{G}$ restricted to G.

14.6 THEOREM. *Let* $[\Sigma, \mathfrak{a}]$ *be a natural system and* G *an open subset of* $\Sigma\setminus\partial[\Sigma, \mathfrak{a}]$. *Then for each* $a \in \mathfrak{a}$, $|a|_G = \limsup_{\sigma \xrightarrow[G]{} \infty} |a(\sigma)|$.

Proof. Set

$$\rho = \limsup_{\sigma \xrightarrow[G]{} \infty} |a(\sigma)|.$$

Then by definition

$$\rho = \inf_{K \subset\subset G} |a|_{G\setminus K}.$$

Hence for arbitrary $\varepsilon > 0$ there exists $K_\varepsilon \subset\subset G$ such that $\rho \le |a|_{G\setminus K_\varepsilon} \le \rho + \varepsilon$. Note that ρ may be infinite. Since always $\rho \le |a|_G$ the desired result follows trivially if $\rho = \infty$. In any case suppose $\rho < |a|_G$. Then if $\rho + \varepsilon < |a|_G$ there exists $K_\varepsilon \subset\subset G$ such that $|a|_{G\setminus K_\varepsilon} < |a|_G$. This implies that $|a|_{K_\varepsilon} = |a|_G$. In particular $\sigma_0 \in K_\varepsilon$ exists such that $|a(\sigma_0)| = |a|_G$, so the set $\{\sigma \in G : a(\sigma) = a(\sigma_0)\}$ is a *compact* peak set for $[G, \mathfrak{a}]$ (i.e. a compact local peak set for $[\Sigma, \mathfrak{a}]$) and accordingly must contain an independent point for $[\Sigma, \mathfrak{a}]$. Since this contradicts the assumption that $G \subseteq \Sigma\setminus\partial[\Sigma, \mathfrak{a}]$ the theorem follows. ◊

14.7 COROLLARY. *If* $\bar{G}$ *is compact then* $|a|_{\bar{G}} = |a|_{\mathrm{bd}G}$ *for every* $a \in \mathfrak{a}$. *If* $\Sigma\setminus\partial[\Sigma, \mathfrak{a}]$ *is locally compact then* $\partial[\Sigma, \mathfrak{a}]$ *is equal to the intersection of all those closed sets* $B \subseteq \Sigma$ *such that* $\hat{B} = \Sigma$.

§15. APPLICATIONS OF THE LOCAL MAXIMUM PRINCIPLE

The example $[\mathbb{C}^n, \mathcal{P}]$, for which $\partial[\mathbb{C}^n, \mathcal{P}]$ is vacuous, suggests that those pairs $[\Sigma, \mathfrak{a}]$ with $\partial[\Sigma, \mathfrak{a}] = \emptyset$ might be rather special. However the next theorem shows that we can in a sense always reduce to the special case.

15.1 THEOREM. *Let* $[\Sigma, \mathfrak{a}]$ *be an arbitrary system and set* $\Sigma_0 = \Sigma\setminus\partial_0[\Sigma, \mathfrak{a}]$. *Then* Σ_0 *is* $\mathfrak{a}$-*convex. If* $[\Sigma, \mathfrak{a}]$ *is natural then* $[\Sigma_0, \mathfrak{a}]$ *is natural and* $\partial[\Sigma_0, \mathfrak{a}] = \emptyset$.

Proof. It is immediate from the definition of an independent point that the set Σ_0 is $\mathfrak{a}$-convex. Therefore by Theorem 7.1, the system $[\Sigma_0, \mathfrak{a}]$ will be natural if $[\Sigma, \mathfrak{G}]$ is natural. Now assume that $[\Sigma, \mathfrak{G}]$ is natural and let δ be an arbitrary point of Σ_0. Since $\delta \notin \partial_0[\Sigma, \mathfrak{G}]$ it is not independent for $[\Sigma, \mathfrak{G}]$, so there exists a compact set $K \subset\subset \Sigma$ such that $\delta \in \hat{K}\setminus K$. Moreover, points of $\hat{K}\setminus K$ are obviously not independent points for $[\Sigma, \mathfrak{G}]$, so $\hat{K}\setminus K \subset \Sigma_0$. Since K is compact there exists an open set G such that $K \subset G$ and $\delta \notin \bar{G}$. Now consider the set $U = \hat{K}\setminus\bar{G}$. Then U is relatively open in $\hat{K}$ and since $\partial_{\mathfrak{G}}\hat{K} \subseteq K$ we have $\delta \in U \subset \hat{K}\setminus\partial_{\mathfrak{G}}\hat{K}$. Therefore by the local maximum principle it follows that $\delta \in \widehat{bd_{\hat{K}}U}$. But since $\bar{U} \subseteq \hat{K}\setminus G$ we have $bd_{\hat{K}}U \subseteq \hat{K}\setminus K$, so $bd_{\hat{K}}U \subset \Sigma_0$ and δ is not independent for $[\Sigma_0, \mathfrak{G}]$. Therefore $\partial_0[\Sigma_0, \mathfrak{G}] = \emptyset$. ◆

The next theorem contains a generalization of a result due to Wilken [W 3].

15.2 THEOREM. *Let* $[\Sigma, \mathfrak{G}]$ *be a natural system and* Ω *a closed subset of* Σ *such that* $\Sigma\setminus\Omega = \Sigma_0 \cup \Sigma_1$, *where* $\bar{\Sigma}_0 \cap \Sigma_1 = \Sigma_0 \cap \bar{\Sigma}_1 = \emptyset$. *If* $\Omega^{\#} \subseteq \Sigma_0 \cup \Omega$ *then* $\Sigma_0 \cup \Omega$ *is* $\mathfrak{G}$*-convex.*

Proof. Define $\Sigma_0' = \Sigma_0 \cup \Omega$ and observe that Σ_0' is a closed set containing Σ_0. Let K be an arbitrary compact subset of Σ_0' and set $U = \hat{K}\setminus\Sigma_0'$. Then U is a relatively open subset of $\hat{K}$ contained in Σ_1. Therefore $bd_{\hat{K}}U \subseteq \hat{K} \cap \bar{\Sigma}_1 \cap \Sigma_0' \subseteq \hat{K} \cap \Omega$. Since $\Omega^{\#} \subseteq \Sigma_0'$ we have

$$\widehat{bd_{\hat{K}}U} \subseteq \widehat{\hat{K} \cap \Omega} \subseteq \Sigma_0'.$$

Moreover since $U \subseteq \hat{K}\setminus K \subseteq \hat{K}\setminus\partial_{\mathfrak{G}}\hat{K}$ it follows by Theorem 14.2 that

$$U \subseteq \widehat{bd_{\hat{K}}U} \subseteq \Sigma_0'.$$

Therefore $\hat{K} \subseteq \Sigma_0'$, which proves that Σ_0' is $\mathfrak{G}$-convex. ◆

15.3 COROLLARY. *If* Ω *is* $\mathfrak{G}$*-convex then* $\Sigma_0 \cup \Omega$ *and* $\Sigma_1 \cup \Omega$ *are also* $\mathfrak{G}$*-convex.*

15.4 COROLLARY. *Assume that* Σ *admits a decomposition* $\Sigma = \Sigma_1 \cup \Sigma_2$, *where* $\bar{\Sigma}_1 \cup \Sigma_2 = \Sigma_1 \cup \bar{\Sigma}_2 = \emptyset$. *Then* Σ_1 *and* Σ_2 *are* $\mathfrak{G}$*-convex.*

15.5 COROLLARY. *Let* G *be an open subset of* Σ *and let* $K \subset\subset G$. *Then either* $\hat{K} \subset G$ *or* $\hat{K} \cap bdG \neq \emptyset$.

15.6 COROLLARY. *Let* G *be an open subset of* Σ *such that* bdG *is* $\mathfrak{G}$*-convex. Then* $\bar{G}$ *is* $\mathfrak{G}$*-convex.*

15.7 COROLLARY. *Let* X *be an arbitrary subset of* Σ. *If the closed* $\mathfrak{G}$-*convex set* Ω *separates* $\hat{X}$ *then it also separates* X.

Proof. Set $\Omega_0 = \Omega \cap \hat{X}$. Then Ω_0 is a closed $\mathfrak{G}|\hat{X}$-convex subset of $\hat{X}$ and $[\hat{X}, \mathfrak{G}]$ is natural. Assume that $\hat{X}\setminus\Omega_0 = Y_1 \cup Y_2$, where $\bar{Y}_1 \cap Y_2 = Y_1 \cap \bar{Y}_2 = \emptyset$. If Ω does not separate X then either $X \subseteq Y_1$ or $X \subseteq Y_2$. But $Y_1 \cup \Omega_0$ and $Y_2 \cup \Omega_0$ are $\mathfrak{G}$-convex by the theorem. Therefore either $Y_1 \cup \Omega_0$ or $Y_2 \cup \Omega_0$ contains $\hat{X}$ which implies that either Y_2 or Y_1 is empty. In other words Ω does not separate $\hat{X}$. ♦

15.8 COROLLARY. *An open set* $G \subset \Sigma$ *will be* $\mathfrak{G}$-*convex iff each point of* bd$G$ *is independent relative to* G.

Proof. Note that δ will be independent relative to G iff it is not dominated by any compact set $K \subset\subset G$. Hence each point of bd$G$ will be independent relative to G iff $\hat{K} \cap \text{bd}G = \emptyset$ for every $K \subset\subset G$, and the latter condition is equivalent to $\mathfrak{G}$-convexity of G since G cannot separate $\hat{K}$ for any $K \subset\subset G$. ♦

The next theorem gives another less immediate criterion for $\mathfrak{G}$-convexity of a subset of Σ in terms of relative independence.

15.9 THEOREM. *Let* $[\Sigma, \mathfrak{G}]$ *be a natural system and* G *an open subset of* Σ. *Then* G *will be* $\mathfrak{G}$-*convex iff, for each* $\delta \in \text{bd}G$ *and* $K \subset\subset G$, *there exists an open set* H *with* $K \subset H \subset \bar{H} \subset G$ *such that* δ *is independent relative to* $G\setminus\bar{H}$.

Proof. That $\mathfrak{G}$-convexity implies the condition in the theorem is immediate from Corollary 15.7. Therefore assume that the condition is satisfied. Let δ be an arbitrary point of bd$G$ and K an arbitrary compact subset of G. Let H be the open neighborhood of K prescribed by the assumed condition. The theorem will follow if we show that $\hat{K}$ cannot contain δ. Consider the (possibly vacuous) intersection $\hat{K} \cap \text{bd}H$. Since this set and the set K are compact there exists an open neighborhood V of the set $\hat{K} \cap \text{bd}H$ such that $\bar{V} \cap K = \emptyset$ and $\hat{K} \cap \text{bd}H \subseteq V \subseteq \bar{V} \subset G$. Now set $U = \hat{K}\setminus(\overline{H \cup V})$. Then U is a relatively open subset of $\hat{K}\setminus K$ and hence also of $\hat{K}\setminus\partial_{\mathfrak{G}}\hat{K}$. Let $L = \text{bd}_{\hat{K}}U$. Then L is a compact subset of $G\setminus\bar{H}$. Therefore since δ is independent relative to $G\setminus\bar{H}$ we have $\delta \notin \hat{L}$. Furthermore by the local maximum principle applied to $[\hat{K}, \mathfrak{G}]$ it follows that $\bar{U} \subseteq \hat{L}$. Since $\hat{K} \subseteq (\overline{H \cup V}) \cup U$ we conclude that $\delta \notin \hat{K}$. ◆

We close this section with some structure properties of supports of points of Σ, where $[\Sigma, \mathfrak{G}]$ is a natural system, that parallel results obtained by Ryff [R13] for supports of representing measures in the case of the disc algebra.

15.10 THEOREM. *Let* $[\Sigma, \mathfrak{G}]$ *be a natural system and* K *a compact support for a point* $\delta \in \Sigma$. *Then* K *satisfies the following properties:*

(i) $\hat{K}$ *is connected.*

(ii) $K = \mathrm{bd}_{\hat{K}}(\hat{K} \setminus K)$. *In particular the interior of* K *relative to* $\hat{K}$ *is empty.*

(iii) If $\hat{K}$ *is locally connected and* U_0 *denotes the largest connected subset of* $\hat{K} \setminus K$ *that contains* δ *then* $K = \mathrm{bd}_{\hat{K}} U_0$.

Proof. Suppose that $\hat{K} = K_1 \cup K_2$, where $\bar{K}_1 \cap K_2 = K_1 \cap \bar{K}_2 = \emptyset$. Then since $\hat{K}$ is closed both K_1 and K_2 must also be closed. Assume that $\delta \in K_1$. Since K_1 is open relative to $\hat{K}$ an independent point for $[K_1, \mathfrak{G}]$ must also be independent for $[\hat{K}, \mathfrak{G}]$. Therefore

$$\partial_{\mathfrak{G}} K_1 \subseteq K_1 \cap \partial_{\mathfrak{G}} \hat{K} \subseteq K_1 \cap K$$

so $\delta \in K_1 \subseteq \widehat{K_1 \cap K}$. Since K is a support for δ it follows that $K_1 \cap K = K$. Moreover, K_1 is $\mathfrak{G}$-convex by Corollary 15.4, so $K_1 = \hat{K}$. In other words $K_2 = \emptyset$ and $\hat{K}$ is connected, proving (i).

Set $\Gamma = \mathrm{bd}_{\hat{K}}(\hat{K} \setminus K)$. Then $\Gamma \subseteq K$ and by Theorem 14.2 $\delta \in \hat{\Gamma}$. Therefore since K is a support for δ we must have $\Gamma = K$, proving (ii).

Finally set $U = \hat{K} \setminus K$. Then U is open relative to $\hat{K}$ and $\delta \in U$. If $\hat{K}$ is locally connected then the largest connected subset of U that contains δ is also open in $\hat{K}$. Denote it by U_0 and set $\Gamma_0 = \mathrm{bd}_{\hat{K}} U_0$. Again by Theorem 14.2 we have $\delta \in \hat{\Gamma}_0$. We prove that $\Gamma_0 \subseteq K$. Suppose on the contrary that $\Gamma_0 \not\subseteq K$. Then $\Gamma_0 \cap U \neq \emptyset$ and by the local connectedness there exists a connected open set $U_1 \subseteq U$ with $U_1 \cap \Gamma_0 \neq \emptyset$. But $U_1 \cap \Gamma_0 \neq \emptyset$ implies $U_1 \cap U_0 \neq \emptyset$. Therefore $U_1 \cup U_0$ is a connected open subset of U that contains δ. By the maximality of U_0 it follows that $U_1 \cup U_0 = U_0$, so $U_1 \subseteq U_0$. On the other hand $U_0 \cap \Gamma_0 = \emptyset$, contradicting $U_1 \cap \Gamma_0 \neq \emptyset$. Therefore it must be true that $\Gamma_0 \subseteq K$ and hence that $\Gamma_0 = K$ is a support for δ, proving (iii). ◆

CHAPTER IV
HOLOMORPHIC FUNCTIONS

§16. PRESHEAVES OF CONTINUOUS FUNCTIONS

Every holomorphic function defined on a domain in $\mathbb{C}^n$ is locally representable by a power series and is therefore a local uniform limit of polynomials in n variables. In the case of an arbitrary pair $[\Sigma, \mathfrak{a}]$ the analogy with $[\mathbb{C}^n, \mathfrak{P}]$ suggests consideration of functions that are defined on subsets of Σ and are local uniform limits of elements from the algebra $\mathfrak{a}$. Such functions turn out to have many nice properties. On the other hand, as might be expected in the general situation, there are complications not present in the case of $[\mathbb{C}^n, \mathfrak{P}]$. For this reason it will be convenient to begin with a somewhat more general setup involving certain presheaves of continuous functions. At this point the presheaf terminology is used primarily to simplify the discussion.

Since we frequently must restrict to subsets of Σ that are not necessarily open it is desirable to admit functions defined on arbitrary subsets of Σ though continuous with respect to the given topology in Σ. Therefore let us denote by $\mathcal{C}$ the family of all continuous complex-valued functions defined on *arbitrary* subsets of the Hausdorff space Σ. Let $\mathfrak{F}$ be a subset of $\mathcal{C}$ and for arbitrary $E \subseteq \Sigma$ denote by $\mathfrak{F}_E$ the subset of $\mathfrak{F}$ consisting of those functions with domain of definition equal to E. Assume that $\mathfrak{F}$ satisfies the condition that $E \subseteq F \subseteq \Sigma$ imply $\mathfrak{F}_F | E \subseteq \mathfrak{F}_E$. In other words $\mathfrak{F}$ is a presheaf of continuous functions over the space Σ with the discrete topology. By considering the family $\mathfrak{F}^o$ consisting of those elements of $\mathfrak{F}$ that are defined on *open* subsets of Σ we obtain a presheaf of continuous functions over Σ with its given topology. By analogy with the notation for pairs we shall denote these presheaves by $\{\Sigma, \mathfrak{F}\}$ and $\{\Sigma, \mathfrak{F}^o\}$ respectively. The ambiguity with reference to the topology in Σ will cause no trouble. The various notational conventions analogous to those developed for pairs will be followed here when it is convenient to do so. For example if X is an arbitrary subset of Σ with the topology induced by Σ, then $\{X, \mathfrak{F}\}$ will denote the presheaf obtained by considering the subset of $\mathfrak{F}$ consisting of those functions defined on subsets of X. Because $\mathfrak{F}$ satisfies the presheaf condition for *arbitrary* subsets of Σ it follows that $\{X, \mathfrak{F}\}$ is always equal to $\{X, \mathfrak{F}|X\}$, where $\mathfrak{F}|X$ denotes the set of functions in X obtained by restricting elements of $\mathfrak{F}$ to the intersections of their domains of definition with the set X. On the other hand if X is not an open set then the restriction $\mathfrak{F}^o|X$ will in general be properly contained in the set $(\mathfrak{F}|X)^o$ consisting of elements of $\mathfrak{F}$ defined on relatively open subsets of X. We shall also denote by $\{\Sigma, \mathfrak{A}\}$ the presheaf obtained by taking $\mathfrak{F}$ to be the family of all functions obtained by restricting elements of $\mathfrak{A}$ to arbitrary subsets of Σ. We will have occasion later to impose algebra conditions on $\mathfrak{F}$. Thus if $\mathfrak{F}_E$ is closed under multiplication for each $E \subseteq \Sigma$ then the presheaf $\{\Sigma, \mathfrak{F}\}$ is said to be *closed under multiplication*. Similarly if each $\mathfrak{F}_E$ is an algebra of functions then $\{\Sigma, \mathfrak{F}\}$ is called a *presheaf of function algebras*.

In the category of presheaves of continuous functions a *morphism* of $\{\Sigma, \mathfrak{F}\}$ and $\{\Sigma', \mathfrak{F}'\}$ is defined by a *continuous* map $\rho : \Sigma \to \Sigma'$ such that $\mathfrak{F}' \circ \rho \subseteq \mathfrak{F}$. As in the case of pairs we write

$$\rho : \{\Sigma, \mathfrak{F}\} \to \{\Sigma', \mathfrak{F}'\}.$$

If $\mathfrak{F}' \circ \rho = \mathfrak{F}$ then the morphism is an *extension* and is denoted by

$$\rho : \{\Sigma, \mathfrak{F}\} \Rightarrow \{\Sigma', \mathfrak{F}'\}.$$

§17. LOCAL EXTENSIONS. $\mathfrak{F}$-HOLOMORPHIC FUNCTIONS

Now let $\mathcal{E}$ be any set of functions (continuous or not) defined on subsets of Σ. Then a function f defined on a set $F \subseteq \Sigma$ is said to *belong locally* to $\mathcal{E}$ if for each $\sigma \in F$ there exists a neighborhood U_σ in Σ such that $f|(U_\sigma \cap F) \in \mathcal{E}$. The set $\mathcal{E}$ is said to be *local* if every function that belongs locally to $\mathcal{E}$ is already an element of $\mathcal{E}$. In particular a presheaf $\{\Sigma, \mathfrak{F}\}$ is said to be *local* if $\mathfrak{F}$ is local. This is equivalent to requiring that if a set $E \subseteq \Sigma$ is an arbitrary union of a family $\{E_\lambda\}$ of relatively open sets and if f is a function defined on E such that $f|E_\lambda \in \mathfrak{F}_{E_\lambda}$ for each λ then $f \in \mathfrak{F}_E$. Thus $\{\Sigma, \mathfrak{F}^o\}$ will be local iff it is isomorphic with the presheaf of sections of its associated sheaf over Σ [G7, p. 121].

The function f (defined on $F \subseteq \Sigma$) is said to be *locally approximable* by elements of $\mathcal{E}$ if it is a local uniform limit of elements of $\mathcal{E}$. This means precisely that for each $\sigma \in F$ there exists a neighborhood U_σ such that f is a uniform limit on $U_\sigma \cap F$ of a sequence of elements of $\mathcal{E}$ (depending on U_σ) whose domains of definition contain $U_\sigma \cap F$. The set of all functions that are locally approximable by elements of $\mathcal{E}$ is called the *local extension* of $\mathcal{E}$ and denoted by $\mathcal{E}^{1c}$. It is obvious that "1c" is a closure operation in the set of all functions in Σ as defined in §8. In the present context $\mathcal{E}$ is said to be *locally closed* if it is 1c-closed (i.e., if $\mathcal{E} = \mathcal{E}^{1c}$). Also the 1c-closure of $\mathcal{E}$ (i.e., the smallest locally closed set of functions that contains $\mathcal{E}$) is called the *local closure* of $\mathcal{E}$ and denoted by $\mathcal{E}^{(1c)}$. The restriction of the closure operation "1c" to the totality of functions defined on *open* (rather than arbitrary) subsets of Σ will be denoted by "loc". It is obvious from the definition that the local extension $\mathcal{E}^{1c}$ and $\mathcal{E}^{loc}$ are always local. It will be convenient to denote by $\mathcal{E}^a$ the set of all functions obtained by restricting elements of $\mathcal{E}$ to *arbitrary* subsets of Σ and by $\mathcal{E}^o$ the subset of $\mathcal{E}$ consisting of those functions defined on *open* subsets of Σ. The following lemma enables us to pass readily from 1c-closure to loc-closure.

17.1 LEMMA. *Let* $\mathcal{E}$ *be an arbitrary set of functions defined on subsets of* Σ. *If* $\mathcal{E}^a = \mathcal{E}$ *then* $(\mathcal{E}^o)^{loc} = (\mathcal{E}^{1c})^o$ *and* $(\mathcal{E}^o)^{(loc)} = (\mathcal{E}^{(1c)})^o$.

Proof. Since $\mathcal{E}^a = \mathcal{E}$ the first equality is immediate. Also, it is obvious that $(\mathcal{E}^o)^{(loc)} \subseteq (\mathcal{E}^{(1c)})^o$. The opposite inclusion is proved by induction on the 1c-resolution $\{\mathcal{E}_\nu\}$ of $\mathcal{E}^{(1c)}$ (with $\mathcal{E}_0 = \mathcal{E}$) given by Proposition 8.1. It is trivial that $\mathcal{E}_0^o \subseteq (\mathcal{E}^o)^{(loc)}$, so assume that $\mathcal{E}_\alpha^o \subseteq (\mathcal{E}^o)^{(loc)}$ for all $\alpha < \nu$. Then

$$(\bigcup_{\alpha<\nu} \mathcal{E}_\alpha)^o = \bigcup_{\alpha<\nu} \mathcal{E}_\alpha^o \subseteq (\mathcal{E}^o)^{(loc)}.$$

Moreover since

$$\mathcal{E}_\alpha = (\bigcup_{\beta<\alpha} \mathcal{E}_\beta)^{1c}$$

it follows that $\mathcal{E}_\alpha^a = \mathcal{E}_\alpha$ for all α. Hence

$$(\bigcup_{\alpha<\nu} \mathcal{E}_\alpha)^a = \bigcup_{\alpha<\nu} \mathcal{E}_\alpha$$

so we have

$$\mathcal{E}_\nu^o = ((\bigcup_{\alpha<\nu} \mathcal{E}_\alpha)^{1c})^o = ((\bigcup_{\alpha<\nu} \mathcal{E}_\alpha)^o)^{loc}$$
$$\subseteq ((\mathcal{E}^o)^{(loc)})^{loc} = (\mathcal{E}^o)^{(loc)}.$$

Therefore $\mathcal{E}_\nu^o \subset (\mathcal{E}^o)^{(loc)}$ for all ν. ♦

17.2 PROPOSITION. *Let* $\{\Sigma, \mathfrak{F}\}$ *be presheaf of continuous functions over* Σ. *Then* $\{\Sigma, \mathfrak{F}^{(1c)}\}$ *and* $\{\Sigma, \mathfrak{F}^{(loc)}\}$ *are local presheaves of continuous functions over* Σ.

Proof. That $\mathfrak{F}^{(1c)}$ and $\mathfrak{F}^{(loc)}$ are local is immediate from the fact that these sets are locally closed. Also, from the fact that local approximation is uniform it follows that $C^{1c} = C$ and $(C^o)^{loc} = C^o$. Therefore since $\mathfrak{F} \subseteq C$ and $\mathfrak{F}^o \subseteq C^o$ we have $\mathfrak{F}^{(1c)} \subseteq C$ and $\mathfrak{F}^{(loc)} \subseteq C^o$, so all of the functions are continuous. ♦

17.3 DEFINITION. *Let* $\{\Sigma, \mathfrak{F}\}$ *be an arbitrary presheaf of continuous functions over* Σ *and let* $\{\mathfrak{F}_\nu\}$ *be the* 1c*-resolution of* $\mathfrak{F}^{(1c)}$. *Then a function* f *is said to be*

(i) $\mathfrak{F}$*-holomorphic, if* $f \in \mathfrak{F}^{(1c)}$, *and* $\mathfrak{F}$*-holomorphic of order* ν, *if* $f \in \mathfrak{F}_\nu$,

(ii) almost $\mathfrak{F}$*-holomorphic if it is continuous and* $\mathfrak{F}$*-holomorphic on the subset of its domain of definition where it is different from zero.*

The set $\mathfrak{F}^{(1c)}$ of all $\mathfrak{F}$-holomorphic functions will also be denoted by the symbol ${}_{\mathfrak{F}}\mathcal{O}$ which is more suggestive of conventional notation. If G is an open subset of Σ then the subset of $C(\bar{G})$ consisting of those functions that are $\mathfrak{F}$-holomorphic on G will be denoted by ${}_{\mathfrak{F}}\mathcal{O}^*_G$. A function is said to be $\mathfrak{F}$-*holomorphic at a point* σ if it is $\mathfrak{F}$-holomorphic on some neighborhood of σ in its domain, and is said to be *locally* $\mathfrak{F}$-*holomorphic* if it is $\mathfrak{F}$-holomorphic at each point of its domain. A function which is locally $\mathfrak{F}$-holomorphic belongs locally to $\mathfrak{F}^{(1c)}$ and so must actually belong to $\mathfrak{F}^{(1c)}$. It follows that a function will be $\mathfrak{F}$-holomorphic iff it is locally $\mathfrak{F}$-holomorphic.

It is obvious that closure under the elementary algebra operations is preserved by local extension, so by the usual induction argument, algebra properties of $\mathfrak{F}$ will carry over to the $\mathfrak{F}$-holomorphic functions. In particular if $\{\Sigma, \mathfrak{F}\}$ is a presheaf of function algebras then the same is true for $\{\Sigma, \mathfrak{F}^{(1c)}\}$. The next theorem shows that in the function algebra case ordinary holomorphic functions of several variables operate on the $\mathfrak{F}$-holomorphic functions.

17.4 THEOREM. *Assume that* $\{\Sigma, \mathfrak{F}\}$ *is a presheaf of function algebras and let* $h_1,\ldots,h_n$ *be* $\mathfrak{F}$*-holomorphic functions with a common domain of definition* X. *Set* $R = \{(h_1(\sigma),\ldots,h_n(\sigma)) : \sigma \in X\}$ *and let* F *be a holomorphic function of* n *complex variables defined on an open set* G *in* $\mathbb{C}^n$ *that contains* R. *Then the function* h, *where* $h(\sigma) = F(h_1(\sigma),\ldots,h_n(\sigma))$, $\sigma \in X$, *is* $\mathfrak{F}$*-holomorphic on* X.

Proof. Let $\delta \in X$ and choose an open polydisc Δ with center

$$(h_1(\delta),\ldots,h_n(\delta))$$

such that $\bar{\Delta} \subseteq G$. Since the functions $h_1,\ldots,h_n$ are continuous we can choose a neighborhood U of δ such that $(h_1(\sigma),\ldots,h_n(\sigma)) \in \Delta$ for all $\sigma \in U \cap X$. Also, the series expansion of F about the point $(h_1(\delta),\ldots,h_n(\delta))$ converges uniformly to F on $\bar{\Delta}$. In particular there exists a sequence $\{P_k\}$ of polynomials such that $P_k \to F$ uniformly on $\bar{\Delta}$. Therefore

$$\lim_{k\to\infty} P_k(h_1(\sigma),\ldots,h_n(\sigma)) = h(\sigma)$$

uniformly for $\sigma \in U$. Since $\{\Sigma, \mathfrak{F}^{(1c)}\}$ is a presheaf of function algebras

$$P_k(h_1,\ldots,h_n) \in \mathfrak{F}^{(1c)}$$

for all k. Therefore it follows that $h \in (\mathfrak{F}^{(1c)})^{1c} = \mathfrak{F}^{(1c)}$. ♦

17.5 COROLLARY. *If a function* h *is* $\mathfrak{F}$*-holomorphic on a set* $X \subseteq \Sigma$ *and* $h(\sigma) \neq 0$ *for each* $\sigma \in X$ *then* $1/h$ *is also* $\mathfrak{F}$*-holomorphic on* X.

We are primarily interested in holomorphic functions associated with a given pair $[\Sigma, \mathfrak{G}]$ *via* the presheaf $\{\Sigma, \mathfrak{G}\}$. The latter, it will be recalled, consists of the functions obtained by restriction of each element of $\mathfrak{G}$ to arbitrary subsets of Σ. It will accordingly be convenient to introduce special terminology and notations for this case. Note first that $\mathfrak{G}^{1c}$ already contains the presheaf determined by $\mathfrak{G}$ so $\mathfrak{G}^{(1c)}$ is equal to the local closure of the presheaf. Elements of $\mathfrak{G}^{(1c)}$ will be called $\mathfrak{G}$-*holomorphic* functions and the set of all $\mathfrak{G}$-holomorphic functions will be denoted by ${}_{\mathfrak{G}}\mathfrak{O}$. If the pair in question is obvious from the context then we simplify this notation further by dropping the subscript $\mathfrak{G}$. Thus we have the presheaf $\{\Sigma, \mathfrak{O}\}$ of function algebras over Σ. If $E \subseteq \Sigma$ then $\mathfrak{O}_E$ consists of all $\mathfrak{G}$-holomorphic functions defined on E. Again if the pair is obvious we shall often refer to $\mathfrak{G}$-holomorphic functions simply as "holomorphic" functions.

§18. HOLOMORPHIC MAPS

Recall from §2 that if $[\Sigma', \mathfrak{G}']$ and $[\Sigma'', \mathfrak{G}'']$ are arbitrary pairs then a continuous map $\mu : \Sigma' \to \Sigma''$ such that $\mathfrak{G}''\circ\mu \subseteq \mathfrak{G}'$ defines a pair morphism

$$\mu : [\Sigma', \mathfrak{G}'] \to [\Sigma'', \mathfrak{G}''].$$

We introduce in the following definition a weaker condition on μ.

18.1 DEFINITION. *A continuous map* $\mu : \Sigma' \to \Sigma''$ *is called a* ***holomorphic map*** *if* $\mathfrak{G}''\circ\mu \subseteq \mathfrak{O}_{\Sigma'}$; *i.e. if it defines a morphism* $\mu : [\Sigma', \mathfrak{O}_{\Sigma'}] \to [\Sigma'', \mathfrak{G}'']$. *This is a* ***holomorphic extension*** *if* $\mathfrak{O}_{\Sigma''}\circ\mu = \mathfrak{O}_{\Sigma'}$.

It is obvious that if μ defines a pair morphism $\mu : [\Sigma', \mathfrak{G}'] \to [\Sigma'', \mathfrak{G}'']$ then it is automatically holomorphic. For notational convenience we shall denote by $\mathfrak{O}'$ and $\mathfrak{O}''$ the sets of all $\mathfrak{G}'$-holomorphic and $\mathfrak{G}''$-holomorphic functions respectively.

18.2 THEOREM. *If* $\mu : \Sigma' \to \Sigma''$ *is a holomorphic map then it defines presheaf morphisms* $\mu : \{\Sigma', \mathfrak{O}'\} \to \{\Sigma'', \mathfrak{O}''\}$ *and* $\mu : \{\Sigma', (\mathfrak{O}')^{\mathfrak{o}}\} \to \{\Sigma'', (\mathfrak{O}'')^{\mathfrak{o}}\}$.

Proof. Observe first that if h is an element of $\mathfrak{O}''$ defined on the open set $V \subseteq \Sigma''$ then $h\circ\mu$ is a function defined on the open set $\mu^{-1}(V) \subseteq \Sigma'$. Therefore it will be sufficient to prove that μ defines the first morphism. Denote by $\{\mathfrak{O}''_\nu\}$ the lc-resolution of $\mathfrak{O}'' = (\mathfrak{G}'')^{(1c)}$, where $\mathfrak{O}''_0 = \mathfrak{G}''$. Then by the definition of holomorphic map we have $\mathfrak{O}''_0\circ\mu \subseteq \mathfrak{O}'_{\Sigma'} \subseteq \mathfrak{O}'$. Therefore assume that $\mathfrak{O}''_\alpha\circ\mu \subseteq \mathfrak{O}'$ for each $\alpha < \nu$. Then obviously

$$\Big(\bigcup_{\alpha<\nu} \mathfrak{O}''_\alpha\Big)\circ\mu \subseteq \mathfrak{O}'$$

so the desired result will follow by induction if we show that $\mathcal{E} \subseteq \mathfrak{G}''$ and $\mathcal{E}\circ\mu \subseteq \mathfrak{G}'$ imply $\mathcal{E}^{lc}\circ\mu \subseteq \mathfrak{G}'$. Let f be an element of $\mathcal{E}^{lc}$ defined on a set $F \subseteq \Sigma''$. Then the function $f\circ\mu$ is defined on the set $E = \mu^{-1}(F) \subseteq \Sigma'$. For arbitrary $\delta' \in E$ set $\delta'' = \mu(\delta')$. Then $\delta'' \in F$, so there exists a neighborhood V of δ'' in Σ'' such that f is a uniform limit on $V \cap F$ of elements of $\mathcal{E}$. Let $V = \mu^{-1}(V)$. Then U is a neighborhood of δ' in Σ' and $f\circ\mu$ is a uniform limit on $U \cap E$ of elements of $\mathcal{E}\circ\mu$. Since δ' was an arbitrary point of E it follows that $f\circ\mu \in (\mathcal{E}\circ\mu)^{lc}$. Therefore $\mathcal{E}^{lc}\circ\mu \subseteq (\mathcal{E}\circ\mu)^{lc}$. Now since $\mathcal{E}\circ\mu \subseteq \mathfrak{G}'$ and $\mathfrak{G}'$ is locally closed it follows that $(\mathcal{E}\circ\mu)^{lc} \subseteq \mathfrak{G}'$ and hence $\mathcal{E}^{lc}\circ\mu \subseteq \mathfrak{G}'$. ♦

18.3 COROLLARY. *If* $\mu : \Sigma' \to \Sigma''$ *is holomorphic then it defines a pair morphism* $\mu : [\Sigma', \mathfrak{G}_{\Sigma'}] \to [\Sigma'', \mathfrak{G}_{\Sigma''}]$.

The next result concerns preservation of the ranges of functions in holomorphic extensions. It depends only on the fact that inverses of holomorphic functions are holomorphic.

18.4 PROPOSITION. *Let* $\rho : [\Sigma', \mathfrak{G}_{\Sigma'}] \Rightarrow [\Sigma'', \mathfrak{G}_{\Sigma''}]$ *be a faithful extension and let* $g = h\circ\rho$, *where* $g \in \mathfrak{G}_{\Sigma'}$ *and* $h \in \mathfrak{G}_{\Sigma''}$. *Then the function* g *and its extension* h *have the same range of values.*

Proof. It is obvious that $g(\Sigma') \subseteq h(\Sigma'')$. Therefore suppose that there exists $\zeta_0 = h(\omega_0) \in h(\Sigma'')\setminus g(\Sigma')$. Then $(g-\zeta_0)^{-1} \in \mathfrak{G}_{\Sigma'}$. Since ρ is an extension $\mathfrak{G}_{\Sigma''}\circ\rho = \mathfrak{G}_{\Sigma'}$, so there exists $k \in \mathfrak{G}_{\Sigma''}$ such that $k\circ\rho = (g-\zeta_0)^{-1}$. Hence

$$1 = (k\circ\rho)(g-\zeta_0) = (k\circ\rho)(h\circ\rho - \zeta_0) = (k(h-\zeta_0))\circ\rho.$$

Since ρ is faithful it follows that $1 = k(h-\zeta_0)$. But this is impossible since $h(\omega_0) = \zeta_0$. Therefore $g(\Sigma') = h(\Sigma'')$. ♦

§19. EXAMPLES AND REMARKS

In the remainder of this chapter we discuss briefly a number of examples that should help to clarify some of the more obvious questions concerning $\mathfrak{A}$-holomorphic functions that have no doubt already suggested themselves to the reader.

19.1 HOLOMORPHIC FUNCTIONS IN THE ORDINARY SENSE. THE STOLZENBERG EXAMPLE. In the case of the pair $[\mathbb{C}^n, \mathfrak{P}]$ a function defined on an open subset of $\mathbb{C}^n$ will be $\mathfrak{P}$-holomorphic iff it is holomorphic in the classical sense. In fact since the classical functions admit local power series representations it follows that the presheaf ${}_n\mathcal{O}$ of all such functions is contained in $({}_{\mathfrak{P}}\mathcal{O})^{\mathcal{O}}$. Also since uniform limits of holomorphic functions are holomorphic it follows that $(\mathfrak{P}^{loc})^{loc} = \mathfrak{P}^{loc} \subseteq {}_n\mathcal{O}$, so ${}_n\mathcal{O} = \mathfrak{P}^{loc} = ({}_{\mathfrak{P}}\mathcal{O})^{\mathcal{O}}$. In particular $({}_{\mathfrak{P}}\mathcal{O})^{\mathcal{O}}$ consists of $\mathfrak{P}$-holomorphic functions of order one.

A natural suggestion for a definition of holomorphic functions for an arbitrary pair $[\Sigma, \mathfrak{A}]$ is in terms of ordinary holomorphic functions in $\mathbb{C}^n$ *via* holomorphic maps of $\mathbb{C}^n$ into Σ. Actually for this purpose one may as well restrict to the case $n = 1$ and the system $[D, \mathfrak{P}]$, where D is the open unit disc in $\mathbb{C}$. Thus if h is a function defined on an open set $U \subseteq \Sigma$ one might call h *holomorphic in the ordinary sense* if for each holomorphic map (Definition 18.1) $\mu : [D, \mathcal{O}_D] \to [U, \mathfrak{A}]$ the function $h \circ \mu$ is an ordinary holomorphic function in D. By Theorem 18.2 every $\mathfrak{A}$-holomorphic function defined on an open subset of Σ is holomorphic in the ordinary sense. Although in many important examples the converse is also true it is not true in general. A counter example is provided by Stolzenberg's construction [S8] of a compact polynomially convex hull Ω in $\mathbb{C}^3$ which admits no analytic structure; that is, *every* holomorphic map of D into Ω is a constant. This means that *every* continuous function on Ω is holomorphic in the ordinary sense. On the other hand since Ω is $\mathfrak{P}$-convex $[\Omega, \mathfrak{P}]$ is a natural system and Ω is so constructed that $\partial[\Omega, \mathfrak{P}] \neq \Omega$. Since by Theorem 20.1 in the next section $\partial[\Omega, {}_{\mathfrak{P}}\mathcal{O}_\Omega] = \partial[\Omega, \mathfrak{P}]$ and always $\partial[\Omega, C(\Omega)] = \Omega$ it follows that not all continuous functions in Ω are $\mathfrak{P}$-holomorphic. This example shows that for our purposes the notion of ordinary holomorphic function is not restrictive enough to be of any use in the general setting.

There is another definition of holomorphic function which was introduced by R. Arens [A2] for functions defined on open subsets of a linear topological space E. He defines a function h to be holomorphic if there exists a continuous linear mapping $\rho : E \to \mathbb{C}^n$ (for some n) such that h is of the form $f \circ \rho$, where f is holomorphic in the usual sense on an open set $W \subseteq \mathbb{C}^n$, so h is defined on the open set $\rho^{-1}(W) \subseteq E$. This definition may be applied to a system $[\Sigma, \mathfrak{G}]$ *via* the canonical embedding $\tau : \Sigma \to \mathfrak{G}'$ of Σ in the space $\mathfrak{G}'$ of continuous linear functionals on $\mathfrak{G}$ (regarded as a linear topological space), where τ maps each point of Σ to the corresponding point evaluation functional on $\mathfrak{G}$. Note that $< \mathfrak{G}', \mathfrak{G} >$ is a "dual pair" (see §53) of vector spaces and, with the $\mathfrak{G}$-topology on $\mathfrak{G}'$, the embedding $\tau : \Sigma \to \mathfrak{G}'$ is a homeomorphism. Thus a function defined on a subset of Σ may be defined to be *Arens-holomorphic* if it is the restriction of a function holomorphic in $\mathfrak{G}'$ according to the Arens definition. This approach to holomorphic functions will be examined in more detail later. (See Theorem 55.1 in Chapter XI.)

Holomorphic functions may also be defined more traditionally in a linear space in terms of differentiation. This is the usual approach by workers in the field of "Infinite Dimensional Holomorphy" which has an extensive and growing literature. We consider briefly in Chapter XI some aspects of this approach that bear directly on our investigations.

19.2 HOLOMORPHIC FUNCTIONS IN $\mathbb{C}^\Lambda$. Consider the natural system $[\mathbb{C}^\Lambda, \mathfrak{P}]$ for an arbitrary index set Λ. Denote by π any finite subset of Λ and by $\tilde{\pi}$ the canonical projection of $\mathbb{C}^\Lambda$ onto $\mathbb{C}^\pi$; i.e.

$$\tilde{\pi} : \{\zeta_\lambda : \lambda \in \Lambda\} \mapsto \{\zeta_\lambda : \lambda \in \pi\}, \quad \check{\zeta} \in \mathbb{C}^\Lambda.$$

Note that $\tilde{\pi}$ is an open continuous map of $\mathbb{C}^\Lambda$ onto $\mathbb{C}^\pi$. A set $X \subseteq \mathbb{C}^\Lambda$ is said to be *determined by* π if $X = \tilde{\pi}^{-1}(\tilde{\pi}X)$. Similarly a function f defined on a set $F \subseteq \mathbb{C}^\Lambda$ is said to be *determined by* π *on* F if $\check{\xi}, \check{\eta} \in F$ and $\tilde{\pi}\check{\xi} = \tilde{\pi}\check{\eta}$ imply $f(\check{\xi}) = f(\check{\eta})$. In this case f does not "involve" explicitly any of the variables ζ_λ for $\lambda \notin \pi$, and the map

$$f_\pi : \tilde{\pi}\check{\zeta} \mapsto f(\check{\zeta}), \quad \check{\zeta} \in F$$

defines unambiguously a function f_π on the set $\tilde{\pi}F \subseteq \mathbb{C}^\pi$. We call f_π the *projection of* f *into* $\mathbb{C}^\pi$. A function f is said to be *determined by* π *at a point* $\check{\xi} \in F$ if there exists a neighborhood of $\check{\xi}$ in F on which f is determined by π. If f is determined by π at each point of its domain F then it is said to be *determined locally by* π. If g is any function defined on a set $G \subseteq \mathbb{C}^\pi$ then the function

$$_\pi g : \check{\zeta} \mapsto g(\tilde{\pi}\check{\zeta}), \quad \check{\zeta} \in \tilde{\pi}^{-1}(G)$$

is clearly determined by π and $({}_\pi g)_\pi = g$. If f is determined by π on $F \subseteq \mathbb{C}^\Lambda$ then ${}_\pi(f_\pi)$ is an extension of f to the set $\tilde{\pi}^{-1}(\tilde{\pi}F)$. Note that each $P \in \wp$, being a polynomial in a finite number of the variables ζ_λ, is obviously determined on $\mathbb{C}^\Lambda$ by the corresponding finite subset of Λ.

We sketch here a few basic properties of $\wp$-holomorphic functions in $\mathbb{C}^\Lambda$. Proofs will be omitted since they follow from more general results obtained later in the case of dual pairs (Chapter XI). We have given a more complete treatment of the case $[\mathbb{C}^\Lambda, \wp]$ in [R6]. (Cf. also [M1, M2, H3].)

Consider a function h defined on an open set $H \subseteq \mathbb{C}^\Lambda$. Then h will be $\wp$-holomorphic on H iff it is locally finitely determined in H and the associated local projections of h are holomorphic in the usual sense [H3, R6]. It follows from this fact that h is of order at most equal to 1. Also if H is connected then h will vanish identically on H if it vanishes on an open subset of H. Furthermore if H is connected then there exists a finite subset π_h of Λ, *depending only on* h, that determines h locally in H [H3]. Note however that it does not follow necessarily that h is determined by π_h (or any finite subset of Λ) on *all* of H. This is shown by the following example which is a slight modification of an example due to A. Hirschowitz [H3, p. 222].

Denote by K_0 the subset of the complex plane consisting of the semi-circle $\{e^{i\theta} : 0 \leq \theta \leq \pi\}$ plus the two closed intervals $[-1, 1/4]$ and $[1/2, 1]$. (See Figure 1 below.) For each $m \geq 1$ set

$$K_m = \rho^m(K_0), \quad \rho : \zeta \mapsto 2e^{2\pi i/3}\zeta.$$

Observe that for each $m \geq 0$ the sets K_m and K_{m+1} intersect in the single point $\xi_m = 2^{-1}\rho^{m+1} = \rho^{m+1}e^{2\pi i/3}$, the sets K_m and K_{m+2} intersect in the single point $\eta_m = -4^{-1}\rho^{m+2} = \rho^m e^{\pi i/3}$ while K_m and K_{m+k} are disjoint for $k > 2$. Now let

$$U_m = \{\zeta \in C : d(\zeta, K_m) < 4^{-1}\}.$$

Then $U_m \cap U_{m+1}$ is a connected neighborhood of the point ξ_m, $U_m \cap U_{m+2}$ is a connected neighborhood of η_m, and $U_m \cap U_{m+k} = \emptyset$ for $k > 2$.

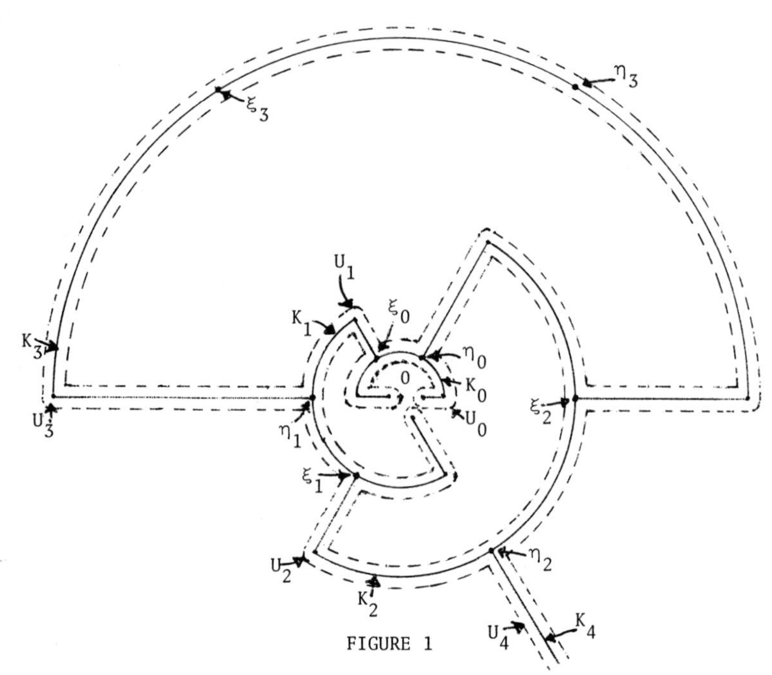

FIGURE 1

Next let D_0 denote the open disc with center at the point 0 and radius equal to 1/2. For $m \geq 1$ let D_m denote the open disc with center at the point $(-1)^m + mi$ and radius equal to $\sqrt{2}$. (Figure 2). Then $D_m \cap D_{m+1} \neq \emptyset$ for $m \geq 0$, and $D_m \cap D_{m+1} \cap D_{m+2} \neq \emptyset$ for $m \geq 1$, while $D_0 \cap D_2 = \emptyset$ and $D_m \cap D_{m+k} = \emptyset$ for $m \geq 0$, $k > 2$.

Now consider in the product space $\mathbb{C}^{\mathbb{N}}$, $\mathbb{N} = \{0,1,2,\ldots\}$, the open set $G = \bigcup_{m=0}^{\infty} G_m$, where

$$G_m = U_m \times D_m \times D_{m-1} \times \ldots \times D_1 \times D_0 \times \mathbb{C}^{\mathbb{N} \setminus \{0,1,\ldots,m+2\}}.$$

Since each of the sets U_m, D_m is connected and both $U_m \cap U_{m+1} \neq \emptyset$ and $D_m \cap D_{m+1} \neq \emptyset$, it follows that $G_m \cap G_{m+1} \neq \emptyset$ for all m, so the set G is obviously connected. Also, since $D_0 \cap D_k = \emptyset$ for all $k > 1$ we note that $G_m \cap G_{m+k} = \emptyset$ for all $k > 1$.

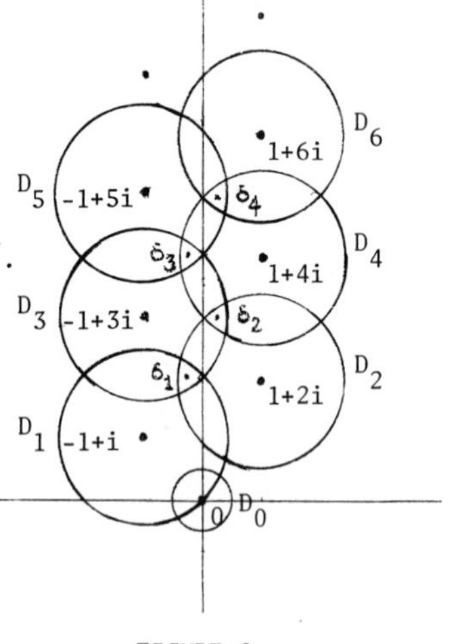

FIGURE 2

Finally let ℓ_0 be a fixed determination of the logarithm defined on the set U_0, and define by induction for each $m \geq 1$ a branch ℓ_m of the logarithm on U_m

such that $\ell_{m-1}(\zeta) = \ell_m(\zeta)$, for $\zeta \in U_{m-1} \cap U_m$. Observe that $\ell_{m-2}(\zeta) \neq \ell_m(\zeta)$, for $\zeta \in U_{m-2} \cap U_m$ and $m \geq 2$. We define a function h on the set $G \subset \mathbb{C}^{\mathbb{N}}$ by setting

$$h(\check{\zeta}) = \ell_m(\zeta_0), \quad \check{\zeta} = \{\zeta_n\} \in G_m.$$

Note that h is well-defined since $\ell_m = \ell_{m+1}$ on $U_m \cap U_{m+1}$ and $G_m \cap G_{m+k} = \emptyset$ for $k > 1$. Also, h is clearly $\wp$-holomorphic and is locally determined by the single index $0 \in \mathbb{N}$ at each point of G. On the other hand, h is not globally determined on G for any finite subset of $\mathbb{N}$. In fact, let n_0 be an arbitrary positive integer and choose $m_0 \geq n_0$. Also choose points $\gamma_0 \in D_0$, $\gamma_1 \in D_1$, $\gamma_2 \in D_2$ and $\delta_m \in D_m \cap D_{m+1} \cap D_{m+2}$ for $m = 1,2,\ldots,m_0 + 2$. Note that since $D_0 \cap D_2 = \emptyset$ the points γ_0 and γ_2 are distinct. Now define

$$\zeta_n' = \begin{cases} \eta_{m_0} & , n = 0 \\ \delta_{m_0 - n+1} & , 1 \leq n \leq m_0 \\ \gamma_2 & , n = m_0 + 1 \\ \gamma_1 & , n = m_0 + 2 \\ \gamma_0 & , n = m_0 + 3 \\ 0 & , n > m_0 + 3 \end{cases}, \qquad \zeta_n'' = \begin{cases} \eta_{m_0} & , n = 0 \\ \delta_{m_0 - n+1} & , 1 \leq n \leq m_0 \\ \gamma_0 & , n = m_0 + 1 \\ 0 & , n > m_0 + 1 \end{cases}.$$

Then $\check{\zeta}' = \{\zeta_n'\} \in G_{m_0+2}$ and $\check{\zeta}'' \in G_{m_0}$, so $\check{\zeta}' \neq \check{\zeta}''$. Also

$$h(\check{\zeta}') = \ell_{m_0+2}(\eta_{m_0}) \neq \ell_{m_0}(\eta_{m_0}) = h(\check{\zeta}'').$$

However, $\zeta_n' = \zeta_n''$ for $n = 0,1,\ldots,n_0$ since $n_0 \leq m_0$. Therefore h is not determined on G by $\{0,1,\ldots,n_0\}$.

The above example shows that a $\wp$-holomorphic function h defined on a connected open set $G \subseteq \mathbb{C}^\Lambda$ need not be globally finitely determined although it is locally finitely determined. However if the domain G is finitely determined then it can be shown that h is finitely determined on all of G.

19.3 HOLOMORPHIC FUNCTIONS OF HIGH ORDER. For the pairs $[\mathbb{C}^n, \wp]$ and $[\mathbb{C}^\Lambda, \wp]$ we have observed that the $\wp$-holomorphic functions are of order at most equal to 1. Although this is also true for other important examples, which will be considered later, it is not always the case. We outline briefly the general situation with respect to this question. It will be sufficient to consider uniform algebras. Therefore let $[\Omega, \mathfrak{B}]$ be a system with a compact space Ω and with $\mathfrak{B}$ a closed subalgebra of $C(\Omega)$.

Denote by $L(\mathfrak{B})$ the uniform closure on Ω of those functions defined on Ω that *belong locally* to $\mathfrak{B}$ and by $H(\mathfrak{B})$ the uniform closure of those functions defined on Ω that are *locally approximable* by elements of $\mathfrak{B}$. Then

$$\mathfrak{B} \subseteq L(\mathfrak{B}) \subseteq H(\mathfrak{B}) \subseteq C(\Omega)$$

and the inclusions are in general proper. If $[\Omega, \mathfrak{B}]$ is natural then it turns out that $[\Omega, L(\mathfrak{B})]$ and $[\Omega, H(\mathfrak{B})]$ are also natural [S7, R3, R4]. This also follows from more general results which will be proved later. (see Theorem 40.3). The proof, which depends on the local maximum principle, is non-trivial. Now set $L^0(\mathfrak{B}) = H^0(\mathfrak{B}) = \mathfrak{B}$ and define inductively for each ordinal ν the algebras

$$L^\nu(\mathfrak{B}) = L(\bigcup_{\alpha<\nu} L^\alpha(\mathfrak{B})), \quad H^\nu(\mathfrak{B}) = H(\bigcup_{\alpha<\nu} H^\alpha(\mathfrak{B})).$$

Again we have

$$\mathfrak{B} \subseteq L^\nu(\mathfrak{B}) \subseteq H^\nu(\mathfrak{B}) \subseteq C(\Omega).$$

Note that $H^\nu(\mathfrak{B})$ is contained in the algebra of all $\mathfrak{B}$-holomorphic functions defined on Ω. If $[\Omega, \mathfrak{B}]$ is natural then, as above, so are the systems $[\Omega, L^\nu(\mathfrak{B})]$ and $[\Omega, H^\nu(\mathfrak{B})]$. If $L(\mathfrak{B}) = \mathfrak{B}$ then $\mathfrak{B}$ is called a *local algebra* and if $H(\mathfrak{B}) = \mathfrak{B}$ then $\mathfrak{B}$ is said to be *holomorphically closed*. If $L^\nu(\mathfrak{B}) = L^{\nu+1}(\mathfrak{B})$, while $L^\alpha(\mathfrak{B}) \subsetneq L^{\alpha+1}(\mathfrak{B})$ for $\alpha < \nu$, then $\mathfrak{B}$ is said to be *nonlocal of order* ν. If $H^\nu(\mathfrak{B}) = H^{\nu+1}(\mathfrak{B})$, while $H^\alpha(\mathfrak{B}) \subsetneq H^{\alpha+1}(\mathfrak{B})$ for $\alpha < \nu$, then $\mathfrak{B}$ is said to be *nonholomorphically closed of order* ν. Clearly nonlocal of order ν implies nonholomorphically closed of order at least equal to ν. The first result here was due to Eva Kallin [K1] who constructed a uniform algebra that is nonlocal and hence nonholomorphically closed of order one. (Cf. also [B4, B5].) The Kallin example is obtained as the uniform closure of the polynomials on a certain compact polynomially convex subset of $\mathbb{C}^4$, so is automatically natural. Next, A. D. Varšavskii [V1] constructed a function algebra nonlocal of order 2. Using the Kallin example as a starting point S. Sidney [S5, S6] constructed for each $\nu \leq \omega$ a natural uniform algebra which is nonlocal of order ν. These examples are also antisymmetric and generated by five or less generators. He also constructed for each $\nu \leq \omega$ a separable antisymmetric natural uniform algebra which is nonholomorphically closed of order ν.

CHAPTER V
MAXIMUM PROPERTIES OF HOLOMORPHIC FUNCTIONS

§20. A LOCAL MAXIMUM PRINCIPLE FOR HOLOMORPHIC FUNCTIONS

We turn now to a study of some of the basic properties of $\mathfrak{G}$-holomorphic functions that derive from the local maximum principle. We shall accordingly assume until further notice that $[\Sigma, \mathfrak{G}]$ is a natural system. The next theorem is an extension of the local maximum principle (given in Theorem 14.2) to almost $\mathfrak{G}$-holomorphic functions (Definition 17.3 (ii)).

20.1 THEOREM. *Assume that the space* Σ *is compact. Let* U *be any open subset of* $\Sigma\backslash\partial[\Sigma, \mathfrak{G}]$ *and* h *a function continuous on* $\bar{U}$ *and almost* $\mathfrak{G}$*-holomorphic on* U. *Then* $|h|_{\bar{U}} = |h|_{bdU}$.

Proof. Observe that h is not required to be defined outside of $\bar{U}$. Also, since we could replace U by the open subset of U on which h is different from zero it will obviously be sufficient to make the proof for a function h that is $\mathfrak{G}$-holomorphic on all of U. Now denote by "$\mathfrak{J}_\nu$" the statement of the theorem with the additional restriction that h be $\mathfrak{G}$-holomorphic of order ν on U. Then "$\mathfrak{J}_0$" is the result in Theorem 14.2, so it is true. Therefore assume that "$\mathfrak{J}_\alpha$" is true for all $\alpha < \nu$ and suppose that "$\mathfrak{J}_\nu$" is false. Then there exists an open set $U \subseteq \Sigma\backslash\partial[\Sigma, \mathfrak{G}]$ and a function f continuous on $\bar{U}$ and $\mathfrak{G}$-holomorphic on U of order ν such that $|f|_{\bar{U}} > |f|_{bdU}$. Consider the multiplicatively closed set

$$\mathfrak{B} = \{af^n : a \in \mathfrak{G}, n = 0,1,\ldots\}$$

generated by $\mathfrak{G}|\bar{U}$ plus f. Then $\partial_{\mathfrak{B}}\bar{U}$ exists, and since $|f|_{\bar{U}} > |f|_{bdU}$ the set U contains a point δ of $\partial_{\mathfrak{B}}\bar{U}$.

Next choose an open neighborhood V of δ such that $\bar{V} \subset U$ and f is a uniform limit on $\bar{V}$ of $\mathfrak{G}$-holomorphic functions of order less than ν. Note that each element of $\mathfrak{B}$ may be so approximated on $\bar{V}$. Now since $\delta \in \partial_{\mathfrak{B}}\bar{U}$ there exists a function $g \in \mathfrak{B}$ such that $|g|_V > |g|_{bdV}$. It follows that there exists by approximation an $\mathfrak{a}$-holomorphic function h of order $\alpha < \nu$ on $\bar{V}$ such that also $|h|_V > |h|_{bdV}$. But this contradicts "$\mathfrak{J}_\alpha$" contrary to the induction hypothesis. Therefore "$\mathfrak{J}_\nu$" must be true for all ν and the theorem follows. ♦

In the next theorem Σ is not assumed to be compact. We denote by $'\mathfrak{G}$ the collection of all almost $\mathfrak{a}$-holomorphic functions defined on subsets of Σ. Then $\{\Sigma, '\mathfrak{G}\}$ is a presheaf of continuous functions but not in general of function algebras since the sets $'\mathfrak{G}_E$ may not be closed under addition although they are closed under multiplication.

20.2 THEOREM. *Let* K *be any compact subset of* Σ. *Then the* $'\mathfrak{G}_{\hat{K}}$*-convex hull of* K *in* $\hat{K}$ *is equal to* $\hat{K}$.

Proof. Suppose that the $'\mathfrak{G}_{\hat{K}}$-convex hull of K were not equal to $\hat{K}$. Then there exists $\delta \in \hat{K}\setminus K$ and $h \in '\mathfrak{G}_{\hat{K}}$ such that $|h(\delta)| > |h|_K$. Set $U = \hat{K}\setminus K$. Then U is a relatively open subset of $\hat{K}$ and $bd_{\hat{K}}U \subseteq K$. Furthermore h is continuous on $\bar{U}$ and almost $\mathfrak{a}|\hat{K}$-holomorphic on U. Hence by Theorem 20.1 we must have $|h|_{\bar{U}} = |h|_{bd_{\hat{K}}U}$ which contradicts the inequality $|h(\delta)| > |h|_K$. ♦

Observe that since the sets $'\mathfrak{G}_E$ are closed under multiplication the Šilov boundary $\partial_{'\mathfrak{G}_K} K$ exists for compact sets $K \subset \Sigma$. [A3].

20.3 COROLLARY. *Let* Ω *be a compact* $\mathfrak{a}$*-convex subset of* Σ. *Then* $\partial_{'\mathfrak{G}_\Omega} \Omega = \partial_{\mathfrak{a}} \Omega$.

If G is an open set that contains $\hat{K}$ then $\mathfrak{a}|G \subseteq '\mathfrak{G}_G$ and $'\mathfrak{G}_G|\hat{K} \subseteq '\mathfrak{G}_{\hat{K}}$. Therefore we have the following corollary.

20.4 COROLLARY. *Let* G *be an open subset of* Σ *and* K *a compact set in* G *with* $\hat{K} \subset G$. *Then the* $'\mathfrak{G}_G$*-convex hull of* K *in* G *is equal to* $\hat{K}$.

20.5 THEOREM. *Let* h *be an almost* $\mathfrak{a}$*-holomorphic function defined on an open set* $G \subseteq \Sigma$ *and let* K *be a compact subset of* $\Sigma\setminus G$ *with* $G \cap \hat{K} \neq \emptyset$. *Then there exists a point* $\delta \in bd_{\hat{K}}(G \cap \hat{K})$ *such that for every neighborhood* V *of* δ, $|h|_{G \cap \hat{K}} = |h|_{V \cap G \cap \hat{K}}$.

Proof. Let $\Gamma = \mathrm{bd}_{\hat{K}}(G \cap \hat{K})$ and note that $\Gamma \neq \emptyset$ by Corollary 14.3. Also set $\beta = |h|_{G \cap \hat{K}}$. Then $0 \leq \beta \leq \infty$ where $\beta = \infty$ is possible since h may be unbounded. The case $\beta = 0$ is trivial so we assume that $\beta > 0$ and let $\{\beta_n\}$ be an increasing positive sequence that converges to β. Let

$$F_n = \overline{\{\sigma \in G \cap \hat{K} : |h(\sigma)| \geq \beta_n\}}.$$

Then F_n is compact and $F_{n+1} \subseteq F_n$. Since $\beta_n < \beta$ also $F_n \neq \emptyset$ for all n. Therefore the intersection F of the sets F_n is not empty. It only remains to prove that $F \cap \Gamma \neq \emptyset$ since any point of this intersection obviously has the required property. Suppose on the contrary that $F \cap \Gamma = \emptyset$. Then $F \subset G \cap \hat{K}$, so $|h|_F = \beta < \infty$ and $F = \{\sigma \in G \cap \hat{K} : |h(\sigma)| = \beta\}$. Now choose a neighborhood U of F in the space $\hat{K}$ such that $\bar{U} \cap \Gamma = \emptyset$. Then h is continuous on $\bar{U}$ and $\mathfrak{G}|\hat{K}$-holomorphic on U. Moreover since $F \cap \mathrm{bd}_{\hat{K}}U = \emptyset$ it follows that $|h|_{\mathrm{bd}_{\hat{K}}U} < |h|_{\bar{U}}$. But this contradicts Theorem 20.1 applied to the natural system $[\hat{K}, \mathfrak{G}]$, so $F \cap \Gamma \neq \emptyset$. ◊

The next result is an extension of Theorem 14.5 to $\mathfrak{G}$-holomorphic functions. The first step is an extension of the notion of a locally independent point.

20.6 DEFINITION. *A point* δ *is called a* ***holomorphically independent point for*** $[\Sigma, \mathfrak{G}]$ *if there exists an open neighborhood* U *of* δ *such that* δ *is an independent point for* $[U, \mathfrak{O}_U]$.

20.7 THEOREM. *Every holomorphically independent point for* $[\Sigma, \mathfrak{G}]$ *is a globally independent point for* $[\Sigma, \mathfrak{G}]$.

Proof. Let U be an open set in Σ and δ an independent point for $[U, \mathfrak{O}_U]$. Choose an $\mathfrak{G}$-convex neighborhood V of δ contained in U. Then δ is also an independent point for $[V, \mathfrak{O}_V]$. Now let K be an arbitrary compact subset of V with $\delta \notin K$. Since $[V, \mathfrak{G}]$ is a natural system $\hat{K} \subset V$ and hence by Corollary 20.4 the $\mathfrak{O}_V$-convex hull of K in V is equal to $\hat{K}$. Therefore by the independence of δ for $[V, \mathfrak{O}_V]$ it follows that $\delta \notin \hat{K}$. In other words δ is also independent for $[V, \mathfrak{G}]$ and hence for $[\Sigma, \mathfrak{G}]$ by Theorem 14.5. ◊

20.8 COROLLARY. *If* G *is an arbitrary open set in* Σ *then*

$$\partial[G, \mathfrak{G}] = \partial[G, \mathfrak{O}_G] = G \cap \partial[\Sigma, \mathfrak{G}]$$

and if $G \subseteq \Sigma \setminus \partial[\Sigma, \mathfrak{G}]$ *then* $\partial[\bar{G}, \mathfrak{G}^*_G] \subseteq \mathrm{bd}\, G$.

The next result is an extension of Theorem 14.6 to $\mathfrak{G}$-holomorphic functions. It does not require that Σ be compact.

20.9 THEOREM. *Let* $[\Sigma, \mathfrak{G}]$ *be an arbitrary natural system and* h *an almost* $\mathfrak{G}$*-holomorphic function defined on an open set* $G \subseteq \Sigma\setminus\partial[\Sigma, \mathfrak{G}]$. *Then*

$$|h|_G = \limsup_{\substack{\sigma\to\infty \\ G}} |h(\sigma)|.$$

Proof. Set $\rho = \limsup_{\substack{\sigma\to\infty \\ G}} h(\sigma)$. Then by definition (see remark preceding Theorem 14.6.) $\rho \leq |h|_G$. Suppose that $\rho < |h|_G$ and choose $\varepsilon > 0$ such that $\rho + \varepsilon < |h|_G$. Then there exists a compact set $K \subset G$ such that $|h|_{G\setminus K} \leq \rho + \varepsilon$. Thus $|h|_{G\setminus K} < |h|_G$, so $|h|_K = |h|_G$. In particular there exists $\sigma_0 \in K$ such that $|h(\sigma_0)| = |h|_G$ and the set $P = \{\sigma \in G : h(\sigma) = h(\sigma_0)\}$ is a compact peak set with peaking function $g = 2^{-1}[h(\sigma_0)^{-1}h + 1]$. Note that $h(\sigma) = 0$ implies $g(\sigma) = 2^{-1}$. Hence there exists an open set U such that $P \subset U \subset G$ on which h is never zero. Therefore h, and hence also g, is $\mathfrak{G}$-holomorphic on U. It follows that P contains an independent point for $[U, \mathfrak{G}_U]$. But such a point is holomorphically independent for $[\Sigma, \mathfrak{G}]$ so must be a globally independent point for $[\Sigma, \mathfrak{G}]$ by Theorem 20.7. This contradicts the assumption that $G \subseteq \Sigma\setminus\partial[\Sigma, \mathfrak{G}]$ and completes the proof of the theorem. ◊

§21. HOLOMORPHIC PEAK SETS

The notions of local peak point and peak set for $[\Sigma, \mathfrak{G}]$ are extended to $\mathfrak{G}$-holomorphic functions in the following definition.

21.1 DEFINITION. *A closed set* $P \subseteq \Sigma$ *is called a* ***holomorphic peak set for*** $[\Sigma, \mathfrak{G}]$ *if there exists a function* f *which is continuous on some neighborhood* U *of* P *in* Σ, *is* $\mathfrak{G}$*-holomorphic on* $U\setminus P$, *and peaks on the set* P *relative to* U. *If* P *consists of a single point then that point is called a* ***holomorphic peak point.***

Our objective here is to obtain some properties of holomorphic peak points and sets and to consider briefly the problem of extending the Rossi theorems to $\mathfrak{G}$-holomorphic functions. For this we are led to consider functions on an open set U that peak at a point $\delta \in U$ and are $\mathfrak{G}$-holomorphic on $U\setminus\{\delta\}$ but not necessarily at δ. This is the reason for requiring f in the above definition to be $\mathfrak{G}$-holomorphic only on the set $U\setminus P$ rather than on all of U.

Just as in the case of peak sets for $[\Sigma, \mathfrak{G}]$ discussed in §14, if g is $\mathfrak{G}$-holomorphic on an open set $U \subseteq \Sigma$ and the maximum set $M = \{\sigma \in U : |g(\sigma)| = |g|_U\}$ is closed in Σ then for each $\sigma_0 \in M$ the set $P = \{\sigma \in U : g(\sigma) = g(\sigma_0)\}$ is a holomorphic peak set for $[\Sigma, \mathfrak{G}]$ with peaking function $2^{-1}[g(\sigma_0)^{-1}g + 1]$. We also have the following result. (Cf. [M4].)

21.2 PROPOSITION. *Let* G *be an open subset of* Σ *and* g *an* $\mathfrak{G}$*-holomorphic function on* G*. For* $\alpha \geq 0$ *set* $M(\alpha) = \{\sigma \in G : |g(\sigma)| \geq \alpha\}$*. If* M_0 *is a non-empty compact subset of* $M(\alpha)$ *such that* $M(\alpha)\setminus M_0$ *is relatively closed in* G *then* $M_0 \cap \partial_0[\Sigma, \mathfrak{G}] \neq \emptyset$.

Proof. Observe that under the conditions on M_0 there exists an open neighborhood U of M_0 contained in G and disjoint from $M(\alpha)\setminus M_0$, so $|g|_U = |g|_{M_0} < \infty$. Choose $\sigma_0 \in M_0$ such that $|g(\sigma_0)| = |g|_{M_0}$. Then the set $P = \{\sigma \in U : g(\sigma) = g(\sigma_0)\}$ is a peak set for $[U, \mathfrak{O}_U]$. Therefore by Proposition 13.6, P contains an independent point for $[U, \mathfrak{O}_U]$ and an application of Theorem 14.5 completes the proof. ◆

The next theorem is a step toward an extension of the Rossi Local Peak Point Theorem to $\mathfrak{G}$-holomorphic functions.

21.3 THEOREM. *Let* $[\Sigma, \mathfrak{G}]$ *be an arbitrary natural system. Then every* $\mathfrak{G}$*-holomorphic peak point for* $[\Sigma, \mathfrak{G}]$ *is a (globally) independent point for* $[\Sigma, \bar{\mathfrak{G}}]$.

Proof. Let δ be an $\mathfrak{G}$-holomorphic peak point for $[\Sigma, \mathfrak{G}]$ with peaking function h defined on a neighborhood U of δ. By Theorem 14.5 it will be sufficient to prove that δ is holomorphically independent for $[\Sigma, \mathfrak{G}]$. Choose an $\mathfrak{G}$-convex neighborhood V of δ contained in U. Let K be an arbitrary compact subset of V and suppose that $\delta \in \hat{K}\setminus K$. Let $G = \hat{K}\setminus K$. Since h peaks at δ and $bd_{\hat{K}}G \subseteq K$, $|h|_{bd_{\hat{K}}G} < 1 = h(\delta)$. Hence there exists a point $\delta' \in G\setminus\{\delta\}$ such that $|h|_{bd_{\hat{K}}G} < |h(\delta')|$. Now choose $u \in \mathfrak{G}$ such that $u(\delta') = 1$ and $u(\delta) = 0$. Also choose an integer m such that

$$(|h|_{bd_{\hat{K}}G})^m |h(\delta')|^{-m} < |u|_{\hat{K}}^{-1}$$

and define $g(\sigma) = h(\delta)^m h(\delta')^{-m} u(\sigma)$, $\sigma \in U$. Then g is almost $\mathfrak{G}$-holomorphic on U and hence on $\hat{K}$. Furthermore $|g|_{bd_{\hat{K}}G} < |g(\delta')| = 1$. Since $[\hat{K}, \mathfrak{G}]$ is natural and $\partial_{\mathfrak{G}}\hat{K} \subseteq K$, this contradicts Theorem 20.1. Therefore δ is an independent point for $[V, \mathfrak{G}_V]$ and hence by Theorem 14.5 is a globally independent point for $[\Sigma, \mathfrak{G}]$.◊

Observe now that, if Σ is compact and $[\Sigma, \mathfrak{G}]$ has the property that independent points are strong boundary points for $[\Sigma, \bar{\mathfrak{G}}]$ (see remarks on p. 34), then the holomorphic peak point is a strong boundary point. Furthermore δ, being a peak point of a continuous function, is a G_δ-set and therefore a global peak point for $[\Sigma, \mathfrak{G}]$. (See [R1], p. 141.) We thus obtain the following extension of the Rossi Peak Point Theorem for uniform algebras. Recall, however, that the Rossi theorems hold for the more general case of Gelfond representations of a commutative Banach algebra.

21.4 COROLLARY. *Let* $[\Sigma, \mathfrak{G}]$ *be a natural system, with compact* Σ *and* $\mathfrak{G} = \bar{\mathfrak{G}}$, *such that* $\partial_0[\Sigma, \mathfrak{G}]$ *consists of strong boundary points. Then every holomorphic peak point is a global peak point for* $[\Sigma, \mathfrak{G}]$.

The peak set problem involves additional complications. We recall first the usual technique for reducing (global) peak sets to peak points. Let P be a local peak set for $[\Sigma, \mathfrak{G}]$. Then P is closed and there exists $u \in \mathfrak{G}$ along with a neighborhood U of P such that $u(\sigma) = 1$ for $\sigma \in P$ and $|u(\sigma)| < 1$ for $\sigma \in U \setminus P$. Let $Q = \{\sigma \in \Sigma : u(\sigma) = 1\}$ and note that Q is closed with respect to the hull-kernel topology in Σ, *i.e.* $Q = \{\sigma \in \Sigma : a|Q = 0 \text{ implies } a(\sigma) = 0\}$. Moreover $Q = P \cup (Q \setminus P)$ and both P and $Q \setminus P$ are closed subsets of Σ. Hence by the Šilov decomposition theorem [R1, p. 169] both P and $Q \setminus P$ are also closed in the hull-kernel topology. Now denote by $\mathfrak{G}_0$ the subalgebra of $\mathfrak{G}$ consisting of those functions that are constant on P, and by Σ_P the compact Hausdorff space obtained from Σ by collapsing the set P to a single point δ. Then $\mathfrak{G}_0$ may be regarded as an algebra of functions on Σ_P. Moreover, since P is hull-kernel closed it is not difficult to prove that $[\Sigma_P, \mathfrak{G}_0]$ is also a natural system (see [R1, p.117] for the compact case) with δ as a local peak point. In order to apply this reduction when P is only a holomorphic peak set, it appears to be necessary to require that P be hull-kernel closed to conclude that $[\Sigma_P, \mathfrak{G}_0]$ is a natural system. Furthermore even if P is hull-kernel closed there remains the problem of showing that the given $\mathfrak{G}$-holomorphic function that peaks on P determines in Σ_P a function that is $\mathfrak{G}_0$-holomorphic near δ. The latter problem is disposed of in the following lemma. Note that the problem remains even if the peaking function for P is assumed to be $\mathfrak{G}$-holomorphic on P as well as on a neighborhood of P.

21.5 LEMMA. *Let P be an arbitrary hull-kernel closed subset of Σ and let $\mathfrak{G}_0$ be the algebra of all functions in $\mathfrak{G}$ that are constant on P. Then a function defined on a subset of $\Sigma \backslash P$ will be $\mathfrak{G}$-holomorphic iff it is $\mathfrak{G}_0$-holomorphic.*

Proof. We have only to prove that $\mathfrak{G}_0$-holomorphic implies $\mathfrak{G}$-holomorphic in $\Sigma \backslash P$. Let σ_0 be an arbitrary point of $\Sigma \backslash P$. Then since P is hull-kernel closed there exists $u \in \mathfrak{G}$ such that $u(\sigma_0) \neq 0$ while $u|P = 0$. Also since hull-kernel closure implies closure in Σ and u is continuous there exists a neighborhood V_0 of the point σ_0 on which u is never zero. Therefore by Corollary 17.5 the function u^{-1} is $\mathfrak{G}_0$-holomorphic on V_0. Now let $\underline{a}$ be an arbitrary element of $\mathfrak{G}$. Then ua vanishes on P so belongs to $\mathfrak{G}_0$. In particular ua is $\mathfrak{G}_0$-holomorphic on V_0. Therefore since $a = u^{-1}(ua)$ on V_0 it follows that $\underline{a}$ is $\mathfrak{G}_0$-holomorphic on V_0. In other words, elements of $\mathfrak{G}$ are locally $\mathfrak{G}_0$-holomorphic in $\Sigma \backslash P$ and hence are $\mathfrak{G}_0$-holomorphic on $\Sigma \backslash P$. The proof of the lemma may now be completed by the usual induction argument. ♦

Observe next that the condition for a point δ to be an independent point for $[\Sigma, \mathfrak{G}]$ is equivalent to requiring that the set $\Sigma \backslash \{\delta\}$ be $\mathfrak{G}$-convex. This suggests calling a subset E of Σ *independent* if $\Sigma \backslash E$ is $\mathfrak{G}$-convex. We note in passing that by Theorem 15.1 the set $\partial_0[\Sigma, \mathfrak{G}]$ is an independent set. Using this terminology, we have the following extension of Theorem 21.3. Observe that points are automatically hull-kernel closed.

21.6 THEOREM. *Let $[\Sigma, \mathfrak{G}]$ be an arbitrary natural system. Then every hull-kernel closed holomorphic peak set for $[\Sigma, \mathfrak{G}]$ is an independent set for $[\Sigma, \mathfrak{G}]$.*

Proof. Let P be an arbitrary holomorphic peak set for $[\Sigma, \mathfrak{G}]$ which is hull-kernel closed. Then the reduced pair $[\Sigma_P, \mathfrak{G}_0]$ is a natural system. Let h be the given peaking function for P, so h is continuous on a neighborhood U of P and $\mathfrak{G}$-holomorphic on $U \backslash P$. Since h is constant on P it determines a function h_0 on the image U_P of U in the space Σ_P. Note that U_P is a neighborhood of the point

δ in Σ_P and the function h_0 is continuous on U_P and peaks at δ. Moreover, by Lemma 21.4 the function h is $\mathfrak{G}_0$-holomorphic on $\Sigma\setminus P$, so h_0 is $\mathfrak{G}_0$-holomorphic on $\Sigma_P\setminus\{\delta\}$. It follows that δ is an independent point for $[\Sigma_P, \mathfrak{G}_0]$. This implies that the set $\Sigma\setminus P$ is $\mathfrak{G}_0$-convex and *a fortiori* is $\mathfrak{G}$-convex. ◊

Whether or not an additional assumption such as that in Corollary 21.4 will yield an extension of the Rossi Peak Set Theorem is not known. It is also not known to what extent the hull-kernel condition might be relaxed. G. R. Allan [A1] has obtained a generalization of the Rossi theorem to holomorphic peak sets where the holomorphic function h is of the form $h(\sigma) = H(a_1(\sigma),\ldots,a_n(\sigma))$ on U where H is an ordinary holomorphic function defined on a neighborhood in $\mathbb{C}^n$ of the set $\{(a_1(\sigma),\ldots,a_n(\sigma)) : \sigma \in U\}$. This means that h is Arens holomorphic [A2] so is a very special kind of $\mathfrak{G}$-holomorphic function. (See 19.1 above and §55, where the Arens holomorphic functions are considered in detail.) Allan's proof, which amounts to a sharpening of one of the proofs of the Rossi theorem (see [S10, p. 90]), also holds for the Banach algebra case and does not require the assumption that the peak set be hull-kernel closed. We are indebted to Allan for uncovering a gap in an earlier "proof" of the result in Corollary 21.4 *without* the assumption on $\partial_0[\Sigma, \mathfrak{G}]$. Whether or not some assumption of this kind is necessary for the result also remains open. Note that the Rossi theorem is not true in general if Σ is not compact (see [M3]).

§22. $\mathfrak{G}$-PRESHEAVES

It will be observed that many of the results for $\mathfrak{G}$-holomorphic functions depend ultimately upon the important property that $\mathfrak{G}$-holomorphic functions preserve $\mathfrak{G}$-convex hulls (Theorem 20.2). Furthermore, as we shall see, the $\mathfrak{G}$-subharmonic functions introduced in the next section also have this property. Therefore it is desirable to introduce a class of presheaves of continuous functions that will cover both types of functions. We shall continue to assume that $[\Sigma, \mathfrak{G}]$ is a given natural system.

22.1 DEFINITION. *A presheaf* $\{\Sigma, \mathfrak{F}\}$ *of continuous functions over* Σ *is called an* $\mathfrak{G}$-*presheaf if it satisfies the following conditions:*

(1) $|\mathfrak{G}| \subseteq |\mathfrak{F}|$.

(2) For each $E \subseteq \Sigma$ the set of functions $\mathfrak{F}_E$ is closed under multiplication.

(3) If K is any compact subset of Σ then the $\mathfrak{F}_{\hat{K}}$-convex hull of K in $\hat{K}$ is equal to $\hat{K}$.

We call $\{\Sigma, \mathfrak{F}\}$ an $\mathfrak{G}$-*presheaf of function algebras* if, in place of (1) and (2), it satisfies the following conditions:

(1)' $\mathfrak{G} \subseteq \mathfrak{F}$.

(2)' For each $E \subseteq \Sigma$ the set $\mathfrak{F}_E$ is an algebra of functions on E.

An equivalent form of condition (3) may be stated as follows:

(3)' If G is an open subset of Σ and K is a compact subset of G such that $\hat{K} \subseteq G$ then the $\mathfrak{F}_G$-convex hull of K in G is equal to $\hat{K}$.

Observe that by condition (1) and the fact that $[\Sigma, \mathfrak{G}]$ is a system, sets of the form

$$U(f_1,\ldots,f_n; \varepsilon) = \{\sigma \in \Sigma : |f_i(\sigma)| < \varepsilon;\ i=1,\ldots,n\}$$

where $\varepsilon > 0$ and $\{f_1,\ldots,f_n\}$ is an arbitrary finite subset of $\mathfrak{F}_\Sigma$ constitute a basis for the topology in Σ. Therefore conditions (1) and (2) ensure the existence of the Šilov boundary of each compact set K relative to the family of functions $\mathfrak{F}_K$ [A3]. Furthermore a point δ will belong to the Šilov boundary iff for every neighborhood V of δ there exists $f \in \mathfrak{F}_K$ such that $|f|_V > |f|_{K\setminus V}$. However these conditions are not sufficient to give the result that strong boundary points are dense in the Šilov boundary. Finally, by condition (3) if Ω is a compact $\mathfrak{G}$-convex subset of Σ then the Šilov boundary of Ω relative to $\mathfrak{F}_\Omega$ is equal to $\partial[\Omega, \mathfrak{G}]$. Note that if $\{\Sigma, \mathfrak{F}\}$ is an $\mathfrak{G}$-presheaf and Ω is any $\mathfrak{G}$-convex subset of Σ then $\{\Omega, \mathfrak{F}\}$ is also an $\mathfrak{G}|\Omega$-presheaf. The presheaf $\{\Sigma, \mathfrak{G}\}$ associated with $[\Sigma, \mathfrak{G}]$ is obviously a minimal $\mathfrak{G}$-presheaf. The presheaf $\{\Sigma, '\mathfrak{H}\}$, where $'\mathfrak{H}$ is the collection of all almost $\mathfrak{G}$-holomorphic functions defined on subsets of Σ, obviously satisfies conditions (1) and (2) of Definition 22.1 and satisfies condition (3) by Theorem 20.2, so is therefore an $\mathfrak{G}$-presheaf. An example involving $\mathfrak{G}$-subharmonic functions will be given in §33.

We observe next that the various maximum properties obtained above for $\mathfrak{G}$-holomorphic functions also hold for the $\mathfrak{F}$-holomorphic functions associated with an $\mathfrak{G}$-presheaf $\{\Sigma, \mathfrak{F}\}$.

22.2 THEOREM. *Let* $\{\Sigma, \mathfrak{F}\}$ *be an arbitrary* $\mathfrak{A}$*-presheaf over* Σ.

(i) If Ω *is a compact* $\mathfrak{A}$*-convex subset of* Σ, G *is a relatively open subset of* $\Omega\backslash\partial[\Omega, \mathfrak{A}]$, *and* f *a continuous function on* $\bar{G}$ *which is almost* $\mathfrak{F}$*-holomorphic on* G. *Then* $|f|_{\bar{G}} = |f|_{bd_{\Omega}G}$.

(ii) $\{\Sigma, {}_{\mathfrak{F}}'\mathfrak{G}\}$ *is an* $\mathfrak{A}$*-presheaf.*

(iii) Every $\mathfrak{F}$*-holomorphically independent point for* $[\Sigma, \mathfrak{A}]$ *is a globally independent point for* $[\Sigma, \mathfrak{A}]$.

Proof. As in the proof of Theorem 20.1 we may assume in (i) that f is $\mathfrak{F}$-holomorphic on all of G. Consider the set $\overset{*}{\mathfrak{F}}_G$ of all continuous functions on $\bar{G}$ whose restrictions to G belong to $\mathfrak{F}_G$. Then $\overset{*}{\mathfrak{F}}_G$ is closed under multiplication, so the Šilov boundary $\partial[\bar{G}, \overset{*}{\mathfrak{F}}_G]$ exists. Now let $\omega \in G$ and choose a neighborhood V of ω in the space Ω such that $\bar{V}$ is $\mathfrak{A}$-convex and contained in G. Then by the local maximum principle for $[\Sigma, \mathfrak{A}]$ (Theorem 14.2) the $\mathfrak{A}$-convex hull of $bd_{\Omega}V$ is equal to $\bar{V}$. Therefore by condition (3) of Definition 22.1 the $\mathfrak{F}_{\bar{V}}$-convex hull of $bd_{\Omega}V$ is equal to $\bar{V}$. This implies that the $\overset{*}{\mathfrak{F}}_G$-convex hull of $bd_{\Omega}V$ is equal to $\bar{V}$ and proves that $\partial[\bar{G}, \overset{*}{\mathfrak{F}}_G] \subseteq bd_{\Omega}G$. It follows immediately that property (i) is true if the restriction of the function f to G belongs to $\mathfrak{F}_G$. We thus have the first step of an induction argument identical with that used in the proof of Theorem 20.1, which completes the proof of (i).

That the presheaf $\{\Sigma, {}_{\mathfrak{F}}'\mathfrak{G}\}$ satisfies condition (1) of Definition 22.1 is trivial and condition (2) follows from the fact that local closure preserves multiplication. Condition (3) may be proved for $\{\Sigma, {}_{\mathfrak{F}}'\mathfrak{G}\}$ using statement (i) of the theorem exactly as Theorem 20.2 was proved using Theorem 20.1, thus establishing statement (ii).

The proof of statement (iii) is given by an argument that parallels the proof of Theorem 20.7 so will be omitted. ◊

§23. A LEMMA OF GLICKSBERG

The result in the next lemma is essentially due to Glicksberg [G5, Lemma 2.1] who obtained a similar result for $\mathfrak{A}$-holomorphic functions (of order 1) on Σ. However the proof given here is different. (Cf. [R2, Lemma 3.1].)

23.1 LEMMA. *Let* $[\Sigma, \mathfrak{G}]$ *be a natural system with compact space* Σ, $\{\Sigma, \mathfrak{F}\}$ *an* $\mathfrak{G}$*-presheaf over* Σ, B *an* $\mathfrak{G}$*-boundary for* Σ, $G = \Sigma \setminus B$ *and* δ *an arbitrary independent point for* $[\bar{G}, {}_{\mathfrak{F}}\overset{*}{\mathcal{O}}_G]$. *Then each neighborhood* U *of* δ *contains another neighborhood* V *such that every almost* $\mathfrak{F}$*-holomorphic function which is defined on* U *and zero on* $U \cap B$ *must be zero throughout* V.

Proof. By Theorem 22.2 (i) it follows that $\partial[\bar{G}, {}_{\mathfrak{F}}\overset{*}{\mathcal{O}}_G] \subseteq \mathrm{bd}\, G$. Therefore $\delta \in \mathrm{bd}\, G = \mathrm{bd}\, B$. Also, by the independence of δ there exists for a given neighborhood U of δ a function $u \in {}_{\mathfrak{F}}\overset{*}{\mathcal{O}}_G$ such that $|u(\delta)| > 1$, $|u|_{\bar{G} \setminus U} < 1/3$. Define $V = \{\sigma \in U \cap \bar{G} : |u(\sigma)| > 1\} \cup (U \setminus \bar{G})$. Then V is an open neighborhood of δ contained in U which, as we shall prove, has the property required by the lemma. Set $W = \{\sigma \in U \cap \bar{G} : |u(\sigma)| > 2/3\} \cup (U \setminus \bar{G})$. Then W is also a neighborhood of δ and $V \subseteq W \subseteq U$. Moreover if $\sigma \in \overline{W \cap G}$ then by the continuity of u we have $|u(\sigma)| \geq 2/3$. Therefore $\overline{W \cap G} \subset U$.

Now let f be an almost $\mathfrak{F}$-holomorphic function which is defined on U and zero on $U \cap B$. Suppose there existed a point $\xi \in V$ with $f(\xi) \neq 0$. Then $\xi \in V \cap G$. Since f is continuous on U it is bounded on $\overline{W \cap G}$. Hence there exists an integer m such that $(2/3)^m |f|_{\overline{W \cap G}} < |f(\xi)|$. Next define

$$g(\sigma) = \begin{cases} u(\sigma)^m f(\sigma), & \sigma \in U \cap \bar{G} \\ 0 & , \ \sigma \in U \setminus \bar{G}. \end{cases}$$

Then g is almost $\mathfrak{F}$-holomorphic on U and is zero on $U \cap B$. By Theorem 22.2 (i) there exists a point $\eta \in \mathrm{bd}(W \cap G)$ such that $|g(\eta)| = |g|_{\overline{W \cap G}}$. Also, since $\xi \in V \cap G$ we have $|u(\xi)| > 1$. Therefore

$$|f(\xi)| < |g(\xi)| \leq |g(\eta)|.$$

On the other hand, $\overline{W \cap G} \subset U$ and g is zero on $U \cap B$, so it follows that $\eta \in U \cap G$. Observe also that since $\eta \in \mathrm{bd}(W \cap G) \subseteq \overline{W \cap G}$ we have $|u(\eta)| \geq 2/3$. Moreover since $\eta \in W \cap G$ the inequality $|u(\eta)| > 2/3$ would imply that $\eta \in W \cap G$, contradicting the fact that $\eta \in \mathrm{bd}(W \cap G)$. Therefore we must have $|u(\eta)| = 2/3$. Thus

$$|g(\eta)| = (2/3)^m |f(\eta)| \leq (2/3)^m |f|_{\overline{W \cap G}} < |f(\xi)|.$$

But this contradicts the preceding inequality involving $|f(\xi)|$ and $|g(\eta)|$, so f must be zero throughout V. ◆

The above lemma provides another example of analytic phenomena for abstract holomorphic functions. It is important in the theory of $\mathfrak{G}$-holomorphic functions and will be used in §34 to establish a fundamental result for "$\mathfrak{F}$-varieties". For the applications we note that the point δ in the lemma could be an independent point for $[\bar{G}, \mathfrak{L}]$, where $\mathfrak{L}$ is any subalgebra of ${}_{\mathfrak{F}}\overset{*}{\mathfrak{O}}_G$ that contains $\mathfrak{G}|\bar{G}$, since such points are *a fortiori* independent for $[\bar{G}, {}_{\mathfrak{F}}\overset{*}{\mathfrak{O}}_G]$. The point δ in the above lemma is also called a *determining point* for $\mathfrak{F}$-holomorphic functions. [R2].

§24. MAXIMAL $\mathfrak{G}$-PRESHEAVES

24.1 PROPOSITION. *For any arbitrary $\mathfrak{G}$-presheaf $\{\Sigma, \mathfrak{F}\}$, there exists a maximal $\mathfrak{G}$-presheaf over Σ that contains it.*

Proof. Since it is obvious that the union of any increasing family of $\mathfrak{G}$-presheaves is an $\mathfrak{G}$-presheaf a routine application of Zorn's Lemma completes the proof. ◆

By Theorem 22.2 (ii) we have the following corollaries the first of which suggests a Rado type result since it implies in particular that an almost holomorphic function is actually holomorphic. (Cf. [G5, Theorem 3.2].)

24.2 COROLLARY. *If $\{\Sigma, \mathfrak{F}\}$ is a maximal $\mathfrak{G}$-presheaf then $\mathfrak{F}$ already contains every almost $\mathfrak{F}$-holomorphic function; i.e. $\mathfrak{F} = {}_{\mathfrak{F}}^{\,\prime}\mathfrak{O}$.*

24.3 COROLLARY. *Every maximal $\mathfrak{G}$-presheaf is locally closed and hence is local.*

Observe that if a function is almost $\mathfrak{G}$-holomorphic then it is almost $\mathfrak{F}$-holomorphic for every $\mathfrak{G}$-presheaf $\{\Sigma, \mathfrak{F}\}$ that contains $\{\Sigma, \mathfrak{G}\}$. Therefore by Corollary 24.2 we have the following result.

24.4 COROLLARY. *The $\mathfrak{G}$-presheaf $\{\Sigma, {}_{\mathfrak{G}}^{\,\prime}\mathfrak{O}\}$ consisting of the almost $\mathfrak{G}$-holomorphic functions in Σ is contained in every maximal $\mathfrak{G}$-presheaf over Σ that contains $\{\Sigma, \mathfrak{G}\}$.*

24.5 PROPOSITION. *If $\{\Sigma, \mathfrak{F}\}$ is a maximal $\mathfrak{G}$-presheaf and Ω is an arbitrary subset of Σ then $\{\Omega, \mathfrak{F}\}$ is a maximal $\mathfrak{G}|\Omega$-presheaf.*

Proof. Let $\{\Omega, \mathfrak{F}'\}$ be any $\mathfrak{G}|\Omega$-presheaf with $\mathfrak{F}|\Omega \subseteq \mathfrak{F}'$. Then $\{\Sigma, \mathfrak{F} \cup \mathfrak{F}'\}$ is obviously a presheaf of continuous functions over Σ such that $|\mathfrak{G}| \subseteq |\mathfrak{F} \cup \mathfrak{F}'|$. Moreover if E is any subset of Σ then

$$(\mathfrak{F} \cup \mathfrak{F}')_E = \begin{cases} \mathfrak{F}'_E , & \text{if } E \subseteq \Omega \\ \mathfrak{F}_E , & \text{if } E \nsubseteq \Omega. \end{cases}$$

Therefore $(\mathfrak{F} \cup \mathfrak{F}')_E$ is closed under multiplication. Similarly if $K \subset\subset \Omega$ then the $(\mathfrak{F} \cup \mathfrak{F}')_{\hat{K}}$-convex hull of K will be equal to the $\mathfrak{F}'_{\hat{K}}$-convex hull of K if $\hat{K} \subseteq \Omega$, and to the $\mathfrak{F}_{\hat{K}}$-convex hull of K if $\hat{K} \nsubseteq \Omega$. In either case the $(\mathfrak{F} \cup \mathfrak{F}')_{\hat{K}}$-convex hull of K is equal to $\hat{K}$, which proves that $\{\Sigma, \mathfrak{F} \cup \mathfrak{F}'\}$ is an $\mathfrak{G}$-presheaf. Finally by the maximality of $\{\Sigma, \mathfrak{F}\}$ it follows that $\mathfrak{F} \cup \mathfrak{F}' = \mathfrak{F}$ and in particular that $\mathfrak{F}|\Omega = \mathfrak{F}'$, so $\{\Omega, \mathfrak{F}\}$ is maximal. ◊

24.6 COROLLARY. *If $\{\Sigma, \mathfrak{F}\}$ is a maximal $\mathfrak{G}$-presheaf and Ω is an arbitrary subset of Σ then for every compact set $K \subset\subset \Omega$ the $\mathfrak{F}_\Omega$-convex hull of K in Ω is equal to $\hat{K} \cap \Omega$.*

We specialize now to $\mathfrak{G}$-presheaves of function algebras and observe first that the proof of Proposition 24.1 carries over without change to this case. Therefore if $\{\Sigma, \mathfrak{F}\}$ is any $\mathfrak{G}$-presheaf of function algebras then there exists a maximal $\mathfrak{G}$-presheaf of function algebras that contains it. Also, if $\{\Sigma, \mathfrak{F}\}$ is maximal and Ω is any subset of Σ then $\{\Omega, \mathfrak{F}\}$ is a maximal $\mathfrak{G}|\Omega$-presheaf of function algebras over Ω. The result in Corollary 24.6 obviously also holds for the function algebras case. The result in Corollary 24.2 (and hence Corollaries 24.3 and 24.4) also carries over but the proof is much more difficult. The problem is that, although $\{\Sigma, {}_{\mathfrak{F}}'\mathfrak{O}\}$ is an $\mathfrak{G}$-presheaf, it in general is not an $\mathfrak{G}$-presheaf of function algebras. We state the function algebra result as a theorem.

24.7 THEOREM. *Let $\{\Sigma, \mathfrak{F}\}$ be a maximal $\mathfrak{G}$-presheaf of function algebras. Then $\mathfrak{F}$ contains the almost $\mathfrak{F}$-holomorphic functions.*

Proof. Let h be an almost $\mathfrak{F}$-holomorphic function defined on a set $H \subseteq \Sigma$. For a set $E \subseteq H$ denote by $\mathfrak{F}'_E$ the algebra of functions on E generated by $\mathfrak{F}_E$ plus the function $h|E$. Also denote by $\mathfrak{F}'$ the union of all of the algebras $\mathfrak{F}'_E$

for arbitrary $E \subseteq H$. Then $\{\Sigma, \mathfrak{F} \cup \mathfrak{F}'\}$ is obviously a presheaf of function algebras over Σ such that $\mathfrak{a} \subseteq \mathfrak{F} \subseteq \mathfrak{F}'$. We wish to prove that $\{\Sigma, \mathfrak{F} \cup \mathfrak{F}'\}$ is actually an $\mathfrak{a}$-presheaf of function algebras, *i.e.* if $K \subset\subset \Sigma$ then the $(\mathfrak{F} \cup \mathfrak{F}')_{\hat{K}}$-convex hull of K is equal to $\hat{K}$. Note that

$$(\mathfrak{F} \cup \mathfrak{F}')_{\hat{K}} = \begin{cases} \mathfrak{F}'_{\hat{K}}, & \text{if } \hat{K} \subseteq H \\ \mathfrak{F}_{\hat{K}}, & \text{if } \hat{K} \nsubseteq H. \end{cases}$$

Since by hypothesis $\mathfrak{F}_{\hat{K}}$-convex hull of K is equal to $\hat{K}$ it remains to prove that the $\mathfrak{F}'_{K}$-convex hull of K is equal to $\hat{K}$ when $\hat{K} \subseteq H$.

Let δ be an independent point for $[\hat{K}, \mathfrak{F}'_{\hat{K}}]$ and assume first that h is $\mathfrak{F}$-holomorphic at δ. Then since each element of $\mathfrak{F}'_{\hat{K}}$ is a polynomial in $h|\hat{K}$ with coefficients from $\mathfrak{F}_{\hat{K}}$ it follows that each element of $\mathfrak{F}'_{\hat{K}}$ is $\mathfrak{F}$-holomorphic at δ. Therefore δ is an $\mathfrak{F}$-holomorphically independent point for $[\hat{K}, \mathfrak{F}_{\hat{K}}]$. Hence by Theorem 22.2 (iii) it is a globally independent point for $[\Sigma, \mathfrak{a}]$ so must belong to K. Thus if $\delta \notin K$ then $\delta \in \mathrm{bd}_{\hat{K}}Z$, where $Z = \{\sigma \in \hat{K} : h(\sigma) = 0\}$. Set $B = K \cup Z$. Then B is an $\mathfrak{a}$-boundary for $\hat{K}$ and if $G = \hat{K}\backslash B$ then $\delta \in \mathrm{bd}_{\hat{K}}G$. Since $h|\bar{G} \in {}_{\mathfrak{F}}\mathcal{O}^{*}_{G}$ it follows that $\mathfrak{F}'_{\bar{U}} \subseteq {}_{\mathfrak{F}}\mathcal{O}^{*}_{G}$, so δ is an independent point for $[\bar{G}, {}_{\mathfrak{F}}\mathcal{O}^{*}_{G}]$. Now choose a neighborhood U of δ in $\hat{K}$ with $U \cap K = \emptyset$ and let V be the neighborhood of δ given by an application of Lemma 23.1 to the system $[\hat{K}, \mathfrak{a}]$ and the $\mathfrak{a}|\hat{K}$-presheaf $\{\hat{K}, \mathfrak{F}\}$. Then since $U \cap B = U \cap Z$ we have $h|(U \cap B) = 0$, so $h|V = 0$ by the lemma. But this contradicts the fact that $\delta \in \mathrm{bd}_{\hat{K}}Z$ and proves that every independent point for $[\hat{K}, \mathfrak{F}'_{\hat{K}}]$ is contained in K. Therefore $\partial[\hat{K}, \mathfrak{F}'_{\hat{K}}] \subseteq K$ and the $\mathfrak{F}_{\hat{K}}$-convex hull of K is equal to $\hat{K}$. Thus it follows that $\{\Sigma, \mathfrak{F} \cup \mathfrak{F}'\}$ is an $\mathfrak{a}$-presheaf of function algebras that contains $\{\Sigma, \mathfrak{F}\}$. Since the latter is maximal we have $\mathfrak{F} \cup \mathfrak{F}' = \mathfrak{F}$ and hence $\mathfrak{F}'_{H} = \mathfrak{F}_{H}$. In particular $h \in \mathfrak{F}_{H}$. ◆

24.8 COROLLARY. *The presheaf of function algebras over* Σ *generated by all almost* $\mathfrak{a}$*-holomorphic functions in* Σ *is an* $\mathfrak{a}$*-presheaf of function algebras and is contained in every maximal* $\mathfrak{a}$*-presheaf of function algebras over* Σ.

As a consequence of this corollary we have the following nontrivial generalization of the local maximum principle for $\mathfrak{a}$-holomorphic function (Theorem 20.1).

24.9 COROLLARY. *Assume that* Σ *is compact and let* U *be an open subset of* $\Sigma \setminus \partial[\Sigma, \mathfrak{G}]$. *Also let* h *be an element of* $C(\bar{U})$ *whose restriction to* U *belongs to the algebra generated by the almost* $\mathfrak{G}$*-holomorphic functions defined on* U. *Then* $|h|_{\bar{U}} = |h|_{\text{bd } U}$.

Remarks. In view of the desirable properties enjoyed by a maximal $\mathfrak{G}$-presheaf of function algebras one might conjecture that the functions belonging to such a presheaf would provide a more appropriate setting for the study of analytic phenomena than the $\mathfrak{G}$-holomorphic function that we have chosen. However, as will be seen in later sections, there are certain powerful techniques that work for $\mathfrak{G}$-holomorphic functions but do not apply to the wider class of functions. An important property of $\mathfrak{G}$-holomorphic functions is that they are locally determined by the structure algebra $\mathfrak{G}$ and this is not true for the wider class of functions even though the maximal presheaf is local. The root of the problem is that there does not appear to be a canonical method of assigning to each natural system a maximal $\mathfrak{G}$-presheaf of function algebras that is relevant to the category of natural systems.

Let X be a compact Hausdorff space and $\mathfrak{L}$ be a proper subalgebra of $C(X)$. Then Glicksberg [G5, p. 923] defines $\mathfrak{L}$ to be *relatively maximal in* $C(X)$ if no subalgebra $\mathfrak{B}$ of $C(X)$ exists, properly containing $\mathfrak{L}$, such that $\partial_{\mathfrak{B}}X = \partial_{\mathfrak{L}}X$. Maximal subalgebras of $C(X)$ are obviously relatively maximal. On the other hand relatively maximal algebras need not be maximal. An example cited by Glicksberg is the algebra $\mathcal{A}(D^n)$ consisting of all functions continuous on the closed polydisc $D^n \subset \mathbb{C}^n$ and holomorphic on the interior of D^n [G5, p. 928]. Glicksberg observes that a subalgebra of $C(X)$ may be embedded in a relatively maximal one with the same Šilov boundary. This in fact suggested our construction of maximal $\mathfrak{G}$-presheaves and many of our results for these presheaves parallel analogous Glicksberg results for relatively maximal algebras. On the other hand if $\{\Sigma, \mathfrak{F}\}$ is a maximal $\mathfrak{G}$-presheaf of function algebras over Σ and X is a compact subspace of Σ then the algebra $\mathfrak{F}_X$ of course has the property that $\partial[X, \mathfrak{F}_X] = \partial[X, \mathfrak{G}]$ by Corollary 24.6, but we cannot prove that it is relatively maximal in $C(X)$. Observe in fact that $\mathfrak{F}_X$ has the additional property that $\partial[K, \mathfrak{F}_X] = \partial[K, \mathfrak{G}]$ for *every* compact set $K \subset\subset X$. We do not have an example of an $\mathfrak{G}$-presheaf $\{\Sigma, \mathfrak{F}\}$ in which $\mathfrak{F}_X$ is not relatively maximal in $C(X)$. The situation is complicated by the fact that $[X, \mathfrak{A}]$ may be natural and $\mathfrak{L}$ a subalgebra of $C(X)$ that contains $\mathfrak{A}|X$ with the property that $\partial[X, \mathfrak{L}] = \partial[X, \mathfrak{A}]$ but $[X, \mathfrak{L}]$ is not natural. (See Example 43.1.)

CHAPTER VI
SUBHARMONIC FUNCTIONS

§25. PLURISUBHARMONIC FUNCTIONS IN $\mathbb{C}^n$

In this section we introduce a class of functions that generalize to the case of an arbitrary system $[\Sigma, \mathfrak{G}]$ the familiar plurisubharmonic functions associated with $[\mathbb{C}^n, \mathcal{P}]$. The functions considered belong to the larger class $\mathcal{U}$ of all upper semicontinuous (usc) functions defined on arbitrary subsets of Σ with values in the extended real numbers $[-\infty, \infty)$. Thus, an element $f \in \mathcal{U}$ may assume the value $-\infty$ but not the value $+\infty$ and, for each point δ in its domain of definition, $\limsup_{\sigma \to \delta} f(\sigma) \leq f(\delta)$. Note that $\mathcal{U}$ is closed under multiplication by positive reals and under restrictions, *i.e.*, if $f \in \mathcal{U}$ and X is a subset of the domain of f then also $f|X \in \mathcal{U}$. Furthermore, if $f, g \in \mathcal{U}$ then $f + g \in \mathcal{U}$ provided $f + g$ exists (i.e. the domains of f and g intersect). With the obvious convention, we accordingly say that $\mathcal{U}$ is "closed under linear combinations with positive coefficients."

One definition of plurisubharmonic (psh) functions in $\mathbb{C}^n$ may be stated as follows: Let G be an open subset of $\mathbb{C}^n$ and f a function defined on G with values in $[-\infty, \infty)$. Then f is *plurisubharmonic on* G provided it is usc and, for every holomorphic map $\eta : D \to G$ of the open unit disc $D \subset \mathbb{C}$ into G, the composition function $f \circ \eta$ is an ordinary subharmonic function on D. [G7, p. 271]. Observe that since subharmonicity is a local property the same is true of psh, *i.e.* f will be psh on G iff each point of G admits a neighborhood on which f is psh. As we saw in the case of $\mathfrak{G}$-holomorphic functions (see 19.1), a definition of this kind is not sharp enough for general systems since nontrivial holomorphic maps of the disc into the space in question may not exist. Therefore we must take as a starting point another characterizing property of psh functions that in the general situation will tie the concept more closely to the structure algebra $\mathfrak{G}$.

An appropriate characterization is based on two well-known but nontrivial approximation theorems for psh functions in $\mathbb{C}^n$. (See for example F1, Theorems 13.9 and 13.10].) The first theorem asserts that if f is psh on an open set $G \subseteq \mathbb{C}^n$ and $H \subset\subset G$ then there exists a sequence $\{f_n\}$ of continuous (in fact, k-times continuously differentiable) psh functions on H such that $f_n \geq f_{n+1} \geq f$ and $\lim_{n\to\infty} f_n = f$ pointwise on H. The second theorem asserts that if g is a continuous psh function on a domain of holomorphy $U \subseteq \mathbb{C}^n$ then for every open set $V \subset\subset U$ the function g is a supremum on V of functions of the form $n^{-1}\log|h|$, where n is a positive integer and h is holomorphic on V. Since each point of $\mathbb{C}^n$ admits arbitrarily small neighborhoods that are domains of holomorphy, it follows that the second approximation theorem always applies locally.

§26. DEFINITIONS. $\mathfrak{G}$-SUBHARMONIC FUNCTIONS

For a general system $[\Sigma, \mathfrak{G}]$, we define generalized psh functions analogously to the way we defined $\mathfrak{G}$-holomorphic functions, *viz* in terms of an appropriate closure operation in the set $\mathcal{U}$ of all usc functions in Σ. Consider first the class $_{\mathfrak{G}}\mathfrak{L}$ of all functions of the form $n^{-1}\log|a|$, where n is a positive integer and $a \in \mathfrak{G}$. Then $_{\mathfrak{G}}\mathfrak{L}$ is obviously contained in $\mathcal{U}$, contains all real constants (including $-\infty$), and is closed under linear combinations with positive *rational* coefficients. The desired generalization of plurisubharmonic functions is obtained by extending the set $_{\mathfrak{G}}\mathfrak{L}$ in $\mathcal{U}$ with respect to the composition of two special closure operations that act on arbitrary subsets of $\mathcal{U}$. (Cf. [L1, L2].)

26.1 DEFINITION. *Let $\mathfrak{F}$ be an arbitrary subset of $\mathcal{U}$.*

(i) The $^{\vee}$-closure of $\mathfrak{F}$ is the set $\mathfrak{F}^{\vee} \subseteq \mathcal{U}$ each of whose elements is locally the supremum of functions belonging to $\mathfrak{F}$.

(ii) The $\downarrow$-closure of $\mathfrak{F}$ is the set $\mathfrak{F}^{\downarrow} \subseteq \mathcal{U}$ each of whose elements is locally the pointwise limit of a nonincreasing sequence of functions belonging to $\mathfrak{F}$.

It is obvious that "$^{\vee}$" and "$\downarrow$" are closure operations in the sense of §8. Note that the requirement that elements of $\mathfrak{F}^{\downarrow}$ belong to $\mathcal{U}$ is automatically satisfied. This is also automatic in the case of $\mathfrak{F}^{\vee}$ if the supremum involves only a finite number of elements of $\mathfrak{F}$. It is obvious that both $\mathfrak{F}^{\vee}$ and $\mathfrak{F}^{\downarrow}$ are closed under arbitrary restrictions. Also, if $\mathfrak{F}$ is closed under linear combinations with positive coefficients then the same is true of $\mathfrak{F}^{\vee}$ and $\mathfrak{F}^{\downarrow}$.

We are actually interested in the compositions of the closure operations "$\vee$" and "$\downarrow$"; *viz* "$\vee\downarrow$" and "$\downarrow\vee$", where $\mathfrak{F}^{\vee\downarrow} = (\mathfrak{F}^{\vee})^{\downarrow}$ and $\mathfrak{F}^{\downarrow\vee} = (\mathfrak{F}^{\downarrow})^{\vee}$. It is easy to verify (for compositions of arbitrary closure operations) that $\mathfrak{F}^{\vee\downarrow} = \mathfrak{F}$ *or* $\mathfrak{F}^{\downarrow\vee} = \mathfrak{F}$ iff both $\mathfrak{F}^{\vee} = \mathfrak{F}$ and $\mathfrak{F}^{\downarrow} = \mathfrak{F}$. Therefore $\mathfrak{F}^{\vee\downarrow} = \mathfrak{F}$ iff $\mathfrak{F}^{\downarrow\vee} = \mathfrak{F}$. It follows in particular that the associated proper closure operations "$(\vee\downarrow)$" and "$(\downarrow\vee)$", as defined in §8, are equal. Because of this fact we may limit our attention to "$\vee\downarrow$".

26.2 DEFINITION. *Let* ${}_{\mathfrak{G}}\mathcal{S} = ({}_{\mathfrak{G}}\mathcal{L})^{(\vee\downarrow)}$, *the* $\vee\downarrow$*-closure of the set* ${}_{\mathfrak{G}}\mathcal{L}$ *in* $\mathcal{U}$. *Then elements of* ${}_{\mathfrak{G}}\mathcal{S}$ *are called* $\mathfrak{G}$*-subharmonic* ($\mathfrak{G}$-sh) *functions. If both* f *and* -f *belong to* ${}_{\mathfrak{G}}\mathcal{S}$, *then* f *is called an* $\mathfrak{G}$*-harmonic function. The set of all* $\mathfrak{G}$*-harmonic functions is denoted by* ${}_{\mathfrak{G}}\mathcal{H}$. *A function is called almost* $\mathfrak{G}$*-subharmonic if it is* usc *and is* $\mathfrak{G}$*-subharmonic on the subset of its domain of definition where it is different from* $-\infty$.

§27. BASIC PROPERTIES OF $\mathfrak{G}$-SUBHARMONIC FUNCTIONS

It will be proved later that in the case of $[\mathbb{C}^n, \mathfrak{P}]$ the set of all $\mathfrak{P}$-subharmonic functions defined on open subsets of $\mathbb{C}^n$ coincides with the set of all psh functions in $\mathbb{C}^n$. When there is no possibility for confusion we shall drop the subscript "$\mathfrak{G}$" and write simply $\mathcal{L}$, $\mathcal{S}$, and $\mathcal{H}$ in place of ${}_{\mathfrak{G}}\mathcal{L}$, ${}_{\mathfrak{G}}\mathcal{S}$, and ${}_{\mathfrak{G}}\mathcal{H}$ respectively. The subset of $\mathcal{S}$ consisting of continuous functions will be denoted by $\mathcal{CS}$. Note that $\mathcal{H} \subseteq \mathcal{CS}$ and elements of $\mathcal{H}$ do not assume infinite values. The subset of $\mathcal{S}$ consisting of functions defined on a given set $\Omega \subset \Sigma$ will be denoted by $\mathcal{S}_{\Omega}$, with similar meanings for $\mathcal{CS}_{\Omega}$ and $\mathcal{H}_{\Omega}$.

Recall that the set $\mathcal{L}$ is closed under linear combinations with positive *rational* coefficients. The next lemma shows that "rational" may be replaced by "real" in passing to the $\vee\downarrow$-closure $\mathcal{S}$. Note that $\mathcal{L}$ contains all real constants.

27.1 LEMMA. *Let* $\mathfrak{F}$ *be a subset of* $\mathcal{U}$ *which is closed under addition of (real) constants and under multiplication by positive rationals. Then* $\mathfrak{F}^{\vee}$ *and* $\mathfrak{F}^{\downarrow}$ *are closed under addition of constants and multiplication by positive reals.*

Proof. The closure under addition of constants is obvious. Consider first an arbitrary element $g \in \check{\mathfrak{F}}$ defined on a set $F \subseteq \Sigma$, and let r be an arbitrary positive real number. Let δ be any point of F and choose t such that $g(\delta) < t$. Set $U_t = \{\sigma \in F : g(\sigma) < t\}$. Since g is usc, U_t is a relatively open neighborhood of the point δ in F. Now define $h = g - t$. Then $h \in \check{\mathfrak{F}}$ and $h(\sigma) < 0$ for $\sigma \in U_t$. Choose a neighborhood $V_\delta \subseteq U_t$ such that for each $\sigma \in V_\delta$

$$h(\sigma) = \sup \{f(\sigma) : f \in \mathfrak{F}, f \le h \text{ on } V_\delta\}.$$

Then for $\sigma \in V_\delta$

$$(rh)(\sigma) = \sup \{(qf)(\sigma) : q \in \mathbb{Q}^+, r \le q ; f \in \mathfrak{F}, f \le h \text{ on } V_\delta\}.$$

Since δ was an arbitrary point of F, it follows that $rh \in \check{\mathfrak{F}}$, so $rg = rh + rt \in \check{\mathfrak{F}}$. This completes the proof for $\check{\mathfrak{F}}$.

Now let $g \in \mathfrak{F}^{\downarrow}$ and proceed exactly as in the above argument to obtain $h = g - t \in \mathfrak{F}^{\downarrow}$. Then choose $\{f_n\} \subset \mathfrak{F}$ such that $f_n \downarrow h$ near the point δ in F. Since $h(\delta) < 0$ there is no loss in assuming that $f_n(\delta) < 0$ for all n. Furthermore, by usc the function f_1, and hence each f_n, is negative on a neighborhood of δ in F. Therefore there exists a neighborhood $V_\delta \subseteq U_t$ such that each of the function f_n is negative on V_δ and $f_n(\sigma) \downarrow h(\sigma)$ for $\sigma \in V_\delta$. Next choose $\{q_n\} \subset \mathbb{Q}^+$ such that $q_n \uparrow r$. Since $f_n(\sigma) < 0$ for $\sigma \in V_\delta$ we thus have $(q_n f_n)(\sigma) \downarrow (rh)(\sigma)$ for $\sigma \in V_\delta$. Therefore it follows that $rh \in \mathfrak{F}^{\downarrow}$. ◆

27.2 COROLLARY. $\mathfrak{S}$ *is closed under linear combinations with positive real coefficients.* $\mathfrak{H}$ *is closed under linear combinations with arbitrary real coefficients.*

By virtue of the local character of the definition of $\mathfrak{a}$-sh functions, the set $\mathfrak{S}$ is obviously local, *i.e.* a function will belong to $\mathfrak{S}$ iff it belongs locally to $\mathfrak{S}$. A more general result is the following.

27.3 THEOREM. $\mathfrak{S}$ *is closed under local uniform limits, i.e.* $\mathfrak{S}^{lc} = \mathfrak{S}$.

Proof. Let f be a function, with domain of definition $F \subseteq \Sigma$, which is locally approximable by elements of $\mathfrak{S}$. Then for each point $\delta \in F$ there exists V_δ in F and a sequence $\{g_n\}$ of elements of $\mathfrak{S}$ defined on V_δ such that $\sigma \in V_\delta$

implies $f(\sigma) - n^{-1} < g_n(\sigma) < f(\sigma) + n^{-1}$, if $f(\sigma) \neq -\infty$, and $g_n(\sigma) < -n$, if $f(\sigma) = -\infty$. We prove first that $f \in \mathcal{U}$. For this it will be sufficient to prove that $f|V_\delta \in \mathcal{U}$. Therefore let $\sigma_0 \in V_\delta$, $t \in \mathbb{R}$, and assume that $f(\sigma_0) < t$. Choose n such that $-n < t - n^{-1}$ and $f(\sigma_0) < t - 2n^{-1}$. Then $g_n(\sigma_0) < t - n^{-1}$ and, since $g_n \in \mathcal{U}$, there exists a neighborhood $U_{\sigma_0} \subseteq V_\delta$ such that $\sigma \in U_{\sigma_0}$ implies $g_n(\sigma) < t - n^{-1}$. Since $f(\sigma) < g_n(\sigma) + n^{-1}$ for all $\sigma \in V_\delta$, it follows that $f(\sigma) < t$ for $\sigma \in U_{\sigma_0}$. Therefore $f \in \mathcal{U}$.

Next let $f_k(\sigma) = \max[f(\sigma), -k]$, $\sigma \in F$. Then $f_k \in \mathcal{U}$ and $f(\sigma) \leq f_{k+1}(\sigma) \leq f_k(\sigma)$ for all k, and $f(\sigma) = \lim_{k\to\infty} f_k(\sigma)$, $\sigma \in F$. Therefore if we prove that $f_k \in \mathcal{S}$ for each k then it will follow that $f \in \mathcal{S}^{\downarrow} = \mathcal{S}$. Now set

$$g_{nk}(\sigma) = \max[g_n(\sigma), -k], \ \sigma \in F.$$

Then $g_{nk} \in \check{\mathcal{S}} = \mathcal{S}$. Also $f(\sigma) = -\infty$ implies $f_k(\sigma) = -k$ and $g_{nk}(\sigma) = -k$ for $n \geq k$. Hence if $n \geq k$ then

$$f_k(\sigma) - n^{-1} < g_{nk}(\sigma) < f_k(\sigma) + n^{-1}, \ \sigma \in V_\delta.$$

Since $g_{nk} - n^{-1} \in \mathcal{S}$ and

$$f_k(\sigma) = \sup_n\{g_{nk}(\sigma) - n^{-1}\}, \ \sigma \in V_\delta$$

it follows that $f_k \in \mathcal{S}^{\vee} = \mathcal{S}$, so $f \in \mathcal{S}$. ◆

Consider next a real-valued function χ defined on an interval $(a, b) \subseteq [-\infty,\infty]$. Assume that χ is nondecreasing and convex. Thus if $a < r < s < b$ then $\chi(r) \leq \chi(s)$ and

$$\chi((1-\theta)r+\theta s) \leq (1-\theta)\chi(r) + \theta\chi(s), \ 0 \leq \theta \leq 1.$$

The function χ is automatically continuous on (a, b) and, for convenience, we extend χ continuously to the half-open interval $[a, b)$ by setting

$$\chi(a) = \lim_{t\to a+} \chi(t).$$

Thus $\chi(a) \leq \chi(t)$ for all $t \in (a, b)$, and if $a = -\infty$ then $\chi(a) = -\infty$ is possible. It follows from the convexity that

$$m_t = \lim_{s\to t+} \frac{\chi(s) - \chi(t)}{s - t}$$

exists for all $t \in (a, b)$ and the graph of χ lies above the line through the point

$(t, \chi(t))$ with slope equal to m_t, *i.e.*

$$\chi(t) + m_t(s-t) \le \chi(s) \ , \ s \in (a, b).$$

Since χ is nondecreasing $m_t \ge 0$ for all t.

27.4 THEOREM. *Let* $f \in \mathcal{S}$ *and assume that the range of* f *is contained in the domain* $[a, b)$ *of the nondecreasing convex function* χ. *Then* $\chi \circ f \in \mathcal{S}$.

Proof. Since χ is continuous and nondecreasing it is immediate that $\chi \circ f \in \mathcal{U}$. Also, since $\mathcal{S}$ contains the constants and $m_t \ge 0$ the function $\chi(t) + n_t(f-t)$ belongs to $\mathcal{S}$ for each $t \in (a, b)$. Therefore if

$$g_t(\sigma) = \max\{\chi(a) \ , \ \chi(t) + m_t(f(\sigma) - t)\}$$

for σ in the domain F of f then also $g_t \in \mathcal{S}$ and for each $t \in (a, b)$, $g_t(\sigma) \le (\chi \circ f)(\sigma)$, $\sigma \in F$. If $f(\sigma) = a$ then $g_t(\sigma) = (\chi \circ f)(\sigma)$ for every t, and if $f(\sigma) \in (a, b)$ then $g_t(\sigma) = (\chi \circ f)(\sigma)$ for $t = f(\sigma)$. Therefore

$$\sup_t g_t(\sigma) = (\chi \circ f)(\sigma) \ , \ \sigma \in F$$

so $\chi \circ f \in \mathcal{S}^{\vee} = \mathcal{S}$. ◊

27.5 COROLLARY.

(i) *If* $f \in \mathcal{S}$ *and* $f \ge 0$ *then* $f^t \in \mathcal{S}$ *for all* $t \ge 1$.

(ii) *If* $f \in \mathcal{S}$ *then* $e^f \in \mathcal{S}$. *In particular if* g *is logarithmically* $\mathcal{A}$*-subharmonic (i.e.,* $g \ge 0$ *and* $\log g \in \mathcal{S}$*) then* $g \in \mathcal{S}$.

(iii) *For arbitrary* $a \in \mathcal{A}$ *both* $\log|a|$ *and* $|a|$ *belong to* $\mathcal{S}$.

The next theorem is an extension of Corollary 27.5 (iii) to $\mathcal{A}$-holomorphic functions.

27.6 THEOREM. *If* h *is an arbitrary* $\mathcal{A}$*-holomorphic function then both* $\log|h|$ *and* $|h|$ *belong to* $\mathcal{S}$.

Proof. Denote by $\mathfrak{F}$ the set of all $\mathcal{A}$-holomorphic function h such that both $\log|h| \in \mathcal{S}$ and $|h| \in \mathcal{S}$. Then $\mathfrak{F}$, as a subset of the set of all $\mathcal{A}$-holomorphic functions $\mathfrak{G}$, is maximal with respect to the property that $\log|\mathfrak{F}| \subseteq \mathcal{S}$ and $|\mathfrak{F}| \subseteq \mathcal{S}$. We must prove that $\mathfrak{F}$ contains all $\mathcal{A}$-holomorphic functions. Since $\mathcal{A} \subseteq \mathfrak{F}$, by

27.5 (iii) it will be sufficient to prove that $\mathfrak{F}^{lc} = \mathfrak{F}$ or, in view of the maximality of $\mathfrak{F}$, that $\log|\mathfrak{F}^{lc}| \subseteq \mathcal{S}$ and $|\mathfrak{F}^{lc}| \subseteq \mathcal{S}$.

Observe first that $|\mathfrak{F}^{lc}| \subseteq |\mathfrak{F}|^{lc} \subseteq \mathcal{S}$, where the second inclusion is given by Theorem 27.3. That $\log|\mathfrak{F}^{lc}| \subseteq \mathcal{S}$ requires more work. Let $h \in \mathfrak{F}^{lc}$ and let H be the domain of h. Then for arbitrary $\delta \in H$ there exists a neighborhood V_δ in H and $\{f_n\} \subset \mathfrak{F}$ such that $\lim_{n\to\infty} f_n = h$ uniformly on V_δ. For an arbitrary positive integer k define

$$|h|_k(\sigma) = \max(e^{-k}, |h(\sigma)|) , \sigma \in H$$

and

$$|f_n|_k(\sigma) = \max(e^{-k}, |f_n(\sigma)|), \sigma \in H.$$

Then $\lim_{n\to\infty}|f_n|_k = |h|_k$ uniformly on V_δ. Moreover since the functions $|f_n|_k$ and $|h|_k$ are uniformly bounded away from zero for fixed k, we also have $\lim_{n\to\infty} \log|f_n|_k = \log|h|_k$ uniformly on V_δ. Observe next that $\log|f_n|_k(\sigma) = \max(-k, \log|f_n(\sigma)|)$, $\sigma \in V_\delta$. Since $\mathcal{S}$ contains the constants and $\log|f_n| \in \mathcal{S}$ it follows that $\log|f_n|_k \in \mathcal{S}^\vee = \mathcal{S}$. This implies that $\log|h|_k \in \mathcal{S}^{lc}$, so $\log|h|_k \in \mathcal{S}$ by Theorem 27.3. Finally observe that $\log|h| \le \log|h|_{k+1} \le \log|h|_k$ for all k, and $\lim_{k\to\infty} \log|h|_k(\sigma) = \log|h(\sigma)|$, $\sigma \in H$. Therefore $\log|h| \in \mathcal{S}^\downarrow = \mathcal{S}$. ◆

27.7 COROLLARY. *Let* h *be an arbitrary* $\mathfrak{A}$*-holomorphic function.*

(i) If h *is never zero then* $\log|h|$ *is* $\mathfrak{A}$*-harmonic.*

(ii) If $h = u + iv$, *where* u *and* v *are real, then* u *and* v *are* $\mathfrak{A}$*-harmonic.*

§28. PLURISUBHARMONICITY

We observed at the beginning of this section that the usual definition of plurisubharmonic functions for $\mathbb{C}^n$ in terms of holomorphic maps was not restrictive enough for general systems. On the other hand it is important to know that $\mathfrak{A}$-subharmonic functions do in fact satisfy that definition. This is given by the next theorem.

28.1 THEOREM. *Let* $\eta : U \to \Sigma$ *be a holomorphic mapping of an open set* $U \subseteq \mathbb{C}$ *into* Σ *and let* f *be an arbitrary* $\mathfrak{A}$*-subharmonic function defined on* $\eta(U)$. *Then* $f\circ\eta$ *is an ordinary subharmonic function on* U.

Proof. Assume first that $f \in \mathcal{L}$, so there exists a positive integer n and an element $a \in \mathfrak{G}$ such that $f = n^{-1}\log|a|$. Then $f\circ\eta = n^{-1}\log|a\circ\eta|$. Since $a\circ\eta$ is holomorphic on U by the definition of holomorphic map, it follows that $f\circ\eta$ is subharmonic on U and the theorem is true for elements of $\mathcal{L}$. Now denote by $\mathcal{F}$ the subset of $\mathcal{S}$ consisting of all those functions for which the theorem is true. In other words if $f \in \mathcal{F}$ and $\eta : U \to \Sigma$ is any holomorphic map of an arbitrary open set $U \subseteq \mathbb{C}$ into Σ such that f is defined on $\eta(U)$ then $f\circ\eta$ is subharmonic in U. We thus have $\mathcal{L} \subseteq \mathcal{F} \subseteq \mathcal{S}$ and must prove that $\mathcal{F} = \mathcal{S}$. It will be sufficient to prove that $\mathcal{F}^{\vee} = \mathcal{F} = \mathcal{F}^{\downarrow}$. Therefore let g be an arbitrary element of either $\mathcal{F}^{\vee}$ or $\mathcal{F}^{\downarrow}$ and let $\eta : U \to \Sigma$ be a holomorphic map such that $\eta(U)$ is contained in the domain G of the function g. Since $g \in \mathcal{U}$ and η is continuous $g\circ\eta$ is usc on U. Let ζ_0 be an arbitrary point of U. Assume first that $g \in \mathcal{F}^{\vee}$. Then there exists a neighborhood V_0 of $\eta(\zeta_0)$ in G such that

$$g(\sigma) = \sup\{f(\sigma) : f \in \mathcal{F} , f \le g \text{ on } V_0\} , \sigma \in V_0.$$

Since η is continuous there exists a neighborhood U_0 of ζ_0 in U such that $\eta(U_0) \subseteq V_0$. By hypothesis, if $f \in \mathcal{F}$ and is defined on V_0 then $f\circ\eta$ is subharmonic on U_0. Also, if $f \le g$ on V_0 then $f\circ\eta \le g\circ\eta$ on U_0. Thus $g\circ\eta$ is the supremum on U_0 of subharmonic functions. Therefore by a well-known property of subharmonic functions [H4, Theorem 1.6.2] $g\circ\eta$ is subharmonic on U_0. Since subharmonicity is a local property it follows that $g\circ\eta$ is subharmonic on U. This completes the proof that $\mathcal{F}^{\vee} = \mathcal{F}$.

Next assume that $g \in \mathcal{F}^{\downarrow}$. Then there exists a neighborhood V_0 of $\eta(\zeta_0)$ in G and $\{f_n\} \subset \mathcal{F}$ such that $g(\sigma) \le f_{n+1}(\sigma) \le f_n(\sigma)$, for all n and $\sigma \in V_0$, and

$$\lim_{n\to\infty} f_n(\sigma) = g(\sigma) , \sigma \in V_0.$$

Choose a neighborhood U_0 of ζ_0 such that $\eta(U_0) \subseteq V_0$. Then $f_n\circ\eta$ is subharmonic on U_0 for each n. Also for $\zeta \in U_0$

$$(g\circ\eta)(\zeta) \le (f_{n+1}\circ\eta)(\zeta) \le (f_n\circ\eta)(\zeta)$$

and

$$\lim_{n\to\infty} (f_n\circ\eta)(\zeta) = (g\circ\eta)(\zeta).$$

Again it follows that $g\circ\eta$ is subharmonic on U_0 [H4, Theorem 1.6.2] and hence on U. Therefore $\mathcal{F}^{\downarrow} = \mathcal{F}$. ◊

The next theorem shows that in the special case $[\mathbb{C}^n, \wp]$ the $\wp$-subharmonic functions reduce to the psh functions.

28.2 THEOREM. *A function defined on an open subset of* $\mathbb{C}^n$ *will be plurisubharmonic iff it is* $\wp$*-subharmonic.*

Proof. That $\wp$-subharmonic implies plurisubharmonic is given by the preceding theorem. For the opposite implication denote by L the set of all functions of the form $n^{-1}\log|h|$, where n is a positive integer and h is holomorphic on an open subset of $\mathbb{C}^n$, by PS the set of all plurisubharmonic functions defined on subsets of $\mathbb{C}^n$, and by CPS the subset of PS consisting of continuous functions. Then by the discussion at the beginning of this section we have $PS \subseteq (CPS)^{\downarrow} \subseteq L^{\vee\downarrow}$. Also, by Theorem 27.6, $L \subseteq {}_{\wp}\mathfrak{S}$. Therefore $PS \subseteq {}_{\wp}\mathfrak{S}^{\vee\downarrow} = {}_{\wp}\mathfrak{S}$ and hence $PS = {}_{\wp}\mathfrak{S}$. ◊

§ 29. MAXIMUM PROPERTIES

We derive next a number of maximum properties, and some of their consequences, for $\mathfrak{A}$-subharmonic functions. Observe first that if f is usc on a compact set $K \subseteq \Sigma$ then f assumes a maximum value on K. If $\mathfrak{F}$ is a set of usc functions defined on K and B is a closed subset of K such that

$$\max_{\sigma\in B} f(\sigma) = \max_{\sigma\in K} f(\sigma)$$

for $f \in \mathfrak{F}$ then we call B an $\mathfrak{F}$-*set*. A simple Zorn's lemma argument shows that there always exists at least one minimal $\mathfrak{F}$-set. If there exists a *unique* minimal $\mathfrak{F}$-set we call it the *Šilov boundary of* K *relative to* $\mathfrak{F}$ and denote it by $\partial_{\mathfrak{F}}K$. The next proposition gives a sufficient condition for the existence of $\partial_{\mathfrak{F}}K$. The proof is a straightforward adaptation of the proof of the existence of a Šilov boundary in the case of an algebra of functions [F1, Theorem 15.1].

29.1 PROPOSITION. *If* $\mathfrak{F}$ *is closed under addition and contains* $\log|\mathfrak{A}|$ *(restricted to* K*), then* $\partial_{\mathfrak{F}}K$ *exists.*

Proof. Consider the set $\exp(\mathfrak{F})$. It consists of positive functions, is closed under multiplication, and contains $|\mathfrak{A}|$. Observe that since the exponential function is increasing, a subset of K will be an $\mathfrak{F}$-set iff it is an $\exp(\mathfrak{F})$-set. Although elements of $\exp(\mathfrak{F})$ are only usc and not necessarily continuous a standard proof for the existence of the Šilov boundary relative to an algebra of continuous functions applies to $\exp(\mathfrak{F})$ completing the proof of the existence of $\partial_{\mathfrak{F}}K$ (see [A3] or [R1; Theorem (3.3.1) and the remark following the proof of 3.3.1].) ◊

It follows from Proposition 29.1 that the Šilov boundary of K relative to $\mathcal{S}_K$ (the $\mathfrak{A}$-sh functions defined on K) exists. We denote it simply by $\partial_{\mathcal{S}}K$. Also, since $\mathcal{S}$ contains $\mathcal{L}$, it follows immediately that $\partial_{\mathfrak{A}}K \subseteq \partial_{\mathcal{S}}K$. If the compact set K is $\mathfrak{A}$-convex then, as in the case of the $\mathfrak{A}$-holomorphic functions (see Corollary 20.3), $\partial_{\mathcal{S}}K = \partial_{\mathfrak{A}}K$. This however is a nontrivial fact which depends on local maximum properties that we shall now develop. As might be expected these properties depend ultimately on the local maximum principle for $\mathfrak{A}$ (Theorem 14.2) and hence involve naturality of the system $[\Sigma, \mathfrak{A}]$.

At this point it is convenient to introduce some definitions which generalize the setting of the local maximum principle expressed in Theorem 14.2 and facilitate the discussion that follows. First we recall that a subset of Σ is said to be *locally closed* if it is the intersection of an open set and a closed set in Σ. Thus every open set and every closed set is locally closed. A locally closed set is obviously relatively open in its closure. In the following, Δ_0 will denote a locally closed set and Δ will denote its closure. Also let $\Gamma = \Delta \setminus \Delta_0$ and note that $\Gamma = \mathrm{bd}_\Delta \Delta_0$ (the boundary of Δ_0 in the space Δ). Now let $\mathfrak{F}$ be a given subset of $\mathcal{U}$. Then $\mathfrak{F}$ is said to *satisfy the local maximum principle in* Δ_0 and Δ_0 is said to be $\mathfrak{F}$-*local* iff for every relatively open set $U \subseteq \Delta_0$ and every $f \in \mathcal{U}$ defined on $\bar{U}$ with $f|U \in \mathfrak{F}|U$ it is true that

$$\sup_{\sigma \in \mathrm{bd}_\Delta U} f(\sigma) = \sup_{\sigma \in \bar{U}} f(\sigma).$$

Note that if Δ_0 is $\mathfrak{F}$-local then every relatively open subset of Δ_0 is $\mathfrak{F}$-local and has a nonempty boundary in Δ. If C is a family of complex-valued functions then Δ_0 is said to be C-local iff it is $|C|$-local. We prove next a useful localization lemma.

29.2 LEMMA. *Let* Δ_0 *be locally closed with compact closure and let* U *be a relatively open subset of* Δ_0. *Also let* g *be an element of* $\mathcal{U}$ *defined on* $\bar{U}$ *and such that*

$$\max_{\sigma \in bd_\Delta U} g(\sigma) < 0 = \max_{\sigma \in \bar{U}} g(\sigma).$$

Then there exists a point $\delta \in U$ *with the following property: For every neighborhood* V *of* δ *there exists a nonnegative integer* m *and a* $u \in \mathcal{G}$ *such that if* $g_V = mg + \log|u|$ *then*

$$\max_{\sigma \in bd_\Delta (U \cap V)} g_V(\sigma) < 0 = g_V(\delta).$$

Proof. Set $Z = \{\sigma \in \bar{U} : g(\sigma) = 0\}$. By hypothesis, $Z \subset U$ and, since g is usc, Z is closed and hence compact. Take δ to be any independent point for $[Z, \mathcal{G}]$ and let V be an arbitrary neighborhood of δ. If $Z \subset V$ then we may take $m = 1$ and $u = 1$ to obtain $g_V = g$ which has the desired property. If $Z \not\subset V$ then since δ is an independent point for $[Z, \mathcal{G}]$ there exists $u \in \mathcal{G}$ such that

$$\max_{\sigma \in Z \setminus V} |u(\sigma)| < e^{-1} < |u(\delta)| = 1$$

so

$$\max_{\sigma \in Z \setminus V} \log|u(\sigma)| < -1 < \log|u(\delta)| = 0.$$

Set $W = \{\sigma \in U : \log|u(\sigma)| < -1\}$. Then W is relatively open in Δ_0 and $(Z\setminus V) \subset W$. Next set $B = [bd_\Delta(U \cap V)]\setminus W$. Then B is compact and $Z \cap B = \emptyset$. If $B = \emptyset$ then $bd_\Delta(U \cap V) \subset W$ and the function $\log|u|$ has the desired property, so $g_V = \log|u|$, obtained by taking $m = 0$. Therefore assume that $B \neq \emptyset$ and set $s = \max_{\sigma \in B} g(\sigma)$. Since $Z \cap B = \emptyset$ it follows that $s < 0$. Hence there exists a positive integer m such that $ms + \log|u|_\Delta < 0$, where $|u|_\Delta = \max_{\sigma \in \Delta} |u(\sigma)|$. Now let $g_V = mg + \log|u|$. Then clearly $g_V(\delta) = 0$. Furthermore for $\sigma \in B$, $g_V(\sigma) \le ms + \log |u|_\Delta < 0$, and for $\sigma \in W \cap bd_\Delta(U \cap V)$ we have $g_V(\sigma) < m\cdot 0 + (-1) = -1$. Therefore

$$\max_{\sigma \in bd_\Delta (U \cap V)} g(\sigma) < 0$$

and the lemma follows. ◊

29.3 THEOREM. *If a locally closed set* Δ_0 *is* $\mathcal{G}$*-local, then it is also* $\mathcal{S}$*-local.*

Proof. Since Δ_0 is assumed to be $\mathcal{G}$-local and the logarithm is an increasing function the set Δ_0 is also $\mathcal{L}$-local. Next let $\mathfrak{F}$ be any set of functions with $\mathcal{L} \subseteq \mathfrak{F} \subseteq \mathcal{U}$, that is closed under addition. Denote by $\mathcal{S}$ either $\mathfrak{F}^{\vee}$ or $\mathfrak{F}^{\downarrow}$. Note that $\mathcal{S}$ is also closed under addition. The desired result will follow by induction on the $^{\vee}{\downarrow}$-resolution of $\mathcal{G}$ (Proposition 8.1) if we show that $\mathfrak{F}$-local implies $\mathcal{S}$-local for the set Δ_0.

Suppose, on the contrary, that Δ_0 is $\mathfrak{F}$-local but not $\mathcal{S}$-local. Then using the fact that $\mathcal{S}$ contains all real constants (since $\mathcal{L} \subseteq \mathcal{S}$) and is closed under addition, we obtain a relatively open set U in Δ_0 and $g \in \mathcal{U}$ defined on $\bar{U}$ such that $g|U \in \mathcal{S}$ and

$$\max_{\sigma \in bd_\Delta U} g(\sigma) < 0 = \max_{\sigma \in \bar{U}} g(\sigma).$$

Hence Lemma 29.2 applies. Let δ be the point in U given by the lemma. Since $g|U \in \mathcal{S}$ and $\mathcal{S}$ is equal to either $\mathfrak{F}^{\vee}$ or $\mathfrak{F}^{\downarrow}$ there exists a neighborhood V of δ such that g is determined in accordance with the definition of $\mathfrak{F}^{\vee}$ or $\mathfrak{F}^{\downarrow}$ on a neighborhood of $\overline{U \cap V}$. Now let $g_V = mg + \log |u|$ be the function given by Lemma 29.2 for the neighborhood V. Choose r so that

$$\max_{\sigma \in bd_\Delta (U \cap V)} g_V(\sigma) < r < 0 = g_V(\delta).$$

Now assume that $\mathcal{S} = \mathfrak{F}^{\vee}$. Then for each $\sigma \in \overline{U \cap V}$

$$g(\sigma) = \sup \{f(\sigma) : f \in \mathfrak{F} ; f \leq g \text{ on } \overline{U \cap V}\}.$$

For arbitrary $f \in \mathfrak{F}$ with $f \leq g$ on $\overline{U \cap V}$ set $f_V = mf + \log|u|$. Then $f_V \leq g_V$ on $\overline{U \cap V}$ and for each $\sigma \in \overline{U \cap V}$

$$g_V(\sigma) = \sup \{f_V(\sigma) : f \in \mathfrak{F} , f \leq g \text{ on } \overline{U \cap V}\}.$$

Therefore we can choose $f \in \mathfrak{F}$, with $f \leq g$ on $\overline{U \cap V}$, such that $r < f_V(\delta)$. Since $f_V \leq g_V$ on $\overline{U \cap V}$, and hence on $bd_\Delta(U \cap V)$, it follows that

$$\max_{\sigma \in bd_\Delta (U \cap V)} f_V(\sigma) < f_V(\delta) \leq \max_{\sigma \in \overline{U \cap V}} f_V(\sigma).$$

But $\mathfrak{F}$ is closed under addition and contains $\mathcal{L}$, so $f_V \in \mathfrak{F}$. We thus have a contradiction of the assumption that Δ_0 is $\mathfrak{F}$-local. This proves that Δ_0 is $\mathcal{S}$-local for $\mathcal{S} = \mathfrak{F}^{\vee}$.

Next assume that $\mathcal{L} = \mathfrak{F}^{\downarrow}$. Then there exists $\{f_n\} \subset \mathfrak{F}$ such that each f_n is defined on a neighborhood $\overline{U \cap V}$, $g \leq f_{n+1} \leq f_n$ for each n, and $\{f_n\}$ converges pointwise to g on $\overline{U \cap V}$. For each n define $f'_n = mf_n + \log|u|$. Then $\{f'_n\} \subset \mathfrak{F}$, $g_V \leq f'_{n+1} \leq f'_n$, and $\{f'_n\}$ converges pointwise to g_V on $\overline{U \cap V}$. Note that $bd_{\Delta}(U \cap V) \neq \emptyset$. Hence for each $\sigma \in bd_{\Delta}(U \cap V)$ there exists n_σ such that $n \geq n_\sigma$ implies $f'_n(\sigma) < r$. Since f'_n is usc there exists a neighborhood N_σ of the point σ such that $f'_{n_\sigma}(\sigma') < r$ for $\sigma' \in N_\sigma \cap (\overline{U \cap V})$. The set $bd_{\Delta}(U \cap V)$ is compact so is covered by a finite number of the neighborhoods N_σ, say $N_{\sigma_1}, \ldots, N_{\sigma_k}$. Let $n_0 = \max\{n_{\sigma_1}, \ldots, n_{\sigma_k}\}$. Then $f'_{n_0} \leq f_{n_{\sigma_i}} < r$ on $N_{\sigma_i} \cap (\overline{U \cap V})$ for each i, so $f'_{n_0}(\sigma) < r$ for $\sigma \in bd_{\Delta}(U \cap V)$. Moreover $0 = g_V(\delta) \leq f'_{n_0}(\delta)$ and $\delta \in U \cap V$, so

$$\max_{bd_{\Delta}(U \cap V)} f'_n(\sigma) < \max_{\sigma \in \overline{U \cap V}} f'_{n_0}(\sigma).$$

This contradicts the assumption that Δ_0 in $\mathfrak{F}$-local and completes the proof of the theorem. ◈

29.4 COROLLARY. *If* Ω *is a compact* $\mathfrak{G}$*-convex subset of* Σ, *then* $\partial_{\mathfrak{G}}\Omega = \partial_{\mathfrak{S}}\Omega$.

The next result extends to $\mathfrak{G}$-subharmonic functions a familiar continuity property for plurisubharmonic functions in $\mathbb{C}^n$. Note that the requirement excluding independent points (Definition 12.2) is automatic in the case of $\mathbb{C}^n$ since the system $[\mathbb{C}^n, \mathfrak{P}]$ has no independent points.

29.5 THEOREM. *Let* f *be* $\mathfrak{G}$*-subharmonic on an open set* $U \subseteq \Sigma$. *If* δ *is any point of* U *which is not an independent point for* $[\Sigma, \mathfrak{G}]$ *then*

$$f(\delta) = \limsup_{\sigma \to \delta} f(\sigma).$$

Proof. Since f is usc it is always true that $f(\delta) \geq \limsup_{\sigma \to \delta} f(\sigma)$. Also, by definition

$$\limsup_{\sigma \to \delta} f(\sigma) = \inf_{V_\delta} \sup_{\sigma \in V'_\delta} f(\sigma)$$

where V_δ ranges over all neighborhoods of δ contained in U and V'_δ denotes the deleted neighborhood $V_\delta \setminus \{\delta\}$. Since δ is not an independent point for $[\Sigma, \mathfrak{G}]$ it cannot be a locally independent point (Theorem 14.5). Recall also that points of Σ admit arbitrarily small $\mathfrak{G}$-convex neighborhoods (see §6). Therefore there exists for each of the neighborhoods V_δ a compact set $K \subset\subset V_\delta$ such that $\hat{K} \subset V_\delta$ and $\delta \in \hat{K}\setminus K$. By Corollary 29.4, $\partial_{\mathcal{S}}\hat{K} = \partial_{\mathfrak{a}}\hat{K} \subseteq K$, so

$$f(\delta) \leq \max_{\sigma\in K} f(\sigma) \leq \sup_{\sigma\in V_\delta'} f(\sigma).$$

Since V_δ may be taken arbitrarily small it follows that $f(\delta) \leq \limsup_{\sigma\to\delta} f(\sigma)$. ◆

29.6 COROLLARY. *Let* Ω *be a compact* $\mathfrak{a}$*-convex subset of* Σ *and let* f *be* $\mathfrak{a}$*-subharmonic on* Ω. *Then for every* $\delta \in \Omega\setminus\partial_{\mathfrak{a}}\Omega$,

$$f(\delta) = \limsup_{\sigma\to\delta} f(\sigma).$$

The next theorem contains a generalization of another familiar result for $[\mathbb{C}^n, \mathcal{P}]$.

29.7 THEOREM. *Let* Δ_0 *be an* $\mathfrak{a}$*-local locally closed subset of* Σ *with compact closure* Δ, *and let* U *be a relatively open subset of* Δ_0. *If* f, h *are functions defined on* $\bar{U}$ *such that* f *is* usc *on* $\bar{U}$ *and* $\mathfrak{a}$*-subharmonic on* U *while* h *is continuous on* $\bar{U}$ *and* $\mathfrak{a}$*-harmonic on* U, *then* $f \leq h$ *on* $\mathrm{bd}_\Delta U$ *implies* $f \leq h$ *on* U.

Proof. By definition both h and $-h$ are continuous on $\bar{U}$ and $\mathfrak{a}$-subharmonic on U. Therefore $f-h$ is usc on $\bar{U}$ and $\mathfrak{a}$-subharmonic on U. By hypothesis $f-h \leq 0$ on $\mathrm{bd}_\Delta U$. Since Δ_0 is $\mathfrak{a}$-local it is also $\mathcal{S}$-local (Theorem 29.3), so $f-h \leq 0$ on $\bar{U}$. In other words $f \leq h$ on $\bar{U}$. ◆

29.8 COROLLARY. *If* f *and* h *are both* $\mathfrak{a}$*-harmonic on* U *then* $f = h$ *on* $\mathrm{bd}_\Delta U$ *implies* $f = h$ *on* U.

§30. INTEGRAL REPRESENTATIONS

We turn next to the construction of integral representations for $\mathfrak{a}$-subharmonic functions. As before let Δ_0 be an $\mathfrak{a}$-local locally closed subset of Σ with compact closure Δ and boundary Γ relative to Δ. Denote by $\mathcal{S}^*_{\Delta_0}$ the set of all functions with values in $[-\infty, \infty)$ that are usc on Δ and $\mathfrak{a}$-subharmonic on Δ_0. Similarly denote by $\mathcal{H}^*_{\Delta_0}$ the set of all real-valued functions that are continuous on Δ and $\mathfrak{a}$-harmonic on Δ_0. Finally denote by $\mathcal{O}^*_{\Delta_0}$ the set of all complex-valued functions that are continuous on Δ and $\mathfrak{a}$-holomorphic on Δ_0. We have $\mathcal{H}^*_{\Delta_0} \subseteq \mathcal{S}^*_{\Delta_0}$, and by Theorem 27.6, $\log |\mathfrak{G}^*_{\Delta_0}| \subseteq \mathcal{S}^*_{\Delta_0}$. The set $\mathcal{S}^*_{\Delta_0}$ is closed under linear combinations with positive coefficients, $\mathcal{H}^*_{\Delta}$ is a closed linear subspace of the Banach space $C_r(\Delta)$ of real continuous functions on Δ, and $\mathfrak{G}^*_{\Delta_0}$ is a closed subalgebra of $C(\Delta)$.

Since Δ_0 is $\mathfrak{a}$-local it is also $\mathcal{S}$-local by Theorem 29.3. Therefore if $\delta \in \Delta$ then

$$g(\delta) \le \max_{\sigma\in\Gamma} g(\sigma),\ g \in \mathcal{S}^*_{\Delta_0}$$

and

$$-\min_{\sigma\in\Gamma} h(\sigma) \le h(\delta) \le \max_{\sigma\in\Gamma} h(\sigma),\ h \in \mathcal{H}^*_{\Delta_0}.$$

In particular

$$\max_{\sigma\in\Gamma} |h(\sigma)| = \max_{\delta\in\Delta} |h(\delta)|,\ h \in \mathcal{H}^*_{\Delta_0}.$$

Hence the map $h \to h|\Gamma$ is a linear isometry of $\mathcal{H}^*_{\Delta_0}$ onto a closed linear subspace $H(\Gamma) \subseteq C_r(\Gamma)$. The space $H(\Gamma)$ consists precisely of those elements of $C_r(\Gamma)$ that admit $\mathfrak{a}$-harmonic extensions to Δ_0, *i.e.* elements of $C_r(\Gamma)$ that admit a solution of a general Dirichlet problem. Note that if a function $f \in C_r(\Gamma)$ admits an $\mathfrak{a}$-harmonic extension $\tilde{f}$ onto Δ_0 then that extension is unique.

Let $\mathcal{S}_0$ denote an arbitrary subset of $\mathcal{S}^*_{\Delta_0}$ which is closed under linear combinations with positive coefficients and contains the constants. We will be interested primarily in the particular cases $\mathcal{S}^*_{\Delta_0}$, $\mathcal{S}^*_{\Delta_0} \cap C_r(\Delta)$, or $\mathcal{H}^*_{\Delta_0}$ for $\mathcal{S}_0$. Consider any extended real-valued function $f : \Gamma \to [-\infty, \infty]$ and define for each $\delta \in \Delta$

$$f^-(\delta) = \sup\{g(\delta) : g \in \mathcal{S}_0,\ g \le f \text{ on } \Gamma\}$$

where we take $f^-(\delta) = -\infty$ if no $g \in \mathcal{S}_0$ exists with $g \le f$. Also define

$$f^+(\delta) = -(-f)^-(\delta),\ \delta \in \Delta.$$

Then $f^+(\delta) = \inf\{-g(\delta) : g \in \mathcal{S}_0,\ f \le -g \text{ on } \Gamma\}$. Note that if $\delta \in \Gamma$ then $f^-(\delta) \le f(\delta) \le f^+(\delta)$. Also, for arbitrary $\delta \in \Delta$, $f^-(\delta) \le f^+(\delta)$. This is obvious if either $f^-(\delta) = -\infty$ or $f^+(\delta) = \infty$. If $f^-(\delta) \ne -\infty$ and $f^+(\delta) \ne \infty$ then, for arbitrary $g', g'' \in \mathcal{S}_0$ such that $g' \le f$ and $g'' \le -f$ on Γ, we have $g' + g'' \le 0$ on Γ. Hence by Theorem 29.3, $g' + g'' \le 0$ on Δ. Therefore $f^-(\delta) + (-f)^-(\delta) \le 0$, so $f^-(\delta) \le f^+(\delta)$, as claimed.

Next we define the set $\mathcal{C}_\delta$ consisting of all $f \in C_r(\Gamma)$ with the property that a $g \in \mathcal{S}_0$ exists such that $g < f$ on Γ and $g(\delta) \ge 0$. Since elements of $\mathcal{S}_0$

are usc they are bounded on Γ, so it is obvious that $\mathcal{C}_\delta$ is not empty. It is also obvious that $\mathcal{C}_\delta$ is an open subset of $C_r(\Gamma)$ which is closed under addition and multiplication by positive reals. Moreover if $f > g'$ and $-f > g''$ on Γ, where g' and g'' belong to $\mathcal{S}_0$, then $g' + g'' < 0$ on Γ. Hence by Theorem 29.3 $g' + g'' < 0$ on Δ. Therefore both f and $-f$ cannot belong to $\mathcal{C}_\delta$. Thus $\mathcal{C}_\delta$ is a proper open cone in $C_r(\Gamma)$. Observe also that, since $0 \in \mathcal{S}_0$, $\mathcal{C}_\delta$ contains the positive cone $C_r^+(\Gamma)$ in $C_r(\Gamma)$ consisting of all strictly positive functions.

Denote by $\mathfrak{m}$ the set of all probability measures on Γ. Let g be any extended real-valued function on Γ which is integrable with respect to m and define $\hat{g}(m) = \int_\Gamma g\,dm$. Observe that if g is bounded above then

$$\hat{g}(m) = \inf\{\hat{f}(m) : f \in C_r(\Gamma),\ g \le f\}$$

and if g is bounded below then

$$\hat{g}(m) = \sup\{\hat{f}(m) : f \in C_r(\Gamma),\ f \le g\}.$$

Thus if g is usc then $\hat{g}(m) > -\infty$ and if g is lsc then $\hat{g}(m) < \infty$.

30.1 LEMMA. *For an arbitrary measure* $m \in \mathfrak{m}$ *the following conditions are equivalent:*

(*) $g(\delta) \le \hat{g}(m)$ *for each* $g \in \mathcal{S}_0$.

(*)' $\hat{f}(m) \ge 0$ *for each* $f \in \mathcal{C}_\delta$.

Proof. Assume first that m satisfies (*) and let $f \in \mathcal{C}_\delta$. Thus there exists $g \in \mathcal{S}_0$, with $g(\delta) \ge 0$, such that $f > g$ on Γ. Hence

$$0 \le g(\delta) \le \hat{g}(m) \le \hat{f}(m)$$

so $\hat{f}(m) \ge 0$. In other words, (*) implies (*)'.

Now assume that m satisfies (*)' and consider an arbitrary $g \in \mathcal{S}_0$. Let $g' = g - g(\delta)$. Then $g' \in \mathcal{S}_0$ and $g'(\delta) = 0$. Therefore, $f \in C_r(\Gamma)$ and $f > g'$ on Γ imply $f \in \mathcal{C}_\delta$. Hence

$$\hat{g}(m) - g(\delta) = \inf\{\hat{f}(m) : f \in C_r(\Gamma),\ f \ge g'\} \ge 0.$$

Therefore (*)' implies (*). ♦

Denote by $\mathfrak{m}_\delta^0$ the subset of $\mathfrak{m}$ consisting of all measures that satisfy condition (*) of Lemma 30.1. Then $\mathfrak{m}_\delta^0$ is a linear convex set which is compact with respect to the $\sigma(C_r(\Gamma)', C_r(\Gamma))$-topology in $C_r(\Gamma)'$. For the special cases in which $\mathcal{S}_0$ is equal to $\mathcal{H}^*_{\Delta_0}$, $\mathcal{S}^*_{\Delta_0} \cap C_r(\Delta)$, or $\mathcal{S}^*_{\Delta_0}$ we shall denote $\mathfrak{m}^0_\delta$ by $\mathfrak{m}^{\mathcal{H}}_\delta$, $\mathfrak{m}^{\mathcal{CS}}_\delta$, or $\mathfrak{m}^{\mathcal{S}}_\delta$ respectively. Observe that since $\mathcal{H}^*_{\Delta_0} \subseteq \mathcal{S}^*_{\Delta_0} \cap C_r(\Delta) \subseteq \mathcal{S}^*_0$ we have $\mathfrak{m}^{\mathcal{H}}_\delta \supseteq \mathfrak{m}^{\mathcal{CS}}_\delta \supseteq \mathfrak{m}^{\mathcal{S}}_\delta$.

The result in the next lemma will be extended to arbitrary (bounded) semi-continuous functions in Theorem 30.5.

30.2 LEMMA. *Let* f_0 *be an arbitrary continuous function defined on* Γ. *Then the range of the function* $\hat{f}_0$ *on the set* $\mathfrak{m}^0_\delta$ *is equal to the interval* $[f^-_0(\delta), f^+_0(\delta)]$.

Proof. Let m be an arbitrary element of $\mathfrak{m}^0_\delta$. If $g', g'' \in \mathcal{S}_0$, with $g' \le f_0$ and $g'' \le -f_0$ on Γ, then $g'(\delta) \le \int g'dm \le \int f_0 dm$ and $g''(\delta) \le \int g''dm \le -\int f_0 dm$. Therefore $g'(\delta) < \hat{f}_0(m) \le -g''(\delta)$ and hence $f^-_0(\delta) \le \hat{f}_0(m) \le f^+_0(\delta)$. This proves that the range of $\hat{f}_0$ on $\mathfrak{m}^0_\delta$ is contained in the interval $[f^-_0(\delta), f^+_0(\delta)]$.

For the proof of the opposite inclusion let $t \in [f^-_0(\delta), f^+_0(\delta)]$ and denote by $\mathcal{F}$ the linear subspace of $C_r(\Gamma)$ spanned by the elements 1 and f_0. We wish to define a linear functional F_0 on $\mathcal{F}$ such that $F_0(1) = 1$, $F_0(f_0) = t$ and $F_0(f) \ge 0$ for $f \in \mathcal{C}_\delta \cap \mathcal{F}$. Note that if $f_0 = c$ (a constant) then $f^-_0(\delta) = c = f^+_0(\delta)$, so $t = c$ and the problem is trivial in this case. Therefore assume that 1 and f_0 are linearly independent so that elements of $\mathcal{F}$ are uniquely of the form $r_0 + rf_0$, where $r_0, r \in \mathbb{R}$. We accordingly define

$$F_0(r_0 + rf_0) = r_0 + rt, \; r_0 + rf_0 \in \mathcal{F}$$

so $F_0(1) = 1$ and $F_0(f_0) = t$. Now suppose that $r_0 + rf_0 \in \mathcal{C}_\delta$. Then there exists $g' \in \mathcal{S}_0$, with $g'(\delta) \ge 0$, such that $r_0 + rf_0 > g'$ on Γ. Note that, if $r = 0$ then $r_0 > g'(\delta) \ge 0$, by Theorem 29.3, so $F_0(r_0 + rf_0) \ge 0$. If $r \ne 0$ then

$$|r|^{-1}(g' - r_0) < \begin{cases} f_0, & \text{if } r > 0 \\ -f_0, & \text{if } r < 0. \end{cases}$$

Since $|r|^{-1}(g' - r_0) \in \mathcal{S}_0$ it follows that

$$|r|^{-1}(g'(\delta) - r_0) \le \begin{cases} f^-_0(\delta), & \text{if } r > 0 \\ (-f_0)^-(\delta) = -f^+_0(\delta), & \text{if } r < 0 \end{cases}$$

so

$$0 \le g'(\delta) \le \begin{cases} r_0 + rf^-_0(\delta), & \text{if } r > 0 \\ r_0 + rf^+_0(\delta), & \text{if } r < 0. \end{cases}$$

But $f_0^-(\delta) \le t \le f^+(\delta)$, so $r_0 + rt \ge 0$ and hence F_0 is nonnegative on $\mathcal{C}_\delta \cap F$. Now, by standard extension techniques for linear functionals, F_0 may be extended to a linear functional F on $C_r(\Gamma)$ such that $F(f) \ge 0$ for $f \in \mathcal{C}_\delta$. Since $C_r^+(\Gamma) \subseteq \mathcal{C}_\delta$ and $F(1) = 1$, F is a positive linear functional of norm 1 on $C_r(\Gamma)$. Therefore there exists a probability measure m^t on Γ such that $F(f) = \int f dm^t$, $f \in C_r(\Gamma)$. By Lemma 30.1 and the definition of $\mathfrak{m}_\delta^0$, $m^t \in \mathfrak{m}_\delta^0$ and $\hat{f}_0(m^t) = t$. Therefore the lemma follows. ◇

30.3 COROLLARY. *For every* $\delta \in \Delta$ *the set* $\mathfrak{m}_\delta^0$ *is nonempty, i.e. there exist probability measures* m_δ *on* Γ *such that* $g(\delta) \le \int g dm_\delta$, $g \in \mathcal{S}_0$.

30.4 LEMMA. *Let* g *be an arbitrary extended real function on* Γ *which is integrable with respect to each* $m \in \mathfrak{m}_\delta^0$ *and bounded above, so*

$$\int g dm = \inf\{\hat{f}(m) : f \in C_r(\Gamma),\ g \le f\}.$$

Then there exists $m_\delta^g \in \mathfrak{m}_\delta^0$ *such that* $g^+(\delta) = \hat{g}(m_\delta^g)$.

Proof. For $f \in C_r(\Gamma)$, with $g \le f$, denote by $\mathfrak{m}_\delta^0(f)$ the $\sigma(C_r(\Gamma)^*, C_r(\Gamma))$-closure of the set of all $m \in \mathfrak{m}_\delta^0$ for which $h \in C_r(\Gamma)$ exists such that $g \le h \le f$ and $\hat{h}(m) = h^+(\delta)$. By Lemma 30.2, the set $\mathfrak{m}_\delta^0(f)$ is nonempty. Since $\mathfrak{m}_\delta^0$ is compact, $\mathfrak{m}_\delta^0(f)$ is a compact subset of $\mathfrak{m}_\delta^0$. Moreover, if $\{f_1,\ldots,f_n\}$ is any finite subset of $C_r(\Gamma)$ such that $g \le f_i$ for each i and $f_0 = \inf(f_1,\ldots,f_n)$ then $f_0 \in C_r(\Gamma)$, $g \le f_0$ and $\mathfrak{m}_\delta^0(f_0) \subseteq \mathfrak{m}_\delta^0(f_1) \cap \cdots \cap \mathfrak{m}_\delta^0(f_n)$. In other words, the sets $\mathfrak{m}_\delta^0(f)$ for $g \le f$ possess the finite intersection property and so have a nonempty intersection. Choose

$$m_\delta^g \in \bigcap_{g \le f} \mathfrak{m}_\delta^0(f).$$

We prove that $\hat{g}(m_\delta^g) = g^+(\delta)$.

Note that, exactly as in the first part of the proof of Lemma 30.2, we always have $\hat{g}(m) \le g^+(\delta)$ for $m \in \mathfrak{m}_\delta^0$. Now let $\hat{g}(m_\delta^g) < t$. Then there exists $f \in C_r(\Gamma)$, with $g \le f$, such that $\hat{g}(m_\delta^g) \le \hat{f}(m_\delta^g) < t$. Since $m_\delta^g \in \mathfrak{m}_\delta^0(f)$, we must have $\hat{f}(m_\delta^g) = f^+(\delta)$, so $\hat{g}(m_\delta^g) \le f^+(\delta) < t$. Now choose $g' \in \mathcal{S}_0$, with $f \le -g'$ on Γ, such that $f^+(\delta) \le -g'(\delta) < t$. Then $\hat{g}(m_\delta^g) \le -g'(\delta) < t$. Since $g \le f \le -g'$, $g^+(\delta) \le -g'(\delta)$ and it follows that $\hat{g}(m_\delta^g) \le g^+(\delta) < t$. From the arbitrariness of t we conclude that $\hat{g}(m_\delta^g) = g^+(\delta)$. ◆

The next theorem extends the result in Lemma 30.2 to semi-continuous functions.

30.5 THEOREM. *Let* g *be an arbitrary bounded semi-continuous function defined on* Γ. *Then the range of the function* $\hat{f}$ *on the set* $\mathfrak{m}_\delta^0$ *is equal to the interval* $[g^-(\delta), g^+(\delta)]$.

Proof. Since g is assumed to be bounded we may apply Lemma 30.4 to both g and $-g$ obtaining measures m_δ^+ and m_δ^- in $\mathfrak{m}_\delta^0$ such that $g^+(\delta) = \hat{g}(m_\delta^+)$ and $g^-(\delta) = \hat{g}(m_\delta^-)$. Note that each t in $[g^-(\delta), g^+(\delta)]$ has the form

$$t = \theta g^-(\delta) + (1-\theta)g^+(\delta),\ 0 \le \theta \le 1.$$

Also, since $\mathfrak{m}_\delta^0$ is convex, it follows that $m = \theta m_\delta^- + (1-\theta)m_\delta^+ \in \mathfrak{m}_\delta^0$. Therefore

$$\hat{g}(m) = \theta\hat{g}(m_\delta^-) + (1-\theta)\hat{g}(m_\delta^+) = \theta g^-(\delta) + (1-\theta)g^+(\delta) = t$$

and hence the range of $\hat{g}$ contains $[g^-(\delta), g^+(\delta)]$. On the other hand, as in the first part of the proof of Lemma 30.2, the range of $\hat{g}$ on $\mathfrak{m}_\delta^0$ is also contained in $[g^-(\delta), g^+(\delta)]$, so we have equality. ◆

30.6 THEOREM. *(i) If* $m \in \mathfrak{m}_\delta^{\mathfrak{H}}$ *then* m *is a representing measure for the point* δ *relative to both* $\mathfrak{H}_{\Delta_0}^*$ *and* $\mathfrak{G}_{\Delta_0}^*$. *(ii) If* $m \in \mathfrak{m}_\delta^{\mathfrak{S}}$ *then* m *is a Jensen representing measure for* δ *relative to* $\mathfrak{G}_{\Delta_0}^*$.

Proof. Observe first that by Corollary 27.7 (ii) the real and imaginary parts of an element of $\mathfrak{G}_{\Delta_0}^*$ both belong to $\mathfrak{H}_{\Delta_0}^*$. Therefore it will be sufficient to prove (i) for $\mathfrak{H}_{\Delta_0}^*$. Now if $h \in \mathfrak{H}_{\Delta_0}^*$ then both h and $-h$ belong to $\mathfrak{S}_{\Delta_0}^*$. Therefore by condition (*) of Lemma 30.1, $h(\delta) \le \int h\,dm$ and $-h(\delta) \le \int(-h)\,dm$. Hence $h(\delta) = \int h\,dm$ so (i) follows. Finally, by Theorem 27.6, $\log|\mathfrak{G}_{\Delta_0}^*| \subseteq \mathfrak{S}_{\Delta_0}^*$, so condition (*) of Lemma 30.1 gives the Jensen inequality $\log|h(\delta)| \le \int \log|h|\,dm$, which proves (ii). ◆

§31. CHARACTERIZATION OF $\mathfrak{G}$-HARMONIC FUNCTIONS

We consider next the problem of characterizing elements of the space $H(\Gamma)$ in $C_r(\Gamma)$. First let $\mathfrak{m}_\delta$ denote the subset $\mathfrak{m}$ consisting of all those measures on Γ

that represent the point δ relative to $\mathcal{H}^*_{\Delta_0}$, *i.e.* $h(\delta) = \int h\,dm = \hat{h}(m)$, $h \in \mathcal{H}^*_{\Delta_0}$, for each $m \in \mathfrak{m}_\delta$. Thus if $f \in H(\Gamma)$ and $\tilde{f}$ is its $\mathfrak{G}$-harmonic extension to Δ_0, so $\tilde{f} \in \mathcal{H}^*_{\Delta_0}$, then $\tilde{f}(\delta) = \hat{f}(m)$ for each $m \in \mathfrak{m}_\delta$. Therefore a trivial necessary condition for an element $f \in C_r(\Gamma)$ to belong to $H(\Gamma)$ is that the function $\hat{f}$ be constant on each of the sets $\mathfrak{m}_\delta$. As we shall see, this condition is also sufficient. On the other hand, it is natural to conjecture a better result, *viz.* that f will belong to $H(\Gamma)$ if $\hat{f}$ is constant on each of the smaller sets $\mathfrak{m}^{\mathfrak{S}}_\delta$. Although we are unable to prove this conjecture, we have a result almost as good.

31.1 THEOREM. *Let* $f \in C_r(\Gamma)$. *Then a necessary and sufficient condition for* $f \in H(\Gamma)$ *is that the function* $\hat{f}$ *be constant on each of the sets* $\mathfrak{m}^{\mathfrak{CS}}_\delta$ *for* $\delta \in \Delta$. *In this case the* $\mathfrak{G}$*-harmonic extension of* f *is given by*

$$\tilde{f}(\delta) = \int f\,dm_\delta, \quad m_\delta \in \mathfrak{m}^{\mathfrak{CS}}_\delta, \quad \delta \in \Delta.$$

Proof. As already observed, the necessity of the condition is trivial. Therefore assume that $\hat{f}$ is constant on each of the sets $\mathfrak{m}^{\mathfrak{CS}}_\delta$. Then the function $\tilde{f}(\delta) = \int f\,dm_\delta$, $\delta \in \Delta$, where m_δ is any element of $\mathfrak{m}^{\mathfrak{CS}}_\delta$, is well-defined on Δ. Moreover, if $\delta \in \Gamma$ then $\tilde{f}(\delta) = f(\delta)$, since $\mathfrak{m}^{\mathfrak{CS}}_\delta$ obviously contains the unit point mass at δ when $\delta \in \Gamma$. In other words, $\tilde{f}$ is an extension of f. Furthermore, by Lemma 30.2, $f^-(\delta) = \tilde{f}(\delta) = f^+(\delta)$, $\delta \in \Delta$. Now since f^- is a supremum of continuous functions (elements of $\mathcal{S}^*_{\Delta_0} \cap C_r(\Delta)$), it is always lsc on Δ. Also, being a supremum of $\mathfrak{G}$-subharmonic functions on Δ_0, it will be $\mathfrak{G}$-subharmonic iff it is usc (and hence continuous). Similarly f^+ is always usc and $-f^+$ will be $\mathfrak{G}$-subharmonic iff f^+ is lsc. Therefore it follows that $\tilde{f}$ is continuous on Δ and $\mathfrak{G}$-harmonic on Δ_0.

31.2 COROLLARY. *A sufficient condition for* $H(\Gamma) = C_r(\Gamma)$ *is that each point of* Δ *have a unique representing measure on* Γ *relative to* $\mathfrak{G}^*_{\Delta_0}$.

§32. HARTOGS FUNCTIONS

For domains in $\mathbb{C}^n$ Bochner and Martin [B6, §6] have given a definition of "Hartogs function" that is closely related to, and in fact suggested, our definition of $\mathfrak{G}$-subharmonic function. Their definition follows.

32.1 DEFINITION (Bochner & Martin). *Let* G *be a domain in* $\mathbb{C}^n$. *Denote by* H_G *the smallest set of real functions defined on* G *that contains* $\log|\mathcal{O}_G|$ *and is closed under the following operations:*

(i) The operation of taking linear combinations of elements of H_G *with nonnegative coefficients.*

(ii) The operation of forming the supremum of any set of elements of H_G *uniformly bounded above on compact subsets of* G.

(iii) The operation of forming the pointwise limit of any nonincreasing sequence of elements of H_G.

(iv) The operation of forming the function $f^*(\zeta) = \limsup_{\xi\to\zeta} f(\xi)$, $\zeta \in G$, *for any* $f \in H_G$.

A function h *defined on* G *is called a* ***Hartogs function*** *if* $h|G' \in H_{G'}$ *for every domain* G' *with* $\overline{G'} \subset G$.

Observe that a Hartogs function is not required to be usc. However, if one limits attention to usc functions (take H_G to be the smallest set of *usc* real functions satisfying the conditions in the definition), then from the fact that usc functions are bounded above on compact sets (see operation (ii)) and Theorem 29.5, it will follow that H_G is contained in the set of all $\mathfrak{P}$-subharmonic functions defined on G and hence, by Theorem 28.2, in the set of all plurisubharmonic functions on G. Therefore every usc Hartogs function defined on G is plurisubharmonic. However, Bremerman [B8] has given an example of a domain G and a plurisubharmonic function on G which is not Hartogs. On the other hand, if G is a domain of holomorphy then a usc function on G will be plurisubharmonic iff it is Hartogs. Since points of $\mathbb{C}^n$ admit arbitrarily small neighborhoods that are domains of holomorphy, the Bremerman example shows that a function may be locally Hartogs without being Hartogs. Although the definition of $\mathfrak{A}$-subharmonic functions is essentially a local version of the definition of Hartogs function, the problem suggested by the Bremerman example does not arise for $\mathfrak{A}$-subharmonic functions precisely because the definition *is* local.

We remark in passing that T. W. Gamelin [G2] has defined a "*Hartogs function*" on the spectrum $\Phi_{\mathfrak{A}}$ of a uniform algebra $\mathfrak{A}$ to be a member of the smallest family H of functions with values in $[-\infty, \infty)$ such that (1) $\mathbb{R}^+\log|\hat{\mathfrak{A}}| \subseteq H$ and (2) H

contains the upper limit of any sequence of its elements that is bounded above. Also, in a recent paper [G3] Gamelin calls a real-valued function defined on a subset of the spectrum of $\mathfrak{A}$ an "$\mathfrak{A}$-subharmonic" function if it is bounded below and is the upper envelope of functions from $\mathbb{R}^+\log|\hat{\mathfrak{A}}|$. Such functions, which we shall call "*Gamelin* $\mathfrak{A}$-*subharmonic*", are automatically lsc and will obviously be $\mathfrak{A}$-subharmonic in our sense if they are usc and hence continuous. Also, the continuous Gamelin $\mathfrak{A}$-subharmonic functions are precisely those functions that are uniform limits of sequences of functions of the form

$$u = \max\{-M, C_1 \log|\hat{a}_1|, \ldots, C_m \log|\hat{a}_m|\}$$

where $M > 0$, $C_j > 0$ $(j = 1,\ldots,m)$ and $a_j \in \mathfrak{A}$ $(j = 1,\ldots,m)$. One of the main results for these functions is the following "localization theorem" which asserts that the property of being Gamelin $\mathfrak{A}$-subharmonic is local.

32.2 THEOREM [G3]. *A bounded* lsc *function* u *on* $\Phi_{\mathfrak{A}}$ *will be Gamelin* $\mathfrak{A}$-*subharmonic iff there exists for each point of* $\Phi_{\mathfrak{A}}$ *a compact neighborhood* N *on which* u *is Gamelin* $\hat{\mathfrak{A}}_N$-*subharmonic, where* $\hat{\mathfrak{A}}_N$ *denotes the closure of* $\hat{\mathfrak{A}}|N$ *in* $C(N)$.

Next let m be a Jensen representing measure on $\Phi_{\mathfrak{A}}$ for a point $\varphi \in \Phi_{\mathfrak{A}}$ relative to $\hat{\mathfrak{A}}$; i.e.

$$\log|\hat{a}(\varphi)| \le \int \log|\hat{a}|\,dm, \quad a \in \mathfrak{A}.$$

If the unit point mass at φ is the only Jensen measure for φ, then φ is called a *Jensen boundary point* and the set of all such points is the *Jensen boundary* of $\Phi_{\mathfrak{A}}$ relative to $\hat{\mathfrak{A}}$. For continuous functions we have the following strengthening of the preceding result.

32.3 THEOREM [G3]. *A continuous function on* $\Phi_{\mathfrak{A}}$ *will be Gamelin* $\mathfrak{A}$-*subharmonic iff it is Gamelin* $\mathfrak{A}$-*subharmonic at each point of* $\Phi_{\mathfrak{A}}$ *outside the Jensen boundary.*

One consequence of the above localization theorem is that the notion of being a Jensen boundary point is local.

32.4 THEOREM [G3]. *If a point* $\varphi \in \Phi_{\mathfrak{A}}$ *admits a compact neighborhood* N *such that* φ *is a Jensen boundary point of* N *(relative to* $\hat{\mathfrak{A}}_N$*), then* φ *is a Jensen boundary point of* $\Phi_{\mathfrak{A}}$.

CHAPTER VII
VARIETIES

§33. VARIETIES ASSOCIATED WITH AN $\mathfrak{G}$-PRESHEAF

The main purpose of this section is the investigation of an abstract analogue for $[\Sigma, \mathfrak{G}]$ of the familiar notion of an analytic variety in finite dimensions. As might be expected, the fundamental idea is to let the $\mathfrak{G}$-holomorphic functions play a role in the abstract situation analogous to that of the ordinary holomorphic functions in the finite dimensional case. However, in the general case it turns out to be desirable to formulate the definition of a variety in terms of an arbitrary presheaf of continuous functions over Σ. (See §16.) In particular, the definition will admit "varieties" associated with $\mathfrak{G}$-subharmonic functions as well as $\mathfrak{G}$-holomorphic functions. We are accordingly interested primarily in the following choices for the presheaf in question:

1^0. The presheaf $\{\Sigma, '\mathfrak{G}\}$, where $'\mathfrak{G}$ denotes the set of all almost $\mathfrak{G}$-holomorphic functions (see Definition 17.3 (ii)) defined on arbitrary subsets of Σ.

2^0. The presheaf $\{\Sigma, \exp('\mathfrak{CS})\}$, where $'\mathfrak{CS}$ denotes the set of all continuous *almost* $\mathfrak{G}$-subharmonic functions (see Definition 26.2) defined on arbitrary subsets of Σ. Thus, $\exp('\mathfrak{CS})$ is the set of all non-negative continuous functions defined on subsets of Σ that are *logarithmically* $\mathfrak{G}$-subharmonic on the set where they differ from zero.

Observe that $|'\mathfrak{G}|$ obviously contains $|\mathfrak{G}|$ and $\exp('\mathfrak{CS})$ contains $|\mathfrak{G}|$ by Corollary 27.5 (iii). Moreover, $'\mathfrak{G}$ and $\exp('\mathfrak{CS})$ are closed under multiplication. Finally, we note that $'\mathfrak{G}$ preserves $\mathfrak{G}$-convex hulls, by Theorem 20.2, and $\exp('\mathfrak{CS})$ preserves $\mathfrak{G}$-convex hulls, by Corollary 29.4. In other words, each of these presheaves is an $\mathfrak{G}$-presheaf in the sense of Definition 22.1. With this fact in mind, we begin our discussion with an arbitrary $\mathfrak{G}$-presheaf $\{\Sigma, \mathfrak{F}\}$ over Σ.

33.1 DEFINITION. *Let* Θ *and* Ω *be subsets of* Σ *with* $\Theta \subseteq \Omega$. *Then* Θ *is called an* $\mathfrak{F}$-*subvariety of* Ω *if for each point* $\omega \in \Omega$ *there exists a neighborhood* U *such that* $U \cap \Theta$ *consists of the common zeros of fucntions from* $\mathfrak{F}_{U\cap\Omega}$. *If this condition is only required at points of* Θ, *then* Θ *is called a local* $\mathfrak{F}$-*subvariety of* Ω.

Observe that Θ will be a local $\mathfrak{F}$-subvariety of Ω iff it is local in the more usual sense that each of its points admits a neighborhood U such that $U \cap \Theta$ is an $\mathfrak{F}$-subvariety of $U \cap \Omega$. Since $|\mathfrak{F}|$ contains $|\mathfrak{G}|$ by hypothesis, $\mathfrak{F}$ contains nonzero constant functions on each subset of Σ. Therefore, if $\omega \in \Omega\backslash\bar{\Theta}$ then the condition at ω for Θ to be an $\mathfrak{F}$-subvariety of Ω is automatically satisfied. It is also automatically satisfied at interior points of Θ relative to Ω. Thus, if Θ is relatively closed in Ω then it will be an $\mathfrak{F}$-subvariety of Ω iff the condition in the definition holds at each point of $\mathrm{bd}_\Omega\,\Theta$. If there exists a neighborhood W of the set Θ such that Θ consists of the common zeros of functions from $\mathfrak{F}_{W\cap\Omega}$, then Θ is an $\mathfrak{F}$-subvariety of Ω and is said to be *globally defined*. A special case here is an $\mathfrak{F}$-*hypersurface* in Ω which is a relatively closed subset of Ω equal to the zero set of a single element of $\mathfrak{F}$ defined on a relatively open subset of Ω. Note that in general the functions from $\mathfrak{F}_{U\cap\Omega}$ required by Definition 33.1 may be finite or infinite in number. For the special case $\mathfrak{F} = {}'\mathfrak{G}$ we shall use the terminology "$\mathfrak{G}$-*subvariety* in place of "'$\mathfrak{G}$-*subvariety*".

It will be convenient to have the following notations. Let $\mathcal{E}$ be an arbitrary set of functions with a common domain of definition E. Then $Z(\mathcal{E})$ will denote the set of common zeros of the elements of $\mathcal{E}$ in the set E. In particular, if f is a single function defined on E then $Z(f)$ is the set of zeros of f in E. Observe that if $\mathcal{E}_1$ and $\mathcal{E}_2$ are any two sets of functions defined on E then

$$Z(\mathcal{E}_1 \cup \mathcal{E}_2) = Z(\mathcal{E}_1) \cap Z(\mathcal{E}_2), \quad Z(\mathcal{E}_1\mathcal{E}_2) = Z(\mathcal{E}_1) \cup Z(\mathcal{E}_2)$$

where $\mathcal{E}_1\mathcal{E}_2 = \{f_1f_2 : f_1 \in \mathcal{E}_1, f_2 \in \mathcal{E}_2\}$. In the next proposition we record a few of the elementary properties of $\mathfrak{F}$-subvarieties.

33.2 PROPOSITION. *(i) The empty set* $\emptyset$ *and the full set* Ω *are* $\mathfrak{F}$-*subvarieties of* Ω.

(ii) If Θ *is an* $\mathfrak{F}$*-subvariety of* Ω *and* $\Omega' \subseteq \Omega$ *then* $\Theta \cap \Omega'$ *is an* $\mathfrak{F}$*-subvariety of* Ω'.

(iii) If G *is a relatively open subset of* Ω *and* Θ *is an* $\mathfrak{F}$*-subvariety of* G *that is relatively closed in* Ω *then* Θ *is an* $\mathfrak{F}$*-subvariety of* Ω.

(iv) If Θ *is an* $\mathfrak{F}$*-subvariety of* Ω *and* $\Theta = \Theta_1 \cup \Theta_2$, *where* $\Omega \cap \Theta_1 \cap \bar{\Theta}_2 = \Omega \cap \bar{\Theta}_1 \cap \Theta_2 = \emptyset$, *then* Θ_1 *and* Θ_2 *are also* $\mathfrak{F}$*-subvarieties of* Ω.

(v) If Θ_1 *and* Θ_2 *are* $\mathfrak{F}$*-subvarieties of* Ω *then* $\Theta_1 \cap \Theta_2$ *and* $\Theta_1 \cup \Theta_2$ *are* $\mathfrak{F}$*-subvarieties of* Ω.

Proof. The proofs of (i)-(iii) are trivial so will be omitted. Therefore consider (v). We prove that Θ_1 is an $\mathfrak{F}$-subvariety of Ω. Observe first that the condition on Θ_1 and Θ_2, plus the fact that Θ is relatively closed in Ω, implies that Θ_1 and Θ_2 are relatively closed in Ω. Also, if $\omega \in \Theta_1$ then there exists a neighborhood U of ω such that $U \cap \Theta_2 = \emptyset$, so $U \cap \Theta = U \cap \Theta_1$. Therefore we may choose U so that $U \cap \Theta_1$ consists of the common zeros of a subset of $\mathfrak{F}_{U\cap\Omega}$, proving that Θ_1 is an $\mathfrak{F}$-subvariety of Ω.

For the proof of (v), let ω be an arbitrary element of Ω and choose a neighborhood U of ω such that $U \cap \Theta_1$ and $U \cap \Theta_2$ consist respectively of the common zeros of subsets $\mathcal{E}_1$ and $\mathcal{E}_2$ in $\mathfrak{F}_{U\cap\Omega}$. Then since

$$Z(\mathcal{E}_1 \cup \mathcal{E}_2) = Z(\mathcal{E}_1) \cap Z(\mathcal{E}_2) = U \cap \Theta_1 \cap \Theta_2$$
$$Z(\mathcal{E}_1\mathcal{E}_2) = Z(\mathcal{E}_1) \cup Z(\mathcal{E}_2) = U \cap (\Theta_1 \cup \Theta_2)$$

and

$$\mathcal{E}_1 \cup \mathcal{E}_2 \subseteq \mathfrak{F}_{U\cap\Omega}, \ \mathcal{E}_1\mathcal{E}_2 \subseteq \mathfrak{F}_{U\cap\Omega}$$

it follows that $\Theta_1 \cap \Theta_2$ and $\Theta_1 \cup \Theta_2$ are $\mathfrak{F}$-subvarieties of Ω. ♦

§34. CONVEXITY PROPERTIES

The next theorem, which generalizes a well-known result for ordinary analytic varieties in finite dimensions (see [G7, p. 219]), is a good example of "analytic phenomena" since the notion of a subvariety is local in character while $\mathfrak{G}$-convexity is a global property.

34.1 THEOREM. *Let* $\{\Sigma, \mathfrak{F}\}$ *be an arbitrary* $\mathfrak{G}$*-presheaf over* Σ *and let* Ω *be an* $\mathfrak{G}$*-convex subset of* Σ*. Then every* $\mathfrak{F}$*-subvariety of* Ω *is* $\mathfrak{G}$*-convex.*

Proof. Let Θ be an $\mathfrak{F}$-subvariety of Ω and let K be any compact subset of Θ. Then $\hat{K} \subseteq \Omega$ and, since Θ is relatively closed in Ω, the set $B = \hat{K} \cap \Theta$ is closed and contains K so is an $\mathfrak{G}$-boundary for $\hat{K}$. Suppose that $\hat{K}$ were not contained in Θ and set $G = \hat{K} \backslash B$. Then G is a nonempty relatively open subset of $\hat{K}$. Let δ be an independent point for $[\bar{G}, \mathfrak{G}]$. Then by the local maximum principle (Theorem 14.2) $\delta \in \mathrm{bd}_{\hat{K}} G$. Now choose a neighborhood U of δ such that $U \cap \Theta$ consists of the common zeros of a set $\mathcal{E} \subseteq \mathfrak{F}_{U \cap \Omega}$. Since $\{\Sigma, \mathfrak{F}\}$ preserves $\mathfrak{G}$-convex hulls, the point δ is also independent for $[\bar{G}, \mathfrak{F}_G^*]$, where $\mathfrak{F}_G^*$ denotes the set of all $g \in C(\bar{G})$ such that $g|G \in \mathfrak{F}_G$. Therefore by Lemma 23.1 there exists a neighborhood V of δ contained in U such that any element of $\mathfrak{F}_{U \cap \hat{K}}$ that vanishes on the set $U \cap B$ must also vanish on $V \cap B$. Since $\mathfrak{F}_{U \cap \Omega}|V \cap \hat{K} \subseteq \mathfrak{F}_{V \cap \hat{K}}$ and elements of $\mathcal{E}$ vanish on $U \cap \Theta$ (and hence on $U \cap B$) it follows that the elements of $\mathcal{E}$ must vanish on $V \cap \hat{K}$. Hence $V \cap \hat{K} \subset \Theta$. But this is impossible since V contains the point δ, which is a boundary point of G, so V must contain points of $\hat{K} \backslash \Theta$. Therefore $\hat{K} \subseteq \Theta$ and Θ is $\mathfrak{G}$-convex. ◊

The method of proof used for the above theorem also gives the following useful result.

34.2 THEOREM. *Let* H *be an arbitrary open subset of* Σ *and* Θ *a local* $\mathfrak{F}$*-subvariety of* H*. Then* Θ *will be* $\mathfrak{G}$*-convex if* $\hat{K} \cap \Theta$ *is closed in* Σ *for every compact set* $K \subseteq \Theta$*.*

Proof. That $\mathfrak{G}$-convexity implies the stated condition is obvious. Therefore let K be an arbitrary compact subset of Θ and proceed as in the proof of Theorem 34.1. In the present case however, the $\mathfrak{G}$-hull $\hat{K}$ *a priori* need not be contained in H. On the other hand, the set $\hat{K} \cap \Theta$, being closed by hypothesis, is an $\mathfrak{G}$-boundary for $\hat{K}$. Therefore if K is not contained in Θ then $\mathrm{bd}_{\hat{K}} G$ contains an independent point for $[\bar{G}, \mathfrak{G}]$, as before. Now, since H is open the neighborhood U may be assumed to be contained in H. This forces $U \cap \hat{K} \subset H$ and the remainder of the argument is identical with that for Theorem 34.1. ◊

§35. GENERALIZATIONS OF SOME RESULTS OF GLICKSBERG

The next several results (35.1-35.5) were suggested by work of Glicksberg [G5, §4] generalizing certain familiar results for the disc algebra. We shall assume that Σ is compact and specialize the general $\mathfrak{G}$-presheaf $\{\Sigma, \mathfrak{F}\}$, involved in the preceding discussion, to the case of a *maximal* $\mathfrak{G}$-presheaf of function algebras. By Corollary 24.4 we automatically have $'\mathfrak{G} \subseteq \mathfrak{F}$.

Let G be any open subset of $\Sigma \setminus \partial[\Sigma, \mathfrak{G}]$ and Θ an arbitrary $\mathfrak{G}$-subvariety (*i.e.* $'\mathfrak{G}$-subvariety) of Σ. Set $H = G \setminus \Theta$ and consider the algebra $\mathfrak{F}_{\bar{H}}$.

Since $H \subseteq \Sigma \setminus \partial[\Sigma, \mathfrak{G}]$ and $\{\Sigma, \mathfrak{F}\}$ is an $\mathfrak{G}$-presheaf, it follows from Theorem 22.2 (iii) that any independent point for $[\bar{H}, \mathfrak{F}_{\bar{H}}]$ is an independent point for $[\Sigma, \mathfrak{G}]$. Therefore $\partial[\bar{H}, \mathfrak{F}_{\bar{H}}] \subseteq \text{bd } H$. When $\bar{H}$ is compact we have the sharper result that $\partial[\bar{H}, \mathfrak{G}_{\bar{H}}] \subseteq \overline{(\text{bd } H) \setminus \Theta}$. This fact, which is not obvious but not so very difficult to prove, is contained in the next much deeper theorem (Cf. [G5, Theorem 4.8]) whose proof is rather long and tedious.

35.1 THEOREM. *Let* Σ *be compact,* G *an open subset of* $\Sigma \setminus \partial[\Sigma, \mathfrak{G}]$, *and* Θ *an* $\mathfrak{G}$-*subvariety of* Σ. *Denote by* $\mathfrak{B}$ *the subalgebra of* $C(\bar{G} \setminus \Theta)$ *consisting of the bounded functions in* $\mathfrak{F}_{\bar{G} \setminus \Theta}$. *Then for each* $b \in \mathfrak{B}$

$$|b|_{\bar{G} \setminus \Theta} = |b|_{(\text{bd } G) \setminus \Theta}.$$

Proof. As before, let $H = G \setminus \Theta$ and note that $H \subseteq \bar{G} \setminus \Theta = \bar{H} \setminus \Theta \subseteq \bar{H}$, so $\mathfrak{F}_{\bar{H}} | \bar{G} \setminus \Theta \subseteq \mathfrak{B}$. Moreover $(\text{bd } G) \setminus \Theta \subseteq (\text{bd } H)$. Therefore the theorem implies that $\partial[\bar{H}, \mathfrak{F}_{\bar{H}}] \subseteq \overline{(\text{bd } H) \setminus \Theta}$, as remarked above.

Elements of $\mathfrak{B}$ will not in general admit continuous extensions to all of $\bar{H}$. However, $\mathfrak{B}$ is a normed algebra under the supnorm $|b|_{\bar{G} \setminus \Theta}$ so has a compact spectrum $\Phi_{\mathfrak{B}}$. We denote by $\mu : \bar{G} \setminus \Theta \to \Phi_{\mathfrak{B}}$ the homeomorphic embedding of $\bar{G} \setminus \Theta$ in $\Phi_{\mathfrak{B}}$, *i.e.* $\hat{b}(\mu(\sigma)) = b(\sigma)$, for $b \in \mathfrak{B}$ and $\sigma \in \bar{G} \setminus \Theta$. Set $\Omega_0 = \mu(H)$ and $\Omega = \mu(\bar{G} \setminus \Theta)$. Note that since $H \subseteq \bar{G} \setminus \Theta \subseteq \bar{H}$ the set Ω_0 is dense in Ω, so $\bar{\Omega}_0 = \bar{\Omega}$. Now each $\varphi \in \Phi_{\mathfrak{B}}$ determines a homomorphism

$$\tau(\varphi) : \mathfrak{G} \to \mathbb{C}, \ a \to a|(\bar{G} \setminus \Theta) \to \widehat{a|(\bar{G} \setminus \Theta)}(\varphi)$$

of the algebra $\mathfrak{G}$ onto $\mathbb{C}$. Moreover, since $\overline{G \setminus \Theta}$ is compact and

$$|\widehat{a|(\bar{G}\setminus\Theta)}(\varphi)| \leq |a|_{\bar{G}\setminus\Theta} = |a|_{\bar{H}}, \quad a \in \mathfrak{G}$$

the homomorphism $\tau(\varphi) : \mathfrak{G} \to \mathbb{C}$ is continuous, so determines a point $\tau(\varphi) \in \Sigma$ such that

$$\widehat{a|(\bar{G}\setminus\Theta)}(\varphi) = a(\tau(\varphi)), \quad a \in \mathfrak{G}.$$

The mapping $\tau : \Phi_{\mathfrak{B}} \to \Sigma$ is obviously continuous. Also observe that τ is equal to the inverse of μ on Ω, *i.e.* $\tau(\mu(\sigma)) = \sigma$ for $\sigma \in \bar{G}\setminus\Theta$, and $\mu(\tau(\varphi)) = \varphi$ for $\varphi \in \Omega$. In particular $\tau(\Omega_0) = G\setminus\Theta$, and since τ is continuous and $\bar{\Omega}$ is compact, it follows that $\tau(\bar{\Omega}) = \overline{G\setminus\Theta} = \bar{H}$. Moreover $\tau(\bar{\Omega}\setminus\Omega) \subseteq \text{bd}\ \Theta$. In fact, let $\bar{\omega} \in \bar{\Omega}\setminus\Omega$ and suppose that $\tau(\bar{\omega}) = \bar{\sigma} \notin \text{bd}\ \Theta$. Then since Θ is closed there exists a neighborhood $N_{\bar{\sigma}}$ of $\bar{\sigma}$ such that $\bar{N}_{\bar{\sigma}} \cap \Theta = \emptyset$. Set $K = \bar{N}_{\bar{\sigma}} \cap (\bar{G}\setminus\Theta)$. Then K is a closed and hence compact subset of $\bar{G}\setminus\Theta$. Therefore $\mu(K)$ is a compact subset of Ω, so there exists a neighborhood $U_{\bar{\omega}}$ of $\bar{\omega}$ disjoint from $\mu(K)$. On the other hand, $\bar{\omega}$ is a limit point of Ω and τ is continuous, so there exists $\omega \in U_{\bar{\omega}}$ such that $\tau(\omega) \in N_{\bar{\sigma}}$. But then $\omega = \mu(\tau(\omega)) \in \mu(K)$, contradicting the fact that $U_{\bar{\omega}}$ and $\mu(K)$ are disjoint. Therefore $\tau(\bar{\Omega}\setminus\Omega) \subseteq \text{bd}\ \Theta$ as claimed.

Now suppose that the theorem were false. Then there exists a function $u \in \mathfrak{B}$ such that $|u|_{(\text{bd}\ G)\setminus\Theta} \leq 3^{-1}$ and $|u|_{\bar{G}\setminus\Theta} > 1$. Hence, if $\Gamma = \mu((\text{bd}\ G)\setminus\Theta)$ then $|\hat{u}|_{\Gamma} \leq 3^{-1}$ and $|\hat{u}|_{\bar{\Omega}} > 1$. It follows that there exists an independent point $\bar{\omega}$ for $[\bar{\Omega}, \hat{\mathfrak{B}}]$ such that $|\hat{u}(\bar{\omega})| > 1$. Note that since $|\hat{u}|_{\Gamma} \leq 3^{-1}$ the point $\bar{\omega}$ cannot belong to $\bar{\Gamma}$. Suppose that $\bar{\omega} \in \Omega_0$. Then there exists $\delta \in G\setminus\Theta$ with $\mu(\delta) = \bar{\omega}$. But this implies that δ is an independent point for $[G\setminus\Theta, \mathfrak{B}]$ and hence, by Theorem 22.2 (iii), is independent for $[\Sigma, \mathfrak{G}]$. Again this is impossible since $G \subseteq \Sigma\setminus\partial[\Sigma, \mathfrak{G}]$, so it follows that $\bar{\omega} \in \bar{\Omega}\setminus\Omega$ and hence that $\tau(\bar{\omega}) = \bar{\sigma} \in \text{bd}\ \Theta$.

Next choose a neighborhood $N_{\bar{\sigma}}$ of $\bar{\sigma}$ such that $N_{\bar{\sigma}} \cap \Theta$ consists of the common zeros of a set of functions almost $\mathfrak{G}$-holomorphic on $\bar{N}_{\bar{\sigma}}$. Since $\bar{\omega} \notin \bar{\Gamma}$ and τ is continuous, we can choose a neighborhood $U_{\bar{\omega}}$ of $\bar{\omega}$ in $\Phi_{\mathfrak{B}}$ with $\bar{U}_{\bar{\omega}} \cap \bar{\Gamma} = \emptyset$ and $\tau(\bar{U}_{\bar{\omega}} \cap \bar{\Omega}) \subset N_{\bar{\sigma}}$. By the independence of $\bar{\omega}$, there exists $v \in \mathfrak{B}$ such that $|v|_{\bar{\Omega}\setminus U_{\bar{\omega}}} < 3^{-1}$ and $\hat{v}(\bar{\omega})| > 1$. Also, since $\bar{\omega} \in \bar{\Omega} = \bar{\Omega}_0$ there exists $\omega_0 \in U_{\bar{\omega}} \cap \Omega_0$ such that $1 < |\hat{v}(\omega_0)|$. Set $\sigma_0 = \tau(\omega_0)$. Then $|v(\sigma_0)| > 1$. Now define $U = U_{\bar{\omega}} \cap \Omega_0$ and $N = \tau(U)$. Then N is an open subset of H and $\bar{N} \cap [(\text{bd}\ G)\setminus\Theta] = \emptyset$, so $\bar{N} \subset H \cap \Theta$. Also $|v|_{\bar{H}\setminus N} \leq |\hat{v}|_{\bar{\Omega}\setminus U_{\bar{\omega}}} < 3^{-1}$. Since $\sigma_0 \in N_{\bar{\sigma}}\setminus\Theta$, there exists a function g almost $\mathfrak{G}$-holomorphic on $\bar{N}_{\bar{\sigma}}$ such that $g(\sigma_0) \neq 0$ and $g|(N_{\bar{\sigma}} \cap \Theta) = 0$. Choose a positive integer m such that $3^{-m}|g|_{\bar{N}_{\bar{\sigma}}} < |g(\sigma_0)|$ and define

$$h(\sigma) = \begin{cases} v(\sigma)^m g(\sigma), & \sigma \in \bar{N}\setminus\Theta = \bar{N} \cap H \\ 0 & , \ \sigma \in \bar{N} \cap \Theta \end{cases} .$$

Since v is bounded on N and $g(\bar{N} \cap \Theta) = 0$, the function h is continuous on $\bar{N}$. Also h is almost $\mathfrak{F}$-holomorphic on N. By Theorem 24.7 it follows that $h \in \mathfrak{F}_{\bar{N}}$. Finally, $\bar{N} \subseteq H \cup \Theta$, so

$$\text{bd } N = \bar{N}\setminus N \subseteq (H \cap \Theta)\setminus N \subseteq H\setminus N$$

and hence

$$|h|_{\text{bd } N} \leq |h|_{H\setminus N} \leq |v^m|_{H\setminus N}|g|_{\bar{N}_{\bar{\sigma}}} < 3^{-m}|g|_{\bar{N}_{\bar{\sigma}}} < |g(\sigma_0)|.$$

Since $|v(\sigma_0)| > 1$, $|g(\sigma_0)| < |v(\sigma_0)^m||g(\sigma_0)| = |h(\sigma_0)|$, and hence $|h|_{\text{bd } N} < |h(\sigma_0)|$. This implies that N contains an independent point for $[\bar{N}, \mathfrak{F}_{\bar{N}}]$ which by Theorem 22.2 (iii) must be an independent point for $[\Sigma, \mathfrak{G}]$. Again this contradicts $G \subseteq \Sigma\setminus\partial[\Sigma, \mathfrak{G}]$ and completes the proof of the theorem. ◊

The following corollary is a generalization of Schwartz's Lemma obtained by taking $\mathfrak{G}$ to be the disc algebra and $g(\zeta) = \zeta$, $\zeta \in D$.

35.2 COROLLARY. *Let* f, g *be almost* $\mathfrak{G}$*-holomorphic functions defined on* Σ. *Suppose that* f/g *is bounded on* $\Sigma\setminus Z(g)$. *Then*

$$|f/g|_{\Sigma\setminus Z(g)} = |f/g|_{\partial[\Sigma, \mathfrak{G}]\setminus Z(g)}.$$

Next let g be a function whose domain of definition includes the set $X \subseteq \Sigma$, and for $\beta > 0$ define

$$X\{|g| < \beta\} = \{\sigma \in X : |g(\sigma)| < \beta\}.$$

The following corollary is obtained by an application of Theorem 35.1 to the reciprocal of the function g.

35.3 COROLLARY. *Let* Θ *be an* $\mathfrak{G}$*-subvariety of* Σ *and* g *a continuous function on* Σ *that is almost* $\mathfrak{G}$*-holomorphic on* $\Sigma\setminus\partial[\Sigma, \mathfrak{G}]$. *Then if* $Z(g)$ *is contained in the interior of* Θ, $\Sigma\{|g| < \beta\} \cap \partial[\Sigma, \mathfrak{G}] \subseteq \Theta$ *implies* $\Sigma\{|g| < \beta\} \subseteq \Theta$.

In the preceding corollary the condition that the zeros of g lie in the interior of Θ implies that $1/g$ is bounded on $\Sigma\setminus\Theta$, so Theorem 35.1 may be applied. Obviously whenever $Z(g) \cap \partial[\Sigma, \mathfrak{G}]$ is contained in the interior of Θ, there exists $\beta > 0$ such that $\Sigma\{|g| < \beta\} \cap \partial[\Sigma, \mathfrak{G}] \subseteq \Theta$. The condition in Corollary 35.3 that $Z(g)$ be contained in the interior of Θ may be weakened to simple inclusion in Θ at the expense, however, of other restrictions. In particular, we assume that the $\mathfrak{G}$-subvariety of Σ is globally defined (§33).

35.4 COROLLARY. *Let* Θ *be a globally defined* $\mathfrak{G}$*-subvariety of* Σ *determined by a family* $\mathcal{F}$ *of almost* $\mathfrak{G}$*-holomorphic function on* Σ. *Let* g *be continuous on* Σ *and almost* $\mathfrak{G}$*-holomorphic on* $\Sigma\setminus\partial[\Sigma, \mathfrak{G}]$. *Also assume that for each* $f \in \mathcal{F}$ *and positive integer* n *the function* f/g^n *is bounded on* $\Sigma\setminus\Theta$. *Then* $\Sigma\{|g| < \beta\} \cap \partial[\Sigma, \mathfrak{G}] \subseteq \Theta$ *implies* $\Sigma\{|g| < \beta\} \subseteq \Theta$. *In particular, if* $Z(g) \cap \partial[\Sigma, \mathfrak{G}] = \emptyset$ *then* $Z(g)$ *is contained in the interior of* Θ.

Proof. Note that the condition $\Sigma\{|g| < \beta\} \cap \partial[\Sigma, \mathfrak{G}] \subseteq \Theta$ implies $\Sigma\{|g| < \beta\} \cap \partial[\Sigma, \mathfrak{G}] \subseteq Z(f)$ for each $f \in \mathcal{F}$, and hence $|g(\sigma)| \geq \beta$ for $\sigma \in \partial[\Sigma, \mathfrak{G}]\setminus Z(f)$. Thus

$$\left|\frac{f(\sigma)}{(g(\sigma))^n}\right| \leq \frac{|f|_\Sigma}{\beta^n}, \quad \sigma \in \partial[\Sigma, \mathfrak{G}]\setminus Z(f).$$

By Theorem 35.1 we have

$$\left|\frac{f(\sigma)}{(g(\sigma))^n}\right| \leq \frac{|f|_\Sigma}{\beta^n}, \quad \sigma \in \Sigma\setminus Z(f).$$

In particular $Z(g) \subseteq Z(f)$. It follows that

$$|g(\sigma)| \leq \left|\frac{f(\sigma)}{(g(\sigma))^n}\right| |g(\sigma)|^n \leq |f|_\Sigma \left|\frac{g(\sigma)}{\beta}\right|^n$$

for all $\sigma \in \Sigma$ and all n. If $|g(\sigma)| < \beta$, then letting $n \to \infty$ gives $f(\sigma) = 0$. In other words $\Sigma\{|g| < \beta\} \subseteq Z(f)$ for each $f \in \mathcal{F}$, so $\Sigma\{|g| < \beta\} \subseteq \Theta$, proving the corollary. ◊

Let f, g be continuous on Σ and almost $\mathfrak{G}$-holomorphic on $\Sigma\setminus\partial[\Sigma, \mathfrak{G}]$. Then g is said to *divide* f if there exists h continuous on Σ and almost $\mathfrak{G}$-holomorphic on $\Sigma\setminus\partial[\Sigma, \mathfrak{G}]$ such that $f = gh$. (Cf. [G5, p. 931].)

35.5 COROLLARY. *Assume that the set* $Z(f)\setminus\partial[\Sigma, \mathfrak{G}]$ *is without interior. Then if* $\emptyset \neq Z(g) \subseteq Z(f)\setminus\partial[\Sigma, \mathfrak{G}]$ *there exists a largest integer* m *such that* g^m *divides* f.

A proper subalgebra of $C(X)$, where the space X is compact, is said to be *maximal* if it is not properly contained in any uniform subalgebra of $C(X)$ except $C(X)$ itself. An example is the boundary value algebra $\mathcal{a}_T$ mentioned in Example 5.1 [W1]. Since the spectrum of $C(X)$ is equal to X, it is obvious that a maximal subalgebra of $C(X)$ will automatically contain any element of $C(X)$ that, along with the maximal algebra, generates a uniform subalgebra of $C(X)$ with spectrum larger than X. Assume that $\mathfrak{C}$ is a subalgebra of $C(\Sigma)$ with $\mathfrak{G} \subseteq \mathfrak{C}$ such that $\partial[\Sigma, \mathfrak{C}] = \partial[\Sigma, \mathfrak{G}] \neq \Sigma$ and $\mathfrak{C}|\partial[\Sigma, \mathfrak{G}]$ is maximal in $C(\partial[\Sigma, \mathfrak{G}])$. Then, by the definition of "$\mathfrak{G}$-presheaf" and the above observation, it is immediate that $\mathfrak{F}_\Sigma \subseteq \mathfrak{C}$. More generally, we have the following result.

35.6 COROLLARY. *Let* Θ *be an* $\mathfrak{G}$*-subvariety of* Σ *that does not exhaust* $\Sigma\setminus\partial[\Sigma, \mathfrak{G}]$. *Then every element of the algebra* $\mathfrak{B}$ *of Theorem 35.1 with* $G = \Sigma\setminus\partial[\Sigma, \mathfrak{G}]$, *that admits a continuous extension to all of* $\partial[\Sigma, \mathfrak{G}]$, *is the restriction to* $\Sigma\setminus\Theta$ *of an element of* $\mathfrak{C}$. *In particular, if* $\Theta \cap \partial[\Sigma, \mathfrak{G}] = \emptyset$ *then* $\mathfrak{B} \subseteq \mathfrak{C}|(\Sigma\setminus\Theta)$.

§36. CONTINUOUS FAMILIES OF HYPERSURFACES

We consider next "continuous families" of hypersurfaces for a natural system $[\Sigma, \mathfrak{G}]$ and some intersection properties of such families with $\mathfrak{G}$-convex subsets of Σ. Unless otherwise indicated, Σ is not assumed to be compact.

Let G be an open subset of Σ and T a given Hausdorff space. Let $F : G \times T \to \mathbb{C}$ be a continuous complex-valued function defined on $G \times T$ such that for fixed $t \in T$ the function

$$h_t : \sigma \mapsto F(\sigma, t), \sigma \in G$$

is almost $\mathfrak{G}$-holomorphic on G. Set $\Theta_t = Z(h_t)$. Then $\{\Theta_t : t \in T\}$ is called a *continuous family of* $\mathfrak{G}$*-hypersurfaces in* G. (See O1, R2, S9, S10 (§28)). Now let Ω be an arbitrary subset of Σ (i.e. independent of G) and define the sets

$$T_0(\Omega) = \{t \in T : \Theta_t \cap \Omega = \emptyset\}, T_1(\Omega) = \{t \in T : \Theta_t \cap \Omega \neq \emptyset\}.$$

If $T_1(\Omega) \neq \emptyset$, *i.e.* $\Theta_t \cap \Omega \neq \emptyset$ for some $t \in T$, then $\{\Theta_t\}$ is said to *intersect* Ω. If $T_1(\Omega)$ reduces to a single point then $\{\Theta_t\}$ is said to *intersect* Ω *minimally*. Furthermore, if $\Theta_t \cap \Omega$ is closed relative to Ω for each $t \in T$ and $\overline{T_0(\Omega)} \cap T_1(\Omega) \neq \emptyset$, then $\{\Theta_t\}$ is said to *intersect* Ω *nontrivially*.

Note that nontrivial intersection implies that both $T_0(\Omega)$ and $T_1(\Omega)$ are nonempty proper subsets of T. Also, if T is connected and $T_1(\Omega)$ is a nonempty proper closed subset of T then automatically $\overline{T_0(\Omega)} \cap T_1(\Omega) \neq \emptyset$. Now assume that $\{\Theta_t\}$ intersects Ω nontrivially and let t^* be any limit point of $T_0(\Omega)$ in $T_1(\Omega)$. Denote by T^* the subspace of T consisting of $T_0(\Omega)$ plus the single point t^*. Then the family $\{\Theta_t : t \in T^*\}$ intersects Ω both nontrivially and minimally.

For the next theorem, recall that a subset of Σ is said to be locally closed if it is the intersection of an open set and a closed set in Σ. As in §29, we denote by Δ_0 a locally closed $\mathfrak{a}$-local subset of Σ with compact closure Δ. We also denote the boundary of Δ_0 in Δ by Γ, so $\Gamma = \Delta \setminus \Delta_0$.

36.1 THEOREM. *Every continuous family of $\mathfrak{a}$-hypersurfaces that intersects Δ nontrivially must intersect Γ.* [*R2*].

Proof. Let $\{\Theta_t : t \in T\}$ be a continuous family of $\mathfrak{a}$-hypersurfaces that intersects Δ nontrivially. By the preceding remarks there is no loss of generality in assuming that the intersection is minimal, so $T_1(\Delta)$ reduces to a single point t^* and $\Theta_{t^*} \cap \Delta$ is a closed subset of Δ. Let G be the open subset of Δ on which the almost $\mathfrak{a}$-holomorphic functions h_t associated with the hypersurfaces Θ_t are defined. Now suppose that Θ_{t^*} did not intersect Γ. Then $\Theta_{t^*} \cap \Delta \subset \Delta_0$. Since Δ is compact there exists an open set U with $\bar{U} \subset \Delta_0 \cap G$ such that $\Theta_{t^*} \cap \Delta \subset U$. Hence $\inf_{\sigma\in\Gamma}|h_{t^*}(\sigma)| = \beta > 0$.

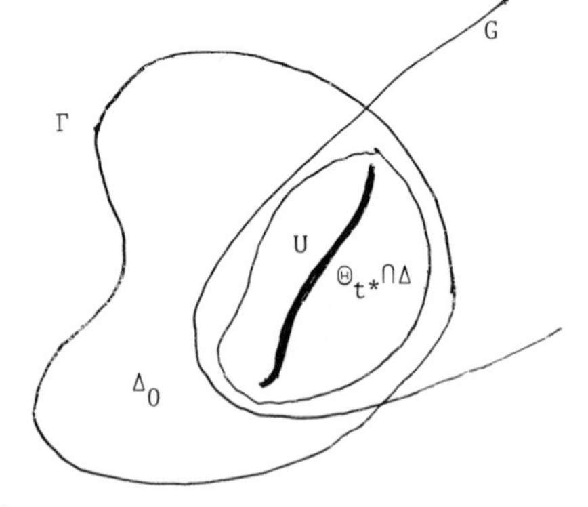

Again, since Γ is compact and t^* is a limit point of $T_0(\Delta)$, continuity of the map $(\sigma, t) \mapsto h_t(\sigma)$ implies the existence of a point $t_0 \in T_0(\Delta)$ such that

$|h_{t_0} - h_{t^*}|_\Delta < \beta/2$. In particular, $|h_{t_0}(\sigma)| > \beta/2$ for $\sigma \in \mathrm{bd}_\Delta U$ and $|h_{t_0}(\sigma)| < \beta/2$ for $\sigma \in \Theta_{t^*} \cap \Delta$. Also, since h_{t_0} is never zero on $\bar{U}$ the function $h_{t_0}^{-1}$ is $\mathfrak{G}$-holomorphic on $\bar{U}$. On the other hand,

$$|h_{t_0}^{-1}|_{\Theta_{t^*}\cap\Delta} > \frac{2}{\beta}, \quad |h_{t_0}^{-1}|_{\mathrm{bd}_\Delta U} < \frac{2}{\beta}$$

which by Theorem 20.1 contradicts the assumption that Δ_0 is $\mathfrak{G}$-local. ◆

36.2 THEOREM. *Let* X *be an arbitrary subset of* Σ *with* $\mathfrak{G}$*-convex hull* $\hat{X}$. *Then every continuous family of* $\mathfrak{G}$*-hypersurfaces which intersects* $\hat{X}$ *nontrivially must intersect* X.

Proof. Let $\{X_\nu\}$ be the #-resolution of the hull $\hat{X}$ given by Proposition 8.1, so $X_0 = X$ and

$$X_\nu = \Big(\bigcup_{\alpha<\nu} X_\alpha\Big)^{\#}.$$

Let $\{\Theta_t : t \in T\}$ be a continuous family of $\mathfrak{G}$-hypersurfaces which intersects $\hat{X}$ nontrivially. As before, we may assume that the intersection is minimal with t^* the unique point of T for which $\Theta_{t^*} \cap X \neq \emptyset$. Suppose that $\{\Theta_t\}$ does not intersect X. Then there exists an ordinal ν such that Θ_{t^*} intersects X_ν but does not intersect X_α for $\alpha < \nu$. Hence there exists a compact set $K \subset\subset \bigcup_{\alpha<\nu} X_\alpha$ such that Θ_{t^*} intersects $\hat{K}$ but does not intersect K. Now set $\Delta_0 = \hat{K}\setminus K$. Then Δ_0 is locally closed and $\mathfrak{G}$-local with compact closure Δ. Since $\{\Theta_t\}$ intersects Δ nontrivially but does not intersect $\Gamma = \mathrm{bd}_\Delta \Delta_0 \subseteq K$, we have a contradiction of Theorem 36.1. ◆

36.3 THEOREM. *Let* Ω *be an arbitrary* $\mathfrak{G}$*-convex subset of* Σ *and let* $\{\Theta_t : t \in T\}$ *be a continuous family of* $\mathfrak{G}$*-hypersurfaces in an open set* $G \subseteq \Sigma$. *If (i)* T *is connected (ii)* G *contains the relative closure of the set* $\bigcup\Theta_t \cap \Omega$ *in* Ω *and (iii) for arbitrary* $K \subset\subset \Omega$ *there exists* $t \in T$ *such that* $\Theta_t \cap K = \emptyset$, *then the set* $\Omega\setminus\bigcup\Theta_t$ *is also* $\mathfrak{G}$*-convex.*

Proof. Let K be an arbitrary compact subset of the set

$$\Omega_0 = \Omega \setminus \bigcup_{t\in T} \Theta_t.$$

Then $\hat{K} \subseteq \Omega$. Hence by condition (ii)

$$F = \overline{\bigcup_{t \in T} \Theta_t} \cap \hat{K} \subseteq \Omega \cap (\overline{\bigcup_{t \in T} \Theta_t \cap \Omega}) \subset G$$

and F is a compact subset of G. By condition (iii) there exists $t \in T$ such that $\Theta_t \cap K = \emptyset$. Since $\Theta_t \cap \hat{K} \subseteq \Theta_t \cap F$ also $\Theta_t \cap \hat{K} = \emptyset$, so $T_0(\hat{K}) \neq \emptyset$. Observe next that for arbitrary $t \in T$, $G \cap \bar{\Theta}_t = \Theta_t$ and $\overline{\Theta_t \cap \hat{K}} \subseteq G$, so $\overline{\Theta_t \cap \hat{K}} \subseteq G \cap \bar{\Theta}_t \cap \hat{K} = \Theta_t \cap \hat{K}$. In other words, $\Theta_t \cap \hat{K}$ is a closed subset of Σ for each $t \in T$.

Now suppose that $T_1(\hat{K}) \neq \emptyset$ and let t_0 be a limit point of $T_1(\hat{K})$ in T. Let V be an arbitrary neighborhood of t_0 in T and define

$$K_V = \overline{\bigcup_{t \in V} \Theta_t \cap \hat{K}}.$$

Then K_V is a nonempty compact subset of G and $V_1 \subseteq V_2$ implies $K_{V_1} \subseteq K_{V_2}$. Therefore there exists a point $\sigma_0 \in \bigcap K_V$. By the continuity of the function $F : (\sigma, t) \mapsto h_t(\sigma)$, there exists for arbitrary $\varepsilon > 0$ neighborhoods U of σ_0 and V of t such that $(\sigma, t) \in U \times V$ implies $|h_t(\sigma) - h_{t_0}(\sigma_0)| < \varepsilon$. Since $\sigma_0 \in K_V$ there exist $t \in V$ and $\sigma \in \Theta_t \cap \hat{K} \cap U$, so $h_t(\sigma) = 0$. This implies that

$$|h_{t_0}(\sigma_0)| = |h_t(\sigma) - h_{t_0}(\sigma_0)| < \varepsilon$$

and hence that $|h_{t_0}(\sigma_0)| < \varepsilon$, for arbitrary $\varepsilon > 0$. Therefore $h_{t_0}(\sigma_0) = 0$, so $\sigma_0 \in \Theta_{t_0} \cap \hat{K}$. In other words $t_0 \in T_1(\hat{K})$ which proves that $T_1(\hat{K})$ is closed. This, together with the fact that T is connected and $T_0(\hat{K}) \neq \emptyset$, implies $\overline{T_0(\hat{K})} \cap T_1(\hat{K}) \neq \emptyset$. It follows that $\{\Theta_t\}$ intersects $\hat{K}$ nontrivially and hence by Theorem 36.1 must intersect K. But this is impossible since $K \subseteq \Omega \setminus \bigcup \Theta_t$. Therefore $T_1(\hat{K}) = \emptyset$ which means that $\hat{K} \subseteq \Omega \setminus \bigcup \Theta_t$ and hence that $\Omega \setminus \bigcup \Theta_t$ is $\mathfrak{G}$-convex. ♦

§37. REMARKS

The result in Theorem 34.1, for the case $\mathfrak{F} = \mathfrak{G}$, was first proved by the author [R4]. As remarked above, the results 35.1 to 35.5 are either suggested by or are straightforward versions of results due to Glicksberg contained in his very interesting paper on *Maximal algebras and a theorem of Rado* [G5]. In this paper, various properties of functions associated with the disc algebra are generalized to the case of a uniform algebra on its spectrum. The varieties considered are either zero sets of elements of the algebra or intersections of such sets (i.e. globally defined varieties).

Except for involving $\mathbb{G}$-holomorphic functions rather than elements of $\mathbb{G}$, Corollary 35.2 is simply Theorem 4.1 of [G5] and Corollaries 35.4 and 35.5 are essentially Corollaries 4.2 and 4.3 of [G5]. Some of the ideas in Theorem 35.1 and its proof are developed further in [R5].

CHAPTER VIII
HOLOMORPHIC AND SUBHARMONIC CONVEXITY

§38. CONVEXITY WITH RESPECT TO AN $\mathfrak{A}$-PRESHEAF

It will be assumed throughout this section that $[\Sigma, \mathfrak{A}]$ is a natural system. Also let $\{\Sigma, \mathfrak{F}\}$ be an $\mathfrak{A}$-presheaf of continuous functions over Σ. (See Definition 22.1). If Ω is any subset of Σ and $K \subset\subset \Omega$ then the $\mathfrak{F}$-*convex hull of* K *in* Ω is the set

$$K_\Omega^{\mathfrak{F}} = \{\sigma' \in \Omega : |f(\sigma')| \leq |f|_K,\ f \in \mathfrak{F}_\Omega\}.$$

Since elements of $\mathfrak{F}$ are continuous the set $K_\Omega^{\mathfrak{F}}$ is always relatively closed in Ω. Also, since $|\mathfrak{A}| \subseteq |\mathfrak{F}|$ it follows that $K_\Omega^{\mathfrak{F}} \subseteq \hat{K} \cap \Omega$, so the closure of $K_\Omega^{\mathfrak{F}}$ is compact.

38.1 DEFINITION. *The set* Ω *is said to be* $\mathfrak{F}$-*convex if* $K_\Omega^{\mathfrak{F}}$ *is compact for every compact set* $K \subset\subset \Omega$. *If* $\Omega_0 \subseteq \Omega$ *and the set* $K_\Omega^{\mathfrak{F}} \cap \Omega_0$ *is compact for every* $K \subset\subset \Omega_0$ *then* Ω_0 *is said to be* $\mathfrak{F}$-*convex within* Ω

Observe that $\mathfrak{F}$-convexity of Ω is the same as convexity of the system $[\Omega, \mathfrak{F}_\Omega]$ according to Definition 7.5. Also, it needs to be emphasized that $\mathfrak{F}$-convexity of a set Ω_0 *within* Ω is *not* the same as $\mathfrak{F}_\Omega$-convexity of Ω_0 (Definition 6.1), since the latter requires that $K_\Omega^{\mathfrak{F}} \subseteq \Omega_0$ for $K \subset\subset \Omega_0$ and says nothing about compactness. The set $K_\Omega^{\mathfrak{F}}$, of course need not be compact. However, since $|\mathfrak{A}| \subseteq |\mathfrak{F}|$ it is obvious that $K_\Omega^{\mathfrak{F}} \subseteq \hat{K}_\Omega$, where $\hat{K}_\Omega = \hat{K} \cap \Omega$ is the $\mathfrak{A}$-convex hull of K in Ω. Therefore, since $[\Sigma, \mathfrak{A}]$ is natural $\hat{K}$ is compact, so $K_\Omega^{\mathfrak{F}}$ has compact closure. Note that the condition in Theorem 34.2 imposed on a local $\mathfrak{F}$-subvariety Θ of an open set $H \subseteq \Sigma$ is equivalent to requiring that Θ be $\mathfrak{A}$-convex within H.

We are primarily interested here in the special cases, $\mathfrak{F} = \mathfrak{O}$ and $\mathfrak{F} = \exp(\mathfrak{CS})$. When $\mathfrak{F} = \mathfrak{O}$ we shall use the special notation $\overset{\frown}{K}_\Omega$ in place of $K_\Omega^{\mathfrak{O}}$ and $\overset{\frown}{K}$ in place of $\overset{\frown}{K}_\Sigma$. Also, when $\mathfrak{F} = \exp(\mathfrak{CS})$ we use $\overset{\sqcap}{K}_\Omega$ in place of $K_\Omega^{\exp(\mathfrak{CS})}$ and $\overset{\sqcap}{K}$ in place of $\bar{K}_\Sigma$. Finally, in place of "$\mathfrak{G}$-convexity" we shall use the terms "$\mathfrak{G}$-holomorphic convexity" or "$\mathfrak{G}$-h convexity" and, in place of "exp($\mathfrak{CS}$)-convexity," the terms "$\mathfrak{G}$-subharmonic convexity" or "$\mathfrak{G}$-sh convexity". Thus, for the system $[\mathbb{C}^n, \mathfrak{P}]$ the notions of $\mathfrak{P}$-h convexity and $\mathfrak{P}$-sh convexity reduce respectively to the familiar notions of holomorphic convexity and p-convexity, as defined, for example, by Gunning and Ross [G7, pp. 208, 276]. For arbitrary $K \subset\subset \Omega$ we always have $\overset{\sqcap}{K}_\Omega \subseteq \overset{\frown}{K}_\Omega \subseteq \hat{K}_\Omega \subseteq \hat{K}$, so it is always true that $\mathfrak{G}$-convexity implies $\mathfrak{G}$-h convexity and $\mathfrak{G}$-h convexity implies $\mathfrak{G}$-sh convexity.

The next theorem is a generalization of Theorem 15.2. The proof will be omitted since it parallels exactly the proof of Theorem 15.2 except for an appeal to Theorem 22.1 in place of Theorem 14.2. Note that if $\{\Sigma, \mathfrak{F}\}$ is an $\mathfrak{G}$-presheaf then Σ is $\mathfrak{F}$-convex (within itself).

38.2 THEOREM. *Let* $\{\Sigma, \mathfrak{F}\}$ *be an* $\mathfrak{G}$*-presheaf and* Ω *a closed subset of* Σ *such that* $\Sigma \setminus \Omega = \Sigma_0 \cup \Sigma_1$, *where* $\bar{\Sigma}_0 \cap \Sigma_1 = \Sigma_0 \cap \bar{\Sigma}_1 = \emptyset$. *If* $K \subset\subset \Omega$ *implies that* $K_\Sigma^{\mathfrak{F}} \subseteq \Sigma_0 \cup \Omega$, *then* $\Sigma_0 \cup \Omega$ *is* $\mathfrak{F}$*-convex within* Σ.

§39. PROPERTIES OF SUBHARMONIC CONVEXITY

We shall now develop some of the properties of $\mathfrak{G}$-sh convexity. First let G be an open subset of Σ, u an arbitrary element of $\mathfrak{CS}_G$, t any real number, and set

$$G\{u < t\} = \{\sigma \in G : u(\sigma) < t\}.$$

Then $G\{u < t\}$ is open, $t_1 < t_2$ implies $G\{u < t_1\} \subseteq G\{u < t_2\}$, and $G = \bigcup_t G\{u < t\}$. Let K be a compact subset of $G\{u < t\}$ and, for convenience, set $K_{u,t} = \overset{\sqcap}{K}_G \cap G\{u < t\}$. Since $\overset{\sqcap}{K}_G$ is relatively closed in G it follows that $\bar{K}_{u,t} \cap G \subseteq \overset{\sqcap}{K}_G$. Moreover, since u is continuous and $K \subset G\{u < t\}$ we have

$$\sup_{\sigma \in \bar{K}_{u,t} \cap G} u(\sigma) = \sup_{\sigma \in K_{u,t}} u(\sigma) = \max_{\sigma \in K} u(\sigma) < t$$

so $\bar{K}_{u,t} \cap G \subseteq G\{u < t\}$. Therefore $\bar{K}_{u,t} \cap G = K_{u,t}$, *i.e.* $K_{u,t}$ is relatively closed in G. The set $K_{u,t}$ will be compact iff it is closed and will be closed iff $\bar{K}_{u,t} \subset G$. Hence, $G\{u < t\}$ will be $\mathfrak{G}$-sh convex within G iff $\bar{K}_{u,t} \subset G$ for every compact $K \subset\subset G\{u < t\}$.

39.1 THEOREM. *(i) If either* G *is* $\mathcal{G}$-sh *convex or* $\overline{G\{u < t\}} \subset G$ *then* $G\{u < t\}$ *is* $\mathcal{G}$-sh *convex within* G.

(ii) If $\overline{G\{u < t\}} \subset G$ *for all real* t *then* G *is* $\mathcal{G}$-sh *convex.*

Proof. Let K be an arbitrary compact subset of $G\{u < t\}$ and observe that $\bar{K}_{u,t} = \overline{\hat{K}_G \cap G\{u < t\}} \subseteq \hat{\bar{K}}_G \cap \overline{G\{u < t\}}$. If G is $\mathcal{G}$-sh convex then $\hat{K}_G$ is compact, so $\hat{\bar{K}}_G = \hat{K}_G \subset G$ and hence $\bar{K}_{u,t} \subset G$. On the other hand, if $\overline{G\{u < t\}} \subset G$ then $\bar{K}_{u,t} \subset G$. Therefore, in either case $G\{u < t\}$ is $\mathcal{G}$-sh convex within G by the preceding remarks. This proves (i).

Now assume that $\overline{G\{u < t\}} \subset G$ for all t and let K be a compact subset of G. Choose $t > \max_{\sigma \in K} u(\sigma)$. Then $\hat{K}_G \subset G\{u < t\} \subset \overline{G\{u < t\}} \subset G$. Since $\hat{K}_G$ is relatively closed in G it follows that $\hat{K}_G$ is in fact closed and therefore compact, proving (ii). ◆

The next series of results require compactness conditions on the open sets involved. The first is a converse to part (ii) of the above theorem. (Cf. [H4, Theorem 2.6.7]).

39.2 THEOREM. *Let* G *be a* σ*-compact open subset of* Σ. *If* G *is* $\mathcal{G}$-sh *convex then there exists* $g \in \mathcal{CS}_G$ *such that for each real* t, $G\{g < t\} \subset\subset G$

Proof. Since G is σ-compact there exists a sequence $\{G_n\}$ of open sets such that $G = \cup G_n$ and $G_n \subset\subset G_{n+1}$ for all n. Since $\bar{G}_n$ is compact and G is $\mathcal{G}$-sh convex the set $(\widehat{\bar{G}_n})_G$ is a compact subset of G. Therefore, we may, by passing to a subsequence if necessary, assume that $(\widehat{\bar{G}_n})_G \subset G_{n+1}$ for all n. Set $F_n = \bar{G}_{n+1} \setminus G_n$. Then $\{F_n\}$ is a sequence of compact sets and

$$G \setminus G_n = \bigcup_{k=n}^{\infty} F_k \,.$$

Note that $(\widehat{\bar{G}_n})_G \cap F_{n+1} = \emptyset$. Hence, for each $\delta \in F_{n+1}$ there exists $u_\delta \in \mathcal{CS}_G$ such that

$$\max_{\sigma \in \bar{G}_n} u_\delta(\sigma) < 0 < u_\delta(\delta).$$

Since u_δ is continuous there exists a neighborhood V_δ of δ, with $V_\delta \subset G$, such that $\inf_{\sigma \in V_\delta} u_\delta(\sigma) > 0$. Since F_{n+1} is compact it is covered by a finite number of such neighborhoods, say $V_{\delta_1}, \ldots, V_{\delta_m}$. Define

$$v_n(\sigma) = \max\{u_{\delta_1}(\sigma),\ldots,u_{\delta_m}(\sigma)\},\ \sigma \in G.$$

Then $v_n \in \mathcal{CS}_G$ and

$$\max_{\sigma\in\bar{G}_n} v_n(\sigma) < 0 < \min_{\sigma\in F_{n+1}} v_n(\sigma).$$

Modification of v_n by a positive multiplicative constant will give the conditions

$$\max_{\sigma\in\bar{G}_n} v_n(\sigma) < -n \log 2,\ \log(n+1) < \min_{\sigma\in F_{n+1}} v_n(\sigma).$$

Now, replacing v_n by $w_n = \exp v_n$, we obtain $w_n \geq 0$, $w_n \in \mathcal{CS}_G$, and

$$\max_{\sigma\in\bar{G}_n} w_n(\sigma) < \frac{1}{2^n},\ n+1 < \min_{\sigma\in F_{n+1}} w_n(\sigma).$$

Finally, set

$$g(\sigma) = \sum_{k=1}^{\infty} w_k(\sigma),\ \sigma \in G.$$

If $k > n$ then $\bar{G}_n \subset G_k$. Therefore, for $\sigma \in G_n$

$$\sum_{k=n+1}^{\infty} w_k(\sigma) < \sum_{k=n+1}^{\infty} \frac{1}{2^k} = \frac{1}{2^n}$$

so the series for g converges uniformly on G. In particular, it converges locally uniformly in G and hence $g \in \mathcal{CS}_G$.

Now, for arbitrary real t choose $n \geq t$. Then, if $\delta \in G\setminus G_n$ there exists $m \geq n$ such that $\delta \in G_m$. Since the functions w_k are nonnegative we have

$$g(\delta) = \sum_{k=1}^{\infty} w_k(\delta) \geq w_{m-1}(\delta) > m \geq t.$$

Therefore, $G\{g < t\} \subseteq G_n$ and, since $G_n \subset\subset G$, it follows that $G\{g < t\} \subset\subset G$. ◆

39.3 COROLLARY. *If* G *is an open,* σ*-compact,* $\mathcal{G}$-sh *convex subset of* Σ, *then there exists a sequence* $\{G_n\}$ *of open sets such that*

$$G_n \subset\subset G_{n+1} \subset\subset G = \bigcup_{k=1}^{\infty} G_k$$

and G_n *is* $\mathcal{G}$-sh *convex within* G *for each* n.

One consequence of the next theorem is a converse to the result in the above corollary. (See Theorem 39.6).

39.4 THEOREM. *Let* G *and* H *be open sets in* Σ, *with* $H \subseteq G$, *where* G *is locally compact and* H *is* $\mathcal{G}$-sh *convex within* G. *Then, for every compact set* $K \subset H$ *the hull* $\hat{K}_G$ *is compact and contained in* H.

Proof. Let K be a compact subset of G and suppose that $\hat{K}_G = F \cup F'$, where F is compact and $F \cap \bar{F}' = \emptyset$. Then, since G is locally compact there exists an open set U with compact closure, and hence a compact (possibly empty) boundary bd U, such that $F \subset U \subset\subset G$, $\bar{F}' \cap \bar{U} = \emptyset$. Note that bd $U \subset G \setminus \hat{K}_G$. If bd $U = \emptyset$ consider the function g such that $g(\sigma) = 0$ for $\sigma \in U$, and $g(\sigma) = 1$ for $\sigma \in G \setminus U$. Then $g \in \mathcal{CS}_G$. Since $g(\sigma) = 0$ for $\sigma \in F$ and $g(\sigma) = 1$ for $\sigma \in F'$ it follows that $\hat{F}_G \cap F' = \emptyset$. On the other hand $\hat{F}_G \subseteq \hat{K}_G$, so we must have $\hat{F}_G = F$. Now assume that bd $U \neq \emptyset$. Then for each $\delta \in$ bd U there exists $f_\delta \in \mathcal{CS}_G$ with

$$\max_{\sigma \in \hat{K}_G} f_\delta(\sigma) = \max_{\sigma \in K} f_\delta(\sigma) < 0 < f_\delta(\delta).$$

Set $N_\delta = \{\sigma \in G : 0 < f_\delta(\sigma)\}$. Then N_δ is an open neighborhood of δ. Since bd U is compact it is covered by a finite collection of these neighborhoods which we denote by $N_1, \ldots, N_n$ with associated functions $f_1, \ldots, f_n$. Let

$$t_0 = \{\max f_i(\sigma) : \sigma \in K, i=1,\ldots,n\}.$$

Then $t_0 < 0$. Choose t such that $t_0 < t < 0$ and set

$$V = \{\sigma \in G \setminus \bar{U} : f_i(\sigma) < t, i=1,\ldots,n\}.$$

Then V is open and contains F'. Observe that for each i, $f_i(\sigma) > 0$ if $\sigma \in (\text{bd } U) \cap N_i$ while $f_i(\sigma) \leq t < 0$ if $\sigma \in \bar{V}$. Therefore $(\text{bd } U) \cap \bar{V} = \emptyset$ and hence $\bar{U} \cap \bar{V} = \emptyset$, so $G = (G \setminus \bar{U}) \cup (G \setminus \bar{V})$. Now define

$$g_1(\sigma) = \max\{f_i(\sigma) : i=1,\ldots,n\}, \ \sigma \in G$$

and

$$g_2(\sigma) = \max\{t, g_1(\sigma)\}, \ \sigma \in G.$$

If $\sigma \in (G \setminus \bar{U}) \cap (G \setminus \bar{V})$ then $f_i(\sigma) \geq t$ for some i, so $g_1(\sigma) = g_2(\sigma)$. Therefore, if we let

$$g(\sigma) = \begin{cases} g_1(\sigma) & , \ \sigma \in G \setminus \bar{V} \\ g_2(\sigma) & , \ \sigma \in G \setminus \bar{U} \end{cases}$$

then g is well-defined in G. Furthermore, since the constant t and each of the functions f_i belong to $\mathcal{CS}_G$ it follows that $g \in \mathcal{CS}_G$. Observe next that $F \subset U \subset G \setminus \bar{V}$ while $F' \subset V \subset G \setminus \bar{U}$. Hence

$$\max_{\sigma\in F} g(\sigma) = \max_{\sigma\in F} g_1(\sigma) \le t_0 < t$$

while, for $\sigma \in F'$, we have $g(\sigma) = g_2(\sigma) \ge t$. Therefore $\hat{F}_G \subset G\setminus F'$, so again we conclude that $\hat{F}_G = F$.

Finally, let K be a compact subset of H. Then the set $\hat{K}_G \cap H$ is compact, by hypothesis. Hence $\hat{K}_G = (\hat{K}_G \cap H) \cup (\hat{K}_G\setminus H)$ is a decomposition of the type considered above. Therefore $\widehat{(\hat{K}_G \cap H)}_G = \hat{K}_G \cap H$ and, since $K \subseteq \hat{K}_G \cap H$, it follows that $\hat{K}_G = \hat{K}_G \cap H$. ◆

39.5 COROLLARY. *Let* G *and* H *be as in the theorem and assume that* $H = H_1 \cup H_2$, *where* $\bar{H}_1 \cap H_2 = H_1 \cap \bar{H}_2 = \emptyset$. *Then* H_1 *and* H_2 *are also* $\mathcal{G}$-sh *convex in* G.

We now prove the converse to Corollary 39.3. Call an arbitrary collection of sets *increasing* if the union of any finite subcollection of its elements is contained in an element of the collection, so the collection is "directed" with respect to inclusion.

39.6 THEOREM. *Let* G *be an open, locally compact subset of* Σ *which is a union of an increasing collection of open sets each of which is* $\mathcal{G}$-sh *convex within* G. *Then* G *is* $\mathcal{G}$-sh *convex.*

Proof. Let K be a compact subset of G. There exists an element H of the increasing collection which contains K. Applying Theorem 39.4, we conclude that $\hat{K}_G$ is compact and hence that G is $\mathcal{G}$-sh convex. ◆

Next is an approximation theorem for $\mathcal{G}$-sh hulls of compact sets.

39.7 THEOREM. *Let* G *be a locally compact subset of* Σ *and let* Ω *be a compact subset of* G *such that* $\hat{\Omega}_G = \Omega$. *Then there exists an open set* H *with* $\Omega \subset H \subseteq G$ *such that* H *is* $\mathcal{G}$-sh *convex within* G.

Proof. Since G is locally compact there exists an open set U such that $\Omega \subset U \subset\subset G$. If $\mathrm{bd}\,\Omega = \emptyset$ then Ω is open in Σ and we may take $H = \Omega$. Otherwise, applying to the set Ω the same construction used with F in the proof of Theorem 39.4, we obtain a function $g_1 \in \mathcal{CS}_G$ such that

$$\max_{\sigma\in\Omega} g_1(\sigma) < 0 < \min_{\sigma\in \text{bd } U} g_1(\sigma).$$

Now choose r such that

$$0 < r < \min_{\sigma\in \text{bd } U} g_1(\sigma).$$

Then $\Omega \subset U\{g_1 < 0\} \subset\subset U\{g_1 < r\} \subset U$, where, as before, $U\{g_1 < t\} = \{\sigma \in U : g_1(\sigma) < t\}$. Observe that

$$G = U\{g_1 < r\} \cup (G\setminus\overline{U\{g_1 < 0\}})$$

and if $\sigma \in U\{g_1 < r\} \cap (G\setminus\overline{U\{g < 0\}})$ then $g_1(\sigma) \geq 0$. Next define

$$g_2(\sigma) = \max\{0, g_1(\sigma)\}, \; \sigma \in G.$$

Then $g_2 \in \mathcal{CS}_G$ and $g_1(\sigma) = g_2(\sigma)$ for

$$\sigma \in U\{g_1 < r\} \cap (G\setminus\overline{U\{g_1 < 0\}}).$$

Therefore, if

$$g(\sigma) = \begin{cases} g_1(\sigma) & , \; \sigma \in U\{g_1 < r\} \\ g_2(\sigma) & , \; \sigma \in G\setminus\overline{U\{g_1 < 0\}} \end{cases}$$

then g is a well-defined element of $\mathcal{CS}_G$. Furthermore, $U\{g_1 < 0\} = G\{g < 0\}$. Since $\overline{G\{g < 0\}} \subset G$ it follows by Theorem 39.1 (i) that $G\{g < 0\}$ is $\mathfrak{a}$-sh convex within G. Therefore the theorem follows with $H = G\{g < 0\}$. ◆

The next theorem is a generalization of Corollary 20.4.

39.8 THEOREM. *Let* G *be an open subset of* Σ *and let* K *be any compact subset of* G *with* $\hat{K} \subset G$. *Then* $\check{K}_G = \hat{K}$.

Proof. By Corollary 29.4 we have $\partial_{\mathcal{S}}\hat{K} = \partial_{\mathfrak{a}}\hat{K} \subseteq K$. Therefore $\hat{K} \subseteq \check{K}_G$. Since it is always true that $\check{K}_G \subseteq \hat{K}$ the theorem follows. ◆

Observe that the inclusion $\partial_{\mathcal{S}}\hat{K} \subseteq K$, obtained in the above proof, implies that the convex hull of K in G with respect to $\mathcal{S}_G$ (rather than just $\mathcal{CS}_G$) is equal to $\hat{K}$. This fact enables us to obtain a generalization of a theorem due to Bremerman [B8, Theorem 4].

39.9 COROLLARY. *Let* G *be an open* $\mathfrak{a}$*-convex subset of* Σ *and let* S,T *be subsets of* G *with* $(S\cup T) \subset\subset G$. *If*

$$\max_{\sigma\in\bar{T}} |a(\sigma)| = \max_{\sigma\in\overline{SUT}} |a(\sigma)|, \ a \in \mathfrak{a}$$

then also

$$\max_{\sigma\in\bar{T}} f(\sigma) = \max_{\sigma\in\overline{SUT}} f(\sigma), \ f \in \mathcal{S}_G.$$

§40. NATURALITY PROPERTIES

We obtain next a naturality result for $\mathfrak{a}$-h convex domains. It is a special case of a much more general theorem which will be proved in the next chapter. Let G be an open subset of Σ and $\mathcal{B}$ an algebra of functions on G with $\mathfrak{a}|G \subseteq \mathcal{B} \subseteq \mathcal{O}_G$. Also, for $K \subset\subset G$ denote by $K_G^{\mathcal{B}}$ the $\mathcal{B}$-convex hull of K in G.

40.1 THEOREM. *Let* G *be an open subset of* Σ *and* K *a compact subset of* G *such that* $K_G^{\mathcal{B}}$ *is also compact. Then the system* $[K_G^{\mathcal{B}}, \mathcal{B}]$ *is natural.*

Proof. Denote by Ψ the spectrum of $[K_G^{\mathcal{B}}, \mathcal{B}]$ and for $\psi \in \Psi$ and $h \in \mathcal{B}$ denote by $\hat{h}(\psi)$ the image of $h|K_G^{\mathcal{B}}$ in $\mathbb{C}$ under ψ. Also, for $\sigma \in K_G^{\mathcal{B}}$ denote by ψ_σ the point evaluation on $\mathcal{B}|K_G^{\mathcal{B}}$ associated with the point σ, so $\tau : K_G^{\mathcal{B}} \to \Psi$, $\sigma \to \psi_\sigma$, is a homeomorphism of $K_G^{\mathcal{B}}$ into Ψ. Now, since $\mathfrak{a}|G \subseteq \mathcal{B}$ the restriction of an element $\psi \in \Psi$ to $\mathfrak{a}|G$ defines a continuous homomorphism of $\mathfrak{a}$ onto $\mathbb{C}$. Since $[\Sigma, \mathfrak{a}]$ is natural there exists $\sigma_\psi \in \hat{K}$ such that

$$\hat{a}(\psi) = \widehat{a|G}(\psi) = a(\sigma_\psi).$$

The mapping $\pi : \Psi \to \hat{K}$, $\psi \to \sigma_\psi$, is obviously continuous and equal to τ^{-1} on the set $\tau(K_G^{\mathcal{B}})$. Each of the functions $\hat{a}$, for $a \in \mathfrak{a}$, is constant on each of the stalks $\pi^{-1}(\sigma)$, $\sigma \in \pi(\Psi)$.

Now, for each $h \in \mathcal{B}$ consider the function $\tilde{h}(\psi) = h(\pi(\psi))$, which is defined on the open set $\Psi_0 = \{\psi \in \Psi : \pi(\psi) \in G\}$ in the space Ψ and is constant on each of the stalks $\pi^{-1}(\sigma)$, $\sigma \in \pi(\Psi) \cap G$. Since h is $\mathfrak{a}$-holomorphic on G and $\hat{\mathfrak{a}} = \mathfrak{a}\circ\pi$, it is immediate that $\tilde{h}$ is $\hat{\mathfrak{a}}$-holomorphic on Ψ_0. Furthermore, since $\hat{\mathfrak{a}} \subseteq \hat{\mathcal{B}}$, it follows that $\tilde{h}$ is $\hat{\mathcal{B}}$-holomorphic on Ψ_0.

Next consider the set

$$\Theta = \{\psi \in \Psi_0 : \hat{h}(\psi) - \tilde{h}(\psi) = 0, \ h \in \mathcal{B}\}.$$

Thus Θ is a globally defined $\mathfrak{D}$-subvariety of Ψ_0. (See the remarks following Definition 33.1). Observe that $\tau(K_G^{\mathcal{B}}) \subseteq \Theta$ and, in particular, that $\tau(K) \subseteq \Theta$, which implies that Θ contains the Šilov boundary of Ψ relative to $\mathfrak{D}$. Now let ψ_0 be an arbitrary point of Θ. Suppose that $\pi(\psi_0) \in G \setminus K_G^{\mathcal{B}}$. Then there exists $h \in \mathcal{B}$ such that $|h(\pi(\psi_0))| > |h|_K$. Since $\hat{h}(\psi_0) = \tilde{h}(\psi_0) = h(\pi(\psi_0))$ and $|h|_K = |\hat{h}|_{\tau(K)}$ we have $|\hat{h}(\psi_0)| > |\hat{h}|_{\tau(K)}$, which is impossible. Therefore $\pi(\Theta) \subseteq K_G^{\mathcal{B}}$. Since π is continuous and Θ is relatively closed in Ψ_0 it follows that Θ is actually closed in Ψ. We may now apply Theorem 34.2 to conclude that Θ is $\mathfrak{D}$-convex. This, with the fact that Θ contains the Šilov boundary of Ψ, implies that $\Theta = \Psi$. Finally, since each of the functions $\tilde{h}$ is constant on the stalks $\pi^{-1}(\pi(\Psi))$ it follows that each stalk is a singleton. Therefore $\Psi = \tau(K_G^{\mathcal{B}})$, so τ defines an isomorphism of the system $[K_G^{\mathcal{B}}, \mathcal{B}]$ with the natural system $[\Psi, \hat{\mathcal{B}}]$. Hence $[K_G^{\mathcal{B}}, \mathcal{B}]$ must also be natural. ♦

40.2 COROLLARY. *For any open set* $G \subseteq \Sigma$ *the system* $[G, \mathcal{B}]$ *will be natural iff it is convex (i.e.* $K_G^{\mathcal{B}}$ *is compact for each* $K \subset\subset G$*).*

Since $[\Sigma, \mathfrak{G}]$ is natural the system $[\Sigma, \mathcal{B}]$ is automatically convex for $\mathfrak{G} \subseteq \mathcal{B} \subseteq \mathfrak{G}_\Sigma$, and therefore $[\Sigma, \mathcal{B}]$ is natural. As a matter of fact, the method of proof used for Theorem 40.1 may be adapted to yield a better result than this. Recall that $\mathfrak{G}_\Sigma^*$ denotes the subalgebra of $C(\Sigma)$ consisting of those functions that are $\mathfrak{G}$-holomorphic on $\Sigma \setminus \partial[\Sigma, \mathfrak{G}]$. Obviously $\mathfrak{G}_\Sigma \subseteq \mathfrak{G}_\Sigma^*$ and the inclusion is in general proper. Furthermore, as we shall see (Example 43.1), $[\Sigma, \mathfrak{G}_\Sigma^*]$ need not be natural. On the other hand, if the Šilov boundary $\partial[\Sigma, \mathfrak{G}]$ is replaced by $\partial_0[\Sigma, \mathfrak{G}]$, the set of independent points for $[\Sigma, \mathfrak{G}]$, then a result of the desired kind is true. For this we denote by $\mathfrak{G}_\Sigma^+$ the subalgebra of $C(\Sigma)$ consisting of those functions that are $\mathfrak{G}$-holomorphic on $\Sigma \setminus \partial_0[\Sigma, \mathfrak{G}]$, so $\mathfrak{G}_\Sigma^+ \subseteq \mathfrak{G}_\Sigma^*$ and the inclusion is generally proper.

40.3 THEOREM. *Let* $[\Sigma, \mathfrak{G}]$ *be natural and* $\mathcal{B}$ *an algebra of functions on* Σ *with* $\mathfrak{G} \subseteq \mathcal{B} \subseteq \mathfrak{G}_\Sigma^+$. *Then* $[\Sigma, \mathcal{B}]$ *is also natural.* [R10].

Proof. Observe first that, if $K \subset\subset \Sigma$ then $\partial_0[\Sigma, \mathfrak{G}] \cap \hat{K} \subseteq \partial_0[\hat{K}, \mathfrak{G}]$. Therefore $\mathfrak{G}_\Sigma^+ | \hat{K} \subseteq \mathfrak{G}_{\hat{K}}^+$ and $[\Sigma, \mathcal{B}]$ will be natural if $[\hat{K}, \mathcal{B}]$ is natural for each $K \subset\subset \Sigma$.

It follows that we may assume Σ to be compact without any loss of generality. There is also no loss in assuming that $\mathcal{B}$ is closed in $C(\Sigma)$, by Proposition 3.4. Also, $\partial_0[\Sigma, \mathcal{G}]$ coincides with the set of strong boundary points of Σ relative to $\bar{\mathcal{G}}$ (§13).

Denote by Ψ the spectrum of $\mathcal{B}$ and by $\tau : \Sigma \to \Psi$, $\sigma \mapsto \psi_\sigma$, the usual homeomorphic embedding of Σ into Ψ *via* the point evaluations. Also, since $[\Sigma, \mathcal{G}]$ is natural we obtain, as usual by restricting elements of Ψ to $\mathcal{G}$, a continuous projection $\pi : \Psi \to \Sigma$, $\psi \mapsto \sigma_\psi$, of Ψ onto Σ. Obviously $\pi(\tau(\sigma)) = \sigma$ for all $\sigma \in \Sigma$. Our problem is to prove that each of the stalks $\pi^{-1}(\sigma)$, for $\sigma \in \Sigma$, reduces to the single point $\tau(\sigma)$.

Consider first a point $\delta \in \partial_0[\Sigma, \mathcal{G}]$ and suppose that ψ exists such that $\pi(\psi) = \delta$ but $\psi \neq \tau(\delta)$. Then there exists $h \in \mathcal{B}$ with $\hat{h}(\psi) = 1$ and $\hat{h}(\tau(\delta)) = h(\delta) = 0$. Choose a neighborhood V of δ in Σ such that $|h|_V < 1$. Then, since δ is a strong boundary point of Σ relative to $\bar{\mathcal{G}}$ there exists $v \in \bar{\mathcal{G}}$ with $|v|_{\Sigma \setminus V} < |v|_\Sigma = v(\delta) = 1$. Choose a positive integer m such that $(|v|_{\Sigma \setminus V})^m < |h|_\Sigma^{-1}$ and set $k = v^m h$. Then $k \in \mathcal{B}$ and $\hat{k}(\psi) = v(\delta)^m \hat{h}(\psi) = 1$. Since $|\hat{k}(\psi)| \leq |k|_\Sigma$ this implies that $|k|_\Sigma \geq 1$. On the other hand, $|k|_V \leq |v|_\Sigma^m |h|_V < 1$ and $|k|_{\Sigma \setminus V} \leq (|v|_{\Sigma \setminus V})^m |h|_\Sigma < 1$, so $|k|_\Sigma < 1$. This is a contradiction, so $\pi^{-1}(\delta)$ must be a singleton for $\delta \in \partial_0[\Sigma, \mathcal{G}]$, *i.e.* $\pi(\psi) = \delta$ implies $\psi = \tau(\delta)$.

Next, as in the proof of Theorem 40.1 we define $\tilde{h}(\psi) = h(\pi(\psi))$, $\psi \in \Psi$, where $h \in \mathcal{B}$. Then $\tilde{h}$ is constant on each of the stalks $\pi^{-1}(\sigma)$, $\sigma \in \Sigma$. Also define $F_h = \hat{h} - \tilde{h}$, $h \in \mathcal{B}$, and set $\Theta = \{\psi \in \Psi : F_h(\psi) = 0,\ h \in \mathcal{B}\}$. Observe that $\psi \in \Theta$ iff $\psi = \tau(\pi(\psi))$. In particular, $\tau(\Sigma) \subseteq \Theta$. Now consider the zero set $Z(F_h)$ of the function F_h in Ψ and let $\psi_0 \in \Psi \setminus Z(F_h)$. Since $Z(F_h)$ is closed there exists a neighborhood U_0 of ψ_0 disjoint from $Z(F_h)$. Observe that $\psi \neq \tau(\pi(\psi))$ for $\psi \notin Z(F_h)$. Therefore in particular, $\pi(U_0) \cap \partial_0[\Sigma, \mathcal{G}] = \emptyset$, so h is $\mathcal{G}$-holomorphic on $\pi(U_0)$. Since $\hat{\mathcal{G}} = \mathcal{G} \circ \pi$ and π is continuous we may conclude, by the usual induction argument, that the function $\tilde{h}$ is $\hat{\mathcal{G}}$-holomorphic, and hence $\hat{\mathcal{B}}$-holomorphic, on $\Psi \setminus Z(F_h)$. Therefore F_h is almost $\hat{\mathcal{B}}$-holomorphic on Ψ. This proves that Θ is an $\hat{\mathcal{B}}$-subvariety of Ψ. Hence by Theorem 34.1, Θ is $\hat{\mathcal{B}}$-convex. Since $\tau(\Sigma) \subseteq \Theta$ it follows that $\Theta = \Psi$. In other words each $\hat{h}$ is constant on stalks, which implies that each stalk is a singleton. ♦

§41. HOLOMORPHIC IMPLIED BY SUBHARMONIC CONVEXITY

We turn next to an examination of some of the connections between $\mathcal{G}$-h and $\mathcal{G}$-sh convexity. As has already been noted, $\mathcal{G}$-h convexity obviously implies $\mathcal{G}$-sh convexity. On the other hand, a fundamental problem in convexity theory is to obtain a converse to this statement. More precisely, if G is an open set in Σ and $H \subseteq G$, under what conditions on G, or on H, will $\mathcal{G}$-sh-convexity of H within G imply $\mathcal{G}$-h convexity within G? The most interesting, and also most difficult, case occurs when $H = G$. This is the problem of determining when $\mathcal{G}$-sh convexity of G implies $\mathcal{G}$-h convexity. In the case of $[\mathbb{C}^n, \mathcal{P}]$, every $\mathcal{P}$-sh convex (i.e. p-convex) domain in $\mathbb{C}^n$ is $\mathcal{P}$-h convex (holomorphically convex). This is also true for Riemann domains [G7; Theorem 4, p. 283], but is not true in general for complex manifolds [G7, p. 276]. Such results are not obvious and the usual proofs depend heavily on special properties of $\mathbb{C}^n$. Nevertheless, we are able to obtain certain general convexity results that are still nontrivial in finite dimensions. The first is an analogue of Theorem 39.8. (Cf. [L2; Proposition 3, p. 56] and [G7; Theorem 15, p. 278].)

41.1 THEOREM. *Let* G *be an open subset of* Σ *and* K *a compact subset of* G *such that* $\widehat{K}_G$ *is also compact. Then* $\overset{\sqcap}{K}_G = \widehat{K}_G$.

Proof. Observe that since $|\mathfrak{G}| \subseteq \mathcal{S}$, $\mathcal{G}$-subharmonicity is equivalent to $\mathfrak{G}$-subharmonicity. Therefore, in view of Theorem 40.1, we may apply Corollary 29.4 and obtain $\partial_{\mathcal{S}} K_G = \partial_{\mathfrak{G}_G} \widehat{K}_G \subseteq K$. Thus $\overset{\sqcap}{K}_G = \widehat{K}_G$, as claimed. ◆

41.2 THEOREM. *Let* G *be an open* $\mathcal{G}$*-h convex subset of* Σ *and let* $H \subseteq G$. *Then* H *will be* $\mathcal{G}$*-h convex within* G *iff it is* $\mathcal{G}$*-sh convex within* G. *(Cf. [G7, Theorem 15, p. 278].)*

Proof. We have only to prove that $\mathcal{G}$-sh convexity implies $\mathcal{G}$-h convexity. Therefore let K be a compact subset of H and assume that $\overset{\sqcap}{K}_G \cap H$ is compact. By Theorem 41.1 we have $\overset{\sqcap}{K}_G = \widehat{K}_G$, so $\widehat{K}_G \cap H = \overset{\sqcap}{K}_G \cap H$. Hence $\widehat{K}_G \cap H$ is compact and the theorem follows. ◆

Theorem 41.2 along with Theorem 39.1 (i), gives the following corollary.

41.3 COROLLARY. *Let* $u \in \mathcal{CS}_G$. *Then for arbitrary real* t *the set* $G\{u < t\}$ *is* $\mathfrak{G}$-h *convex within* G.

We also have an analogue of Theorem 41.2 for $\mathfrak{G}$-convexity in place of $\mathfrak{G}$-h convexity.

41.4 THEOREM. *Let* G *be an open* $\mathfrak{G}$*-convex subset of* Σ *and* H *an open subset of* G. *Then* H *will be* $\mathfrak{G}$*-convex iff it is* $\mathfrak{G}$-sh *convex within* G.

Proof. Let $K \subset\subset H$. Then by Theorem 35.6 we have $\overset{\sqcap}{K}_G = \hat{K}$, so $\hat{K} = (\overset{\sqcap}{K}_G \cap H) \cup (\overset{\sqcap}{K}_G \setminus H)$. By hypothesis, $\overset{\sqcap}{K}_G \cap H$ is compact. Therefore we have a decomposition of $\hat{K}$ into disjoint compact sets. By Corollary 15.4 the set $\overset{\sqcap}{K}_G \cap H$ is $\mathfrak{G}$-convex and hence is equal to $\hat{K}$, since it contains K. In particular, $\hat{K} \subset H$, so H is $\mathfrak{G}$-convex. ◆

41.5 COROLLARY. *Let* $u \in \mathcal{CS}_G$. *Then for arbitrary real* t *the set* $G\{u < t\}$ *is* $\mathfrak{G}$*-convex*.

§42. LOCAL PROPERTIES

The condition on the set G in Theorem 41.2 is considerably more restrictive than we would like. In particular, the theorem gives us no information if $H = G$. Although the situation here remains unclear in the general case, it is possible to obtain certain "local convexity" results without restrictions on G.

A set $H \subseteq \Sigma$ is said to be $\mathfrak{G}$*-convex at a point* $\delta \in \Sigma$ if there exists a neighborhood U of δ such that $U \cap H$ is $\mathfrak{G}$-convex. If H is $\mathfrak{G}$-convex at every point of Σ then it is said to be *locally* $\mathfrak{G}$*-convex*. The notions of $\mathfrak{G}$-h and $\mathfrak{G}$-sh *convexity at a point*, as well as *local* $\mathfrak{G}$-h *convexity* and *local* $\mathfrak{G}$-sh *convexity*, are defined similarly. As a consequence of the naturality of $[\Sigma, \mathfrak{G}]$, each point of Σ admits arbitrarily small $\mathfrak{G}$-convex neighborhoods (§6). Therefore the condition that an open set be locally convex in any of the above senses is actually only a condition on its boundary. It is also easy to see that if an open set is "globally" convex in any of the above senses then it is locally convex in the same sense.

In terms of one set being convex within another, we may formulate the following more restrictive definition of convexity at a point. A set H is said to be *relatively convex* (in one of the above senses) *at a point* $\delta \in \Sigma$ if there exists a neighborhood U of δ such that $U \cap H$ is convex *within* U. Note that by Corollary 15.4, if H is open and $H \cap U$ is $\mathcal{G}$-convex within U then $H \cap U$ is itself $\mathcal{G}$-convex. Therefore, relative $\mathcal{G}$-convexity (at a point) reduces to $\mathcal{G}$-convexity, so there is nothing new. The following theorem is a much stronger result along the same lines.

42.1 THEOREM. *Relative* $\mathcal{G}$-sh *convexity (and hence also relative* $\mathcal{G}$-h *convexity) of an open set* H *at a point* δ *is equivalent to* $\mathcal{G}$*-convexity at* δ.

Proof. Let U be a neighborhood of δ such that $U \cap H$ is $\mathcal{G}$-sh convex within U and let V be an $\mathcal{G}$-convex neighborhood of δ contained in U. If $K \subset\subset V \cap H$, the set $\hat{K}_V \cap H$ is relatively closed in H and contained in the compact set $\hat{K}_U \cap H$. Therefore $\hat{K}_V \cap H$ is compact and it follows that $H \cap V$ is $\mathcal{G}$-sh convex within V. Finally, since V is $\mathcal{G}$-convex it follows that $H \cap V$ is $\mathcal{G}$-convex by Theorem 41.4. In other words, H is $\mathcal{G}$-convex at the point δ. ◆

42.2 COROLLARY. *Let* G *and* H *be open sets in* Σ, *where* $H \subseteq G$ *and* H *is* $\mathcal{G}$-sh *convex within* G. *Then* H *is* $\mathcal{G}$*-convex at each point of* G. *If* $\bar{H} \subset G$ *then* H *is locally* $\mathcal{G}$*-convex.*

Consider any open set G and a function $u \in \mathcal{CS}_G$. For real t and $\delta \in G\{u < t\}$ choose an open $\mathcal{G}$-convex neighborhood U of δ contained in G. Then U is of course also $\mathcal{G}$-sh convex. Therefore by Theorem 39.1 (i) (with U in place of G) $U\{u < t\}$ is $\mathcal{G}$-sh convex within U. Since $U\{u < t\} = U \cap G\{u < t\}$ this means that $G\{u < t\}$ is relatively $\mathcal{G}$-sh convex at δ. Thus we have the following corollary.

42.3 COROLLARY. *Let* $u \in \mathcal{CS}_G$, *where* G *is an arbitrary open subset of* Σ. *Then for each* t *the set* $G\{u < t\}$ *is* $\mathcal{G}$*-convex at each point of* G. *If also* $\overline{G\{u < t\}} \subset G$ *then* $G\{u < t\}$ *is locally* $\mathcal{G}$*-convex.*

We close this section with a convexity property somewhat suggestive of the notion of linear convexity in a linear vector space.

Let Ω be a subset of Σ. A point $\delta \in \Omega$ is called an *inner point* of Ω if it is not a locally independent point for the system $[\Omega, \mathcal{G}]$ (Definition 14.4).

This amounts to saying that if U is any neighborhood of δ then there exists a compact set $K \subset U \cap \Omega$ such that $\delta \in \hat{K} \setminus K$. The point δ is called an *extension point* of Ω if any function, which is $\mathfrak{G}$-holomorphic on a deleted neighborhood of δ within Ω, has an $\mathfrak{G}$-holomorphic extension to the full neighborhood in Ω. These notions, although formulated for an arbitrary set, are primarily of interest for an $\mathfrak{G}$-variety. For example, if Θ is an ordinary variety of $\dim \geq 2$ (say, in $\mathbb{C}^n$) then each of its points is an inner extension point. Observe also that if Θ is an $\mathfrak{G}$-subvariety of an open set $G \subseteq \Sigma$ (Definition 26.1) then it is $\mathfrak{G}$-convex at each of its points. In fact, let $\delta \in \Theta$ and choose an open $\mathfrak{G}$-convex neighborhood V of δ with $V \subseteq G$. Then $V \cap \Theta$ is an $\mathfrak{G}$-subvariety of the $\mathfrak{G}$-convex set V and so, by Theorem 34.1, is itself $\mathfrak{G}$-convex.

42.4 THEOREM. *Let* G *be an open set in* Σ *and* δ *a point of* Σ *at which* G is $\mathfrak{G}$-h-convex. *Also let* Ω *be a set which is* $\mathfrak{G}$-*convex at* δ *and for which* δ *is an inner extension point. Then* $\Omega \setminus \{\delta\} \subseteq G$ *implies* $\Omega \subseteq G$.

Proof. As in the proof of Theorem 41.4, choose an open $\mathfrak{G}$-convex neighborhood V of δ such that $V \cap G$ is $\mathfrak{G}$-h convex and $V \cap \Omega$ is $\mathfrak{G}$-convex. Since δ is an inner point of Ω there exists a compact set $K \subset V \cap \Omega$ such that $\delta \in \hat{K} \setminus K$. Note that $\hat{K} \subset V \cap \Omega$. Assume now that $\Omega \setminus \{\delta\} \subseteq G$. Then $K \subset V \subseteq G$, so $\widehat{K}_{V\cap G}$ is a compact subset of $V \cap G$ that contains K. Suppose that $\delta \notin \widehat{K}_{V\cap G}$. Then $\hat{K} \setminus \widehat{K}_{V\cap G}$ is a relatively open subset of $\hat{K}$ that contains δ. Since $\hat{K}$ cannot contain an isolated point outside of K (Corollary 14.3) it follows that the set $\hat{K} \setminus \widehat{K}_{V\cap G}$ must contain a point δ' different from δ. Then $\delta' \in V \cap (\Omega \setminus \{\delta\}) \subset V \cap G$. Hence there exists $h \in \Theta_{V\cap G}$ such that $|h(\delta')| \geq |h|_{\widehat{K}_{V\cap G}} = |h|_K$. Note that the restriction of h to $(V \cap \Omega) \setminus \{\delta\}$ is $\mathfrak{G}$-holomorphic. Therefore since δ is an extension point of Ω the function h extends to an $\mathfrak{G}$-holomorphic function on $V \cap \Omega$. In particular, h becomes $\mathfrak{G}$-holomorphic on $\hat{K}$. But then the inequality $|h(\delta')| > |h|_K$ contradicts the local maximum principle for $\mathfrak{G}$-holomorphic functions. Thus the assumption $\delta \notin \widehat{K}_{V\cap G}$ leads to a contradiction, so we must have $\delta \in \widehat{K}_{V\cap G}$ and hence $\delta \in G$. ♦

If G is assumed to be $\mathfrak{G}$-convex at δ then the assumption that Ω be $\mathfrak{G}$-convex at δ and that δ be an extension point for Ω in Theorem 42.4 may be dropped.

To see this, proceed as before to obtain an open $\mathcal{G}$-convex neighborhood V of δ such that $V \cap G$ is $\mathcal{G}$-convex, and a compact set $K \subset V \cap \Omega$ such that $\delta \in \hat{K}\setminus K$. If $\Omega\setminus\{\delta\} \subset G$ then $K \subset V \cap G$ so $\hat{K} \subset V \cap G$ and hence $\delta \in G$.

§43. REMARKS AND AN EXAMPLE

Corollary 40.2 is due to the author [R4]. It is a special case of a much more general result [R9; §4] to be discussed in the next chapter (Theorem 47.3). Note that the convexity condition in Corollary 40.2 is automatic if $G = \Sigma$. In particular, the result holds if $\mathscr{B}$ is an $\mathcal{G}$-local algebra on Σ (*i.e.* each element of $\mathscr{B}$ belongs locally to $\mathcal{G}$). For compact Σ this is a result due to Stolzenberg [S7]. F. Quigley (written communication) has also given an elegant proof of the result in Corollary 40.2 for the case $G = \Sigma$ and compact Σ. (See [G1, p. 93]). In fact, the idea for using the function $\tilde{h}$ in the proof of Theorem 40.1 is from the Quigley proof.

In the case of a domain G in $\mathbb{C}^n$, it turns out that local p-convexity (and hence local holomorphic convexity) at the boundary of G implies holomorphic convexity of G [H4; Theorem 2.6.10]. This is a nontrivial result and we do not have an adequate extension for $\mathcal{G}$-h convexity in general. The problem is also open even with $\mathcal{G}$-h convexity replaced by $\mathcal{G}$-convexity.

The following example shows that results such as that in Corollary 40.2 and Theorem 40.3, even in the compact case, are somewhat more delicate than one might think at first. The example is essentially one constructed by S. Sidney in response to a question posed by J. Garnett (oral communication). It shows that one cannot in general strengthen Corollary 40.2 by replacing the algebra by the possibly larger algebra $\mathcal{O}^*_\Sigma$ of functions continuous on Σ and $\mathcal{G}$-holomorphic on $\Sigma\setminus\partial[\Sigma, \mathcal{G}]$. It also answers in the negative a question attributed to Kenneth Hoffman by I. Glicksberg [G5, p. 924 n.], *viz.* if $[\Sigma, \mathcal{G}]$ is natural (with compact Σ) and $\mathcal{G} \subseteq \mathcal{B} \subseteq C(\Sigma)$, where $\partial[\Sigma, \mathcal{B}] = \partial[\Sigma, \mathcal{G}]$, then is $[\Sigma, \mathcal{B}]$ natural? (See also [W3].)

43.1 EXAMPLE. For $r > 0$ denote the open bidisc of radius r in $\mathbb{C}^2$ by

$$\Delta_r = \{(\zeta_1, \zeta_2) \in \mathbb{C}^2 : |\zeta_1| < r, |\zeta_2| < r\}$$

and its closure by $\bar{\Delta}_r$. Let $\{(\xi_1^{(n)}, \xi_2^{(n)})\}$ be a sequence of points dense in $\Delta_{1/2}$

and for each n consider in $\mathbb{C}^2 \times \mathbb{R}$ the closed interval

$$I_n = \{(\xi_1^{(n)}, \xi_2^{(n)}, t) : 0 \le t \le \frac{1}{n}\}.$$

Since the length of I_n converges to zero as $n \to \infty$, the union

$$\Sigma = (\Delta_1 \times (0)) \cup (\bigcup_{n=1}^{\infty} I_n)$$

is a compact subset of the space $\mathbb{C}^2 \times \mathbb{R}$. Next let $\mathfrak{A}$ be the algebra consisting of all functions $f(\zeta_1, \zeta_2, t)$ defined and continuous on Σ such that for $t = 0$ the function $f(\zeta_1, \zeta_2, 0)$ is holomorphic in Δ_1. Then $\mathfrak{A}$ is a uniform algebra on Σ and the system $[\Sigma, \mathfrak{A}]$ is natural (see [R1, p. 130].) Since the elements of $\mathfrak{A}$ are arbitrarily continuous on each of the intervals I_n, it is obvious that the Šilov boundary $\partial[\Sigma, \mathfrak{A}]$ contains each I_n and, being closed, must also contain $\bar{\Delta}_{1/2} \times (0)$. In fact,

$$\partial[\Sigma, \mathfrak{A}] = (\bar{\Delta}_{1/2} \times (0)) \cup (T^2 \times (0))$$

where T^2 denotes the torus

$$T^2 = \{(\zeta_1, \zeta_2) \in \mathbb{C}^2 : |\zeta_1| = |\zeta_2| = 1\}.$$

For arbitrary $h \in \mathcal{O}_\Sigma^*$, set

$$h_1(\zeta_1, \zeta_2) = h(\zeta_1, \zeta_2, 0), \; (\zeta_1, \zeta_2) \in \bar{\Delta}_1 \setminus \bar{\Delta}_{1/2}.$$

Then h_1 is holomorphic on $\Delta_1 \setminus \bar{\Delta}_{1/2}$. Now, by the Hartog's phenomenon h_1 admits a unique holomorphic extension $\tilde{h}_1$ to all of Δ_1. If $h \in \mathfrak{A}$ it is obvious that $\tilde{h}_1(\zeta_1, \zeta_2)$ coincides on $\Delta_{1/2}$ with $h(\zeta_1, \zeta_2, 0)$. On the other hand, functions in $\mathcal{O}_\Sigma^*$, being more-or-less arbitrarily continuous in $\Delta_{1/2} \times (0)$, will generally be such that $\tilde{h}_1(\zeta_1, \zeta_2) \ne h(\zeta_1, \zeta_2, 0)$ for $(\zeta_1, \zeta_2) \in \Delta_{1/2}$. Therefore, mappings of the form

$$h \mapsto \tilde{h}_1(\zeta_1, \zeta_2), \; (\zeta_1, \zeta_2) \in \Delta_{1/2}$$

define homomorphisms of $\mathcal{O}_\Sigma^*$ onto $\mathbb{C}$ that are not point evaluations in Σ. In other words, $[\Sigma, \mathcal{O}_\Sigma^*]$ is not natural in this example. Observe also that $\partial[\Sigma, \mathcal{O}_\Sigma^*] = \partial[\Sigma, \mathfrak{A}]$.

Sidney's original example involved discs in place of the intervals I_n and elements of $\mathfrak{A}$ were required to be holomorphic on the discs, so that $\mathfrak{A}$ had the additional property of being antisymmetric.

CHAPTER IX
$[\Sigma, \mathfrak{G}]$-DOMAINS

§44. DEFINITIONS

In this section we consider a generalization of the classical notion of a Riemann domain (or manifold space over $\mathbb{C}^n$). In our case, the complex space $\mathbb{C}^n$ is replaced by the space Σ of a given system $[\Sigma, \mathfrak{G}]$.

44.1 DEFINITION. *A pair* (Φ, p), *consisting of a Hausdorff space* Φ *and an open local homeomorphism* $p : \Phi \to \Sigma$ *of* Φ *into* Σ, *is called a* $[\Sigma, \mathfrak{G}]$-*domain (or simply a* Σ-*domain).* [R9].

If the space Φ is connected then the $[\Sigma, \mathfrak{G}]$-domain (Φ, p) is also said to be *connected*. Note that Definition 44.1 asserts simply that the space Φ is "spread" by p over the base space Σ. The term "$[\Sigma, \mathfrak{G}]$-domain" is used for emphasis and, when the system involved is obvious, will be replaced by the more cryptic "Σ-domain". Also, in order to avoid uninteresting pathology we shall always assume that the base space Σ is both connected and locally connected. The subset of Φ mapped by p into a given point $\sigma \in \Sigma$ is called the *stalk* of Φ over σ.

The condition that $p : \Phi \to \Sigma$ be a *local homeomorphism* means that each point $\varphi \in \Phi$ admits a neighborhood U_φ such that $p : U_\varphi \to p(U_\varphi)$ is a homeomorphism. Also, the condition that $p : \Phi \to \Sigma$ is *open* means that open sets of Φ map onto open sets in Σ. In particular, $p(\Phi)$ is an open subset of Σ and the neighborhood U_φ maps onto a neighborhood $p(U_\varphi)$ of the base point $p(\varphi)$ in Σ. Neighborhoods U_φ such that $p : U_\varphi \to p(U_\varphi)$ is a homeomorphism are called p-*neighborhoods* and the inverse of the homeomorphism is denoted by $p^{-1}_{U_\varphi}$ or, when it is unnecessary to exhibit the particular p-neighborhood involved, simply by p^{-1}_φ. It is essential to remember, however, that the symbol p^{-1}_φ is ambiguous by itself and the implied p-neighborhood must be determined by the context. If G is any subset of Φ such that $p : G \to p(G)$

is one-to-one (and hence a homeomorphism) then G is called a p-*set*. Thus a p-set G is simply a section of the Σ-domain over $p(G)$. Since Σ is assumed to be locally connected the space Φ is also locally connected. If (Φ, p) is connected then any two points of Φ may be joined by a finite chain of p-neighborhoods. The simplest example of a Σ-domain is of the form (G, ι), where G is an open subset of Σ and $\iota : G \hookrightarrow \Sigma$ is the identity map. Note that a p-set over $p(\Phi)$ is necessarily a connected component of the space Φ.

44.2 LEMMA. *Let* U *and* V *be arbitrary connected open* p-*sets such that* $p(U) \cap p(V)$ *is connected. If* $U \cap V \neq \phi$ *then* $U \cup V$ *is a* p-*set. In particular, if* $p(U) = p(V)$ *and* $U \cap V \neq \phi$ *then* $U = V$. *(Cf. [N2; Lemma 4, p. 18].)*

Proof. Define $U' = p_U^{-1}(p(U) \cap p(V))$ and $V' = p_V^{-1}(p(U) \cap p(V))$. Then $U' \cap V' = U \cap V$, $p(U') = p(V')$, and both U' and V' are connected open p-sets. Hence, if $U' \neq V'$ there exists a point φ_1 of U' on the boundary of V' and a point $\varphi_2 \in V'$ such that $p(\varphi_1) = p(\varphi_2)$. Choose disjoint p-neighborhoods $U_{\varphi_1} \subset U'$ and $V_{\varphi_2} \subset V'$ such that $p(U_{\varphi_1}) = p(V_{\varphi_2})$. Since φ_1 is a limit point of V' the neighborhood U_{φ_1} contains a point $\varphi_3 \in V'$. Therefore V_{φ_2} contains a point φ_4 such that $p(\varphi_3) = p(\varphi_4)$. But this contradicts the fact that V' is a p-set and proves that $U' = V' = U \cap V$. Finally let φ' and φ'' be points of $U \cup V$ such that $p(\varphi') = p(\varphi'')$. Then φ' and φ'' belong to $U' \cap V'$ and hence to $U \cap V$, so $\varphi' = \varphi''$. In other words, $U \cup V$ is a p-set. If in addition $p(U) = p(V)$ then $U \cup V$ can be a p-set only if $U = V$. ♦

§45. "DISTANCE" FUNCTIONS

Now let A be a linearly independent system of generators for the structure algebra $\mathfrak{G}$. Denote by α an arbitrary finite subset of the set A and, for each $\sigma_0 \in \Sigma$ and $0 < r \leq \infty$, set

$$N_{\sigma_0}(\alpha, r) = \{\sigma \in \Sigma : |a(\sigma) - a(\sigma_0)| < r, a \in \alpha\}.$$

Then, since $[\Sigma, \mathfrak{G}]$ is a system and A generates $\mathfrak{G}$ the neighborhoods $N_{\sigma_0}(\alpha, r)$ constitute a basis for the topology of Σ. For $\varphi \in \Phi$ and each α, we denote by

$R_\alpha(\varphi)$ the set consisting of 0 plus all $r \in (0, \infty]$ such that p_φ^{-1} is defined on the neighborhood $N_{p(\varphi)}(\alpha, r)$. For $r \in R_\alpha(\varphi)$ and $r \neq 0$, set

$$W_\varphi(\alpha, r) = p_\varphi^{-1}(N_{p(\varphi)}(\alpha, r)).$$

Then $W_\varphi(\alpha, r)$ is called a *basic* p-*neighborhood* of the point φ.

45.1 DEFINITION. *For arbitrary* $\varphi \in \Phi$ *and* $\alpha \subset A$, *we define*

$$d_\alpha(\varphi) = \sup\{r : r \in R_\alpha(\varphi)\}, \ d_A(\varphi) = \sup_\alpha d_\alpha(\varphi)$$

and, for arbitrary $S \subseteq \Phi$,

$$\delta_\alpha(S) = \inf\{d_\alpha(\varphi) : \varphi \in S\}, \ \delta_A(S) = \sup_\alpha \delta_\alpha(S).$$

Observe that $0 \le d_\alpha(\varphi) \le \infty$ and both of the values 0 and ∞ are possible. Also, $d_\alpha(\varphi) = 0$ iff $R_\alpha(\varphi)$ reduces to the single point 0. On the other hand, since the neighborhoods $N_\sigma(\alpha, r)$ generate the topology of Σ it follows that there exists a finite set $\alpha \subseteq A$ such that $d_\alpha(\varphi) > 0$, so $d_A(\varphi) > 0$ for each $\varphi \in \Phi$. Obviously, if $d_\alpha(\varphi) > 0$ then the basic p-neighborhood $W_\varphi(\alpha, r)$ is defined for every $r \in (0, d_\alpha(\varphi))$. Moreover, if

$$W = \bigcup_{0<r<d_\alpha(\varphi)} W_\varphi(\alpha, r)$$

then, since $r_1 \le r_2$ implies $W(\alpha, r_1) \subseteq W_\varphi(\alpha, r_2)$, the set W is a p-neighborhood of φ. Furthermore.

$$p(W) = \bigcup_{0<r<d_\alpha(\varphi)} N_{p(\varphi)}(\alpha, r) = N_{p(\varphi)}(\alpha, d_\alpha(\varphi)).$$

Therefore $W_\varphi(\alpha, r)$ exists for all $r \in (0, d_\alpha(\varphi)]$ when $d_\alpha(\varphi) > 0$. Note that if $d_\alpha(\varphi) = \infty$ then $N_{p(\varphi)}(\alpha, d_\alpha(\varphi)) = \Sigma$. Some further properties of these "distance" functions are given in the next proposition.

45.2 PROPOSITION. *(i) The set* $\Phi_\alpha^\infty = \{\varphi \in \Phi : d_\alpha(\varphi) = \infty\}$ *is open and closed in* Φ *and each of its components is mapped by* p *homeomorphically onto* Σ. *In particular, if* Φ *is connected then either* $\Phi_\alpha^\infty = \emptyset$ *or* $\Phi_\alpha^\infty = \Phi$.

(ii) For each α *the function* d_α *is continuous on the (open) set* $\{\varphi \in \Phi : d_\alpha(\varphi) > 0\}$.

(iii) For each $\varphi \in \Phi$ *we have* $\lim_{\alpha} d_\alpha(\varphi) = d_A(\varphi) > 0$. *The function* d_A *takes its values in* $(0, \infty]$ *and is lower semi-continuous, so* $-\log d_A \in \mathcal{U}$. *(See §25.)*

(iv) If K *is a compact subset of* Φ *then* $\delta_A(K) = \inf\{d_A(\varphi) : \varphi \in K\} > 0$.

(v) If S *is any subset of* Φ *with* $\delta_A(S) > 0$ *and if* $0 < r < \delta_A(S)$, *then* α *exists such that* $W_\varphi(\alpha, r)$ *is defined for each* $\varphi \in S$.

Proof. Let $\varphi \in \Phi_\alpha^\infty$. Then $W_\varphi(\alpha, \infty)$ is defined and $p(W_\varphi(\alpha, \infty)) = \Sigma$. Moreover, if $\varphi' \in W_\varphi(\alpha, \infty)$ then also $d_\alpha(\varphi') = \infty$ and $W_{p(\varphi')}(\alpha, r) = W_{p(\varphi)}(\alpha, r)$. Since Σ is assumed to be connected, $W_\varphi(\alpha, \infty)$ is an open connected subset of Φ. Furthermore, if φ_0 is a limit point of $W_\varphi(\alpha, \infty)$ in Φ then, since $p(\varphi_0) \in W_\varphi(\alpha, \infty)$, there exists $\varphi_0' \in W_\varphi(\alpha, \infty)$ such that $p(\varphi_0') = p(\varphi_0)$. If $\varphi_0 \neq \varphi_0'$ then there exist disjoint p-neighborhoods U_{φ_0} and $U_{\varphi_0'}$ such that $U_{\varphi_0'} \subseteq W_\varphi(\alpha, \infty)$ and $p(U_{\varphi_0}) = p(U_{\varphi_0'})$. But φ_0 is a limit point of $W_\varphi(\alpha, \infty)$, so there exists $\varphi_1 \in U_{\varphi_0} \cap W_\varphi(\alpha, \infty)$. Hence there also exists $\varphi_1' \in U_{\varphi_0'}$ such that $p(\varphi_1') = p(\varphi_1)$. Since this contradicts the fact that $W_\varphi(\alpha, \infty)$ is a p-neighborhood, it follows that $W_\varphi(\alpha, \infty)$ is both open and closed in Φ, proving (i).

Next let φ_0 be any point of Φ. If $d_\alpha(\varphi_0) = \infty$, then by property (i), $d_\alpha(\varphi) = \infty$ on a neighborhood of φ_0, so d_α is continuous at φ_0. Therefore assume that $0 < d_\alpha(\varphi_0) < \infty$. For arbitrary $\varepsilon > 0$ choose $r < \varepsilon$ so that $W_{\varphi_0}(\alpha, r)$ is defined. Then $d_\alpha(\varphi_0) \geq r$. The continuity of d_α at the point φ_0 will follow immediately from the inequality

$$d_\alpha(\varphi_0) - \varepsilon \leq d_\alpha(\varphi) \leq d_\alpha(\varphi_0) + \varepsilon, \ \varphi \in W_{\varphi_0}(\alpha, r).$$

For the proof, set $\sigma_0 = p(\varphi_0)$ and observe that

$$N_\sigma(\alpha, d_\alpha(\varphi_0) - \varepsilon) \subseteq N_{\sigma_0}(\alpha, d_\alpha(\varphi_0)), \ \sigma \in N_{\sigma_0}(\alpha, r).$$

This implies that $W_\varphi(\alpha, d_\alpha(\varphi_0) - \varepsilon)$ is defined for each $\varphi \in W_{\varphi_0}(\alpha, r)$. Therefore $\varphi \in W_{\varphi_0}(\alpha, r)$ implies that $d_\alpha(\varphi) \geq d_\alpha(\varphi_0) - \varepsilon$, proving the left hand side of the desired inequality. Now suppose that φ' were a point of $W_{\varphi_0}(\alpha, r)$ such that $d_\alpha(\varphi') > d_\alpha(\varphi_0) + \varepsilon$. Then, in particular $d_\alpha(\varphi') > r$ and hence $W_{\varphi'}(\alpha, r)$ is defined. Therefore, as in the preceding argument we have $d_\alpha(\varphi') - \varepsilon \leq d_\alpha(\varphi)$ for all $\varphi \in W_{\varphi'}(\alpha, r)$. On the other hand, since $\varphi' \in W_{\varphi_0}(\alpha, r)$ it follows that $\varphi_0 \in W_{\varphi'}(\alpha, r)$, so

$$d_\alpha(\varphi_0) \geq d_\alpha(\varphi') - \varepsilon > (d_\alpha(\varphi_0) + \varepsilon) - \varepsilon = d_\alpha(\varphi_0).$$

This is a contradiction and means that we must have $d_\alpha(\varphi) \leq d_\alpha(\varphi_0) + \varepsilon$ for all $\varphi \in W_{\varphi_0}(\alpha, r)$. Therefore d_α is continuous on the set $\{\varphi \in \Phi : d_\alpha(\varphi) > 0\}$, proving (ii).

It is obvious that $\alpha_1 \subseteq \alpha_2$ implies $d_{\alpha_1}(\varphi) \leq d_{\alpha_1}(\varphi)$. Therefore, for each $\varphi \in \Phi$, $\lim_\alpha d_\alpha(\varphi) = \sup_\alpha d_\alpha(\varphi) = d_A(\varphi)$, where the limit is taken with respect to the directed system consisting of all finite subsets of the set A partially ordered by inclusion. Also, as already noted in the remark following Definition 45.1, we have $d_A(\varphi) > 0$. The function d_A obviously takes its values in $(0, \infty]$ and, being a supremum of continuous functions, is lower semi-continuous. Hence (iii) follows.

Let K be a compact subset of Φ and set $2s = \inf_{\varphi \in K} d_A(\varphi) - \delta_A(K)$. Since

$$\delta_A(K) = \sup_\alpha \inf_{\varphi \in K} d_\alpha(\varphi) \leq \inf_{\varphi \in K} d_A(\varphi)$$

it follows that $s \geq 0$. Suppose now that $s > 0$. Fix $\varphi_0 \in K$ and choose α_0 such that $d_A(\varphi_0) - s < d_{\alpha_0}(\varphi_0)$. Since $d_A(\varphi_0) \geq 2s + \delta_A(K)$, we have $d_A(\varphi_0) - s > 0$, so $d_{\alpha_0}(\varphi_0) > 0$ and d_{α_0} is continuous at φ_0. Hence there exists a neighborhood $U_{\varphi_0}(\alpha_0)$ of the point φ_0 such that

$$d_A(\varphi_0) - s < d_{\alpha_0}(\varphi),\ \varphi \in U_{\varphi_0}(\alpha_0).$$

Since K is compact, it is covered by a finite number of such neighborhoods, say $U_{\varphi_1}(\alpha_1), \ldots, U_{\varphi_n}(\alpha_n)$, with $d_A(\varphi_i) - s < d_{\alpha_i}(\varphi)$, $\varphi \in U_{\varphi_i}(\alpha_i)$, $(i=1,\ldots,n)$. Now take $\alpha = \alpha_1 \cup \cdots \cup \alpha_n$ and for $\varphi \in K$ choose i such that $\varphi \in U_\varphi(\alpha_i)$. Then

$$\inf_{\varphi' \in K} d_A(\varphi') \leq d_A(\varphi_i) < d_{\alpha_i}(\varphi) + s \leq d_\alpha(\varphi) + s$$

so $\inf_{\varphi' \in K} d_A(\varphi') \leq d_\alpha(\varphi) + s$ for $\varphi \in K$. Hence

$$\inf_{\varphi \in K} d_A(\varphi) \leq \delta_A(K) + s < \delta_A(K) + 2s = \inf_{\varphi \in K} d_A(\varphi).$$

This is a contradiction and proves that $s = 0$. Also, since the function d_A is lower semi-continuous and never zero it assumes a nonzero minimum value on K. Therefore $\delta_A(K) > 0$ and (iv) is proved.

Finally, let S be a subset of Φ with $\delta_A(S) > 0$ and choose r such that $0 < r < \delta_A(S)$. Then there exists α such that $r < \delta_\alpha(S)$, so $r < d_\alpha(\varphi)$ for each $\varphi \in S$. Therefore $W_\varphi(\alpha, r)$ is defined for each $\varphi \in S$ and (v) follows. ◆

§46. HOLOMORPHIC FUNCTIONS

The "distance" functions will come up again later in this chapter (Theorem 47.3) when we consider the problem of "$\mathfrak{G}$-holomorphic" convexity of a Σ-domain. We proceed now to the development of an $\mathfrak{G}$-holomorphy theory for Σ-domains. As might be expected from the example of Rieman domains, the idea here is to lift local properties of the base system $[\Sigma, \mathfrak{G}]$ up to the Σ-domain (Φ, p) *via* the local homeomorphism. We accordingly make the following definition.

46.1 DEFINITION. *Let* h *be a function defined on a set* $X \subseteq \Phi$. *Then* h *is said to be* $\mathfrak{G}$-*holomorphic on* X *if for each* $\varphi \in X$ *and* p-*neighborhood* U_φ *the function* $h \circ p_\varphi^{-1}$ *is* $\mathfrak{G}$-*holomorphic in the usual sense on the set* $p(U_\varphi \cap X)$.

The set of all $\mathfrak{G}$-holomorphic functions defined on subsets of Φ will be denoted by ${}_{(\Phi, p)}\mathfrak{G}$. Thus we have the presheaf $\{\Phi, {}_{(\Phi, p)}\mathfrak{G}\}$ of function algebras over Φ. For simplicity, we shall usually write $\mathfrak{G}_X$ or $\mathfrak{G}_{(X, p)}$ in place of ${}_{(\Phi, p)}\mathfrak{G}_X$ for the algebra of all $\mathfrak{G}$-holomorphic functions defined on a set $X \subseteq \Phi$. Now consider the pair $[\Phi, \mathfrak{G}\circ p]$. Although $[\Phi, \mathfrak{G}\circ p]$ need not be a system (the algebra $\mathfrak{G}\circ p$ will usually not even separate points), we nevertheless have the notion of $\mathfrak{G}\circ p$-holomorphic functions in Φ.

46.2 PROPOSITION. *A function in* Φ *will be* $\mathfrak{G}$-*holomorphic according to Definition 46.1 iff it is* $\mathfrak{G}\circ p$-*holomorphic in the usual sense (Definition 14.3).*

Proof. This result is immediate from the fact that being $\mathfrak{G}$-holomorphic is a local property and that the projection $p : \Phi \to \Sigma$ restricted to an arbitrary p-set $G \subseteq \Phi$ induces a pair isomorphism $p : [G, \mathfrak{G}\circ p] \Rightarrow [p(G), \mathfrak{G}]$. ◊

A major problem in the study of a Σ-domain (Φ, p) is that $[\Phi, \mathfrak{G}_\Phi]$ is generally not a system. In fact, $\mathfrak{G}_\Phi$ need not even separate points. On the other hand, since $\mathfrak{G}\circ p \subseteq \mathfrak{G}_\Phi$ and $\mathfrak{G}$ separates the points of Σ, the algebra $\mathfrak{G}_\Phi$ can fail to separate two points φ_1 and φ_2 only if $p(\varphi_1) = p(\varphi_2)$. The problem is clearly global in character since for any p-neighborhood U_φ the pair $[p(U_\varphi), \mathfrak{G}]$, and hence $[U_\varphi, \mathfrak{G}_\Phi]$, is a system. In certain special cases, for example when we have a concept of a derivative and Taylor expansions in Σ so that a holomorphic function is determined on a neighborhood of a point by its value and the values of its derivatives at the point, it is not difficult to prove that the holomorphic functions do separate points. This enables us to handle (in Chapter XII) the Case of holomorphic functions on a domain spread over a vector space. Note, however, that $[\Phi, \mathfrak{O}_\Phi]$ may still fail to be a system even if $\mathfrak{O}_\Phi$ does separate points. For the time being, we shall sidestep this problem by assuming outright that the pair in question is a system whenever necessary.

§47. RELATIVE COMPLETENESS AND NATURALITY

Consider an arbitrary algebra $\mathfrak{H}$ of functions on Φ such that $\mathfrak{G}\circ p \subseteq \mathfrak{H} \subseteq \mathfrak{O}_\Phi$. The next theorem provides important criteria for $[\Phi, \mathfrak{H}]$ to be a natural system. First, however, we need some definitions. Let K be a compact subset of Φ and denote its $\mathfrak{H}$-convex hull in Φ by $\hat{K}$. Since $\mathfrak{G}$-holomorphic functions are continuous the hull $\hat{K}$ is always closed but will not in general be compact. By Definition 7.5 the pair $[\Phi, \mathfrak{H}]$ will be *convex* iff the $\mathfrak{H}$-convex hull of every compact set $K \subseteq \Phi$ is also compact. In particular, if the pair $[\Phi, \mathfrak{O}_\Phi[$ is convex then the space Φ, or domain (Φ, p), is said to be *$\mathfrak{G}$-holomorphically convex*, or simply $\mathfrak{G}$-h *convex*. Note that the $\mathfrak{H}$-convex hull of a single point φ is obviously contained in the stalk $p^{-1}(\varphi)$ and, since stalks are discrete, the hull of φ will be compact iff it consists of a finite number of points. In particular, if $[\Phi, \mathfrak{H}]$ is convex then $\mathfrak{H}$ can fail to separate at most a finite set of points.

Consider next an arbitrary directed set $\mathfrak{D}$, *i.e.* $\mathfrak{D}$ is partially ordered by a relation "<" such that for d' and d'' in $\mathfrak{D}$ there exists $d \in \mathfrak{D}$ with $d' \le d$ and $d'' \le d$. A set $\{\varphi_d : d \in \mathfrak{D}\} \subset \Phi$ is called a *net* in Φ. Recall that a net $\{\varphi_\alpha\}$ *converges to the point* $\varphi \in \Phi$, written $\lim_d \varphi_d = \varphi$, if for each neighborhood U of the point φ there exists $d_u \in \mathfrak{D}$ such that $d \ge d_u$ implies $\varphi_\alpha \in U$. If $\lim_d \varphi_\alpha$ exists then it is clearly unique. Now let H be a linearly independent system of generators for the algebra $\mathfrak{H}$, so the H-topology and the $\mathfrak{H}$-topology in Φ are equivalent. We assume for later convenience that H contains the set $A\circ p$, where A is the previously chosen system of generators for $\mathfrak{G}$ in Definition 45.1. Denote by η an arbitrary finite subset of H. Then for $r > 0$ the set

$$U_{\varphi_0}(\eta, r) = \{\varphi \in \Phi : |h(\varphi) - h(\varphi_0)| < r,\ h \in \eta\}$$

is a neighborhood of φ_0, and such neighborhoods constitute a basis for the $\mathfrak{H}$-topology in Φ. A given net $\{\varphi_d : d \in \mathfrak{D}\}$ will be called a *Cauchy net* if for arbitrary η and $r > 0$ there exists $d(\eta, r) \in \mathfrak{D}$ such that $d', d'' \geq d(\eta, r)$ implies

$$|h(\varphi_{d'}) - h(\varphi_{d''})| < r,\ h \in \eta.$$

Since the $\mathfrak{H}$-topology is weaker than the given topology in Φ it is obvious that every convergent net in Φ is a Cauchy net. If, conversely, every Cauchy net in Φ is convergent then the pair $[\Phi, \mathfrak{H}]$ is said to be *complete*. It is not difficult to prove that if $[\Phi, \mathfrak{H}]$ is complete then it is automatically a system. However, for our purposes, a weaker notion of completeness is needed.

47.1 DEFINITION. *A net $\{\varphi_d : d \in \mathfrak{D}\}$ is said to be dominated by a compact set $K \subset \Phi$ if $\{\varphi_d\} \subset \hat{K}$. The pair $[\Phi, \mathfrak{H}]$ is said to be relatively complete if every dominated Cauchy net in Φ is convergent.*

By the following lemma we see that the notion of "dominated Cauchy net" depends only on the algebra $\mathfrak{H}$ and not on the particular system of generators H.

47.2 LEMMA. *Let H and H' be any two systems of generators for the algebra $\mathfrak{H}$ and let $\{\varphi_d : d \in \mathfrak{D}\}$ be a dominated net in Φ. If $\{\varphi_d\}$ is a Cauchy net with respect to H then it is also a Cauchy net with respect to H'.*

Proof. Assume that $\{\varphi_d\}$ is a Cauchy net with respect to H dominated by the compact set K. Let $\eta' = (h_1', \ldots, h_m')$ be an arbitrary finite subset of H' and $r' > 0$. Since H is a system of generators for $\mathfrak{H}$ there exists a finite set $\eta = (h_1, \ldots, h_n) \subset H$ and polynomials $P_1, \ldots, P_m$ in n variables such that $h_j' = P_j(h_1, \ldots, h_n)$ for $j=1,\ldots,m$. Next let $\rho = \max(|h_1|_K, \ldots, |h_n|_K)$. Then by uniform continuity there exists $r > 0$ such that $|\zeta_i| \leq \rho$, $|\zeta_i''| \leq \rho$ and $|\zeta_i' - \zeta_i''| < r$ for $i=1,\ldots,n$ imply $|P_j(\zeta_1', \ldots, \zeta_n') - P_j(\zeta_1'', \ldots, \zeta_n'')| < r'$ for $j=1,\ldots,m$. Now choose $d(\eta', r')$ such that $d', d'' \geq d(\eta', r')$ implies $|h_i(\varphi_{d'}) - h_i(\varphi_{d''})| < r$ for $i=1,\ldots,n$. Since $\{\varphi_d\} \subset \hat{K}$ we have $|h_i(\varphi_d)| \leq |h_i|_K \leq \rho$ for each i. Therefore

$$|P_j(h_1(\varphi_{d'}), \ldots, h_n(\varphi_{d'})) - P_j(h_1(\varphi_{d''}), \ldots, h_n(\varphi_{d''}))| < r'$$

so $|h_j'(\varphi_{d'}) - h_j'(\varphi_{d''})| < r'$ for $j=1,\ldots,m$. In other words, $\{\varphi_d\}$ is also a Cauchy net with respect to H'. ◊

We are now ready to prove the main theorem of this chapter.

47.3 THEOREM. *Let* $[\Sigma, \mathfrak{G}]$ *be a natural system,* (Φ, p) *a* Σ*-domain and* $\mathfrak{H}$ *an algebra with* $\mathfrak{G}\circ p \subseteq \mathfrak{H} \subseteq \mathfrak{O}_\Phi$ *such that* $[\Phi, \mathfrak{H}]$ *is a system. Then the following properties are equivalent:*

(i) $[\Phi, \mathfrak{H}]$ *is relatively complete.*

(ii) $\delta_A(\hat{K}) > 0$ *for every compact set* $K \subset \Phi$.

(iii) $[\Phi, \mathfrak{H}]$ *is convex (Definition 7.5).*

(iv) $[\Phi, \mathfrak{H}]$ *is natural.*

Proof. By Theorem 7.3, property (iv) always implies (iii) and, by Proposition 45.2 (iv), (iii) implies (ii). For the proof that (ii) implies (i), let $\{\varphi_d : d \in \mathfrak{D}\}$ be a Cauchy net in Φ dominated by a compact set K. Choose r such that $0 < 2r < \delta_A(\hat{K})$. Then, by Proposition 45.2 (v), there exists $\alpha \subset A$ such that the basic p-neighborhood $W_\varphi(\alpha, 2r)$ is defined for each $\varphi \in \hat{K}$. Now choose $\bar{d} \in \mathfrak{D}$ such that $d', d'' \geq \bar{d}$ implies

$$|(a\circ p)(\varphi_{d'}) - (a\circ p)(\varphi_{d''})| < r, \; a \in \alpha.$$

Then $\varphi_d \in W_{\varphi_{\bar{d}}}(\alpha, r)$ for all $d \geq \bar{d}_v$. For each $d \geq \bar{d}$ set $\sigma_d = p(\varphi_d)$. Then $\{\varphi_d\}$ is a net in Σ. Moreover, $\{\sigma_d\}$ is a Cauchy net with respect to the system of generators A and is dominated by the compact set $p(K)$, *i.e.* $\{\sigma_d\} \subset \widehat{p(K)}$. Since $[\Sigma, \mathfrak{G}]$ is natural the set $\widehat{p(K)}$ is also compact. Next, for each $d \geq \bar{d}$ set $K_d = \overline{\{\sigma_{d'} : d' \geq d\}}$. Then K_d is a nonempty compact subset of $\widehat{p(K)}$. Moreover, since $\mathfrak{D}$ is a directed set the family $\{K_d : d \geq \bar{d}\}$ has the finite intersection property. Hence there exists a point σ_0 common to all of the sets K_d. Since $\{\sigma_d\} \subset N_{\sigma_{\bar{d}}}(\alpha, r)$ it follows that $\sigma_0 \in N_{\sigma_{\bar{d}}}(\alpha, 2r)$. We prove that $\lim_d \sigma_d = \sigma_0$. In fact, let α' be an arbitrary finite subset of A and $\varepsilon > 0$. Then there exists $d_1 \geq \bar{d}$ such that $d', d'' \geq d_1$ implies $|a(\sigma_{d'}) - a(\sigma_{d''})| < \varepsilon$ for $a \in \alpha'$. Since $\sigma_0 \in K_{d_1}$, there exists $d_1' \in \mathfrak{D}$ such that $d_1' \geq d_1$ and $\sigma_{d_1'} \in N_{\sigma_0}(\alpha', \varepsilon)$. Therefore $d \geq d_1$ implies

$$|a(\sigma_d) - a(\sigma_0)| \leq |a(\sigma_d) - a(\sigma_{d_1'})| + |a(\sigma_{d_1'}) - a(\sigma_0)| < 2\varepsilon$$

for each $a \in \alpha'$. In other words, $d \geq d_1$ implies $\sigma_d \in N_{\sigma_0}(\alpha', 2\varepsilon)$, so $\lim_d \sigma_d = \sigma_0$, as claimed. Now since $p : W_{\varphi_{\bar{d}}}(\alpha, 2r) \to N_{\sigma_{\bar{d}}}(\alpha, 2r)$ is a surjective homeomorphism we have $\lim_d \varphi_d = \lim_d p^{-1}_{\varphi_{\bar{d}}}(\sigma_d) = p^{-1}_{\varphi_{\bar{d}}}(\sigma_0)$, so the dominated Cauchy net $\{\varphi_d\}$ converges in Φ proving that (ii) implies (i).

The final step of the proof, that (i) implies (iv), is considerably more difficult than were any of the above. We begin by considering an arbitrary natural system extension (see Proposition 10.2.)

$$\mu : [\Phi, \mathcal{H}] \Rightarrow [\Omega, \mathcal{B}]$$

of the system $[\Phi, \mathcal{H}]$. Thus, $\mu : \Phi \to \Omega$ is a homeomorphism of Φ into Ω such that $\mathcal{B}\circ\mu = \mathcal{H}$. Observe that, since $[\Sigma, \mathcal{C}]$ and $[\Omega, \mathcal{B}]$ are natural systems, the product $[\Sigma \times \Omega, \mathcal{C} \otimes \mathcal{B}]$ is also a natural system (Proposition 4.1). Now consider the mapping

$$\tau : \Phi \to \Sigma \times \Omega,\ \varphi \mapsto (p(\varphi), \mu(\varphi)).$$

It is readily verified that τ maps Φ homeomorphically into $\Sigma \times \Omega$. Furthermore, $(\mathcal{C} \otimes \mathcal{B})\circ\tau = \mathcal{H}$. In fact, recall that $\mathcal{C} \otimes \mathcal{B}$ consists of all functions on $\Sigma \times \Omega$ of the form

$$F(\sigma, \omega) = \sum_{i=1}^{n} a_i(\sigma)b_i(\omega)$$

where $a_i \in \mathcal{C}$ and $b_i \in \mathcal{B}$ for each i. Also

$$F\circ\tau = \sum_{i=1}^{n} (a_i\circ p)(b_i\circ\mu).$$

Since $\mathcal{C}\circ p \subseteq \mathcal{H}$ and $\mathcal{B}\circ\mu = \mathcal{H}$ it follows that $(\mathcal{C} \otimes \mathcal{H})\circ\tau = \mathcal{H}$, so τ defines an isomorphism

$$\tau : [\Phi, \mathcal{H}] \Rightarrow [\tau(\Phi), \mathcal{C} \otimes \mathcal{B}].$$

Therefore the system $[\Phi, \mathcal{H}]$ will be natural iff $[\tau(\Phi), \mathcal{C} \otimes \mathcal{B}]$ is natural. Thus, our problem reduces to showing that condition (i) implies naturality of the system $[\tau(\Phi), \mathcal{C} \otimes \mathcal{B}]$. The proof is resolved into a sequence of five propositions.

(1). *The image* $\tau(\Phi)$ *of* Φ *in* $\Sigma \times \Omega$ *is a local* $\mathcal{C} \otimes \mathcal{B}$*-subvariety of the open set* $p(\Phi) \times \Omega$.

Let φ be an arbitrary point of Φ and U_φ a p-neighborhood of φ. Then $p(U_\varphi)$ is a neighborhood of $p(\varphi)$ in Σ and, since μ is a homeomorphism, there exists a neighborhood $V_{\mu(\varphi)}$ of the point $\mu(\varphi) \in \Omega$ such that

$V_{\mu(\varphi)} \cap \mu(\Phi) = \mu(U_\varphi)$. Set $T_{\tau(\varphi)} = p(U_\varphi) \times V_{\mu(\varphi)}$. Then $T_{\tau(\varphi)}$ is a neighborhood of $\tau(\varphi)$ contained in $p(\Phi) \times \Omega$. We shall prove that $T_{\tau(\varphi)} \cap \tau(\Phi)$ consists of the common zeros of $\mathfrak{G} \otimes \mathfrak{B}$-holomorphic functions defined on $T_{\tau(\varphi)}$, so (1) will follow by Definition 33.1.

For each $h \in \mathfrak{H}$ choose $b_h \in \mathfrak{B}$ such that $b_h \circ \mu = h$, and define

$$H_h(\sigma, \omega) = (h \circ p_\varphi^{-1})(\sigma) - b_h(\omega), \ (\sigma, \omega) \in T_{\tau(\varphi)}.$$

Since $h \circ p_\varphi^{-1}$ is $\mathfrak{G}$-holomorphic on $p(U_\varphi)$ it is immediate that H_h is $\mathfrak{G} \otimes \mathfrak{B}$-holomorphic on $T_{\tau(\varphi)}$. Moreover, if $(\sigma, \omega) \in \tau(\Phi)$ then $\omega = \mu(p_\varphi^{-1}(\sigma))$, so

$$\begin{aligned} H_h(\sigma, \omega) &= h(p_\varphi^{-1}(\sigma)) - b_h(\mu(p_\varphi^{-1}(\sigma))) \\ &= h(p_\varphi^{-1}(\sigma)) - h(p_\varphi^{-1}(\sigma)) = 0. \end{aligned}$$

Hence $T_{\tau(\varphi)} \cap \tau(\Phi) \subseteq Z(\{H_h : h \in \mathfrak{H}\})$, the set of common zeros of the functions $\{H_h, h \in \mathfrak{H}\}$. Now let $(\sigma, \omega) \in T_{\tau(\varphi)} \setminus \tau(\Phi)$. Then $\omega \neq \mu(p_\varphi^{-1}(\sigma))$, so there exists $b \in \mathfrak{B}$ such that $b(\omega) \neq b(\mu(p_\varphi^{-1}(\sigma)))$. Set $h = b \circ \mu$. Then $h \in \mathfrak{H}$ and $H_h(\sigma, \omega) \neq 0$. Therefore $T_{\tau(\varphi)} \cap \tau(\Phi) = Z(\{H_h : h \in \mathfrak{H}\})$ completing the proof of (1).

(2). *Let* K *be a compact subset of* Φ. *Then* $\tau(\hat{K}) = \widehat{\tau(K)} \cap \tau(\Phi)$, *where* "^" *denotes the* $\mathfrak{G} \otimes \mathfrak{B}$-*convex hull in* $\Sigma \times \Omega$.

If $F \in \mathfrak{G} \otimes \mathfrak{B}$ then $|F \circ \tau|_K = |F|_{\tau(K)}$. Therefore, since $(\mathfrak{G} \otimes \mathfrak{B}) \circ \tau = \mathfrak{H}$ it follows that $\varphi \in \hat{K}$ iff $\tau(\varphi) \in \widehat{\tau(K)}$. In other words, $\tau(\hat{K}) = \widehat{\tau(K)} \cap \tau(\Phi)$, as claimed.

(3). *The set* $\tau(\Phi)$ *will be* $\mathfrak{G} \otimes \mathfrak{B}$-*convex in* $\Sigma \times \Omega$ *iff* $\widehat{\tau(K)} \cap \tau(\Phi)$ *is a closed (and hence compact) set for each compact set* $K \subset\subset \Phi$.

Since $\tau : \Phi \to \tau(\Phi)$ is a homeomorphism the image $\tau(K)$ of an arbitrary compact set $K \subset\subset \Phi$ is compact in $\tau(\Phi)$ and every compact subset of $\tau(\Phi)$ is of this form. Therefore (3) follows by Theorem 34.2.

(4). *If* K *is any compact subset of* Φ *and if* (σ_0, ω_0) *is a limit point of* $\widehat{\tau(K)} \cap \tau(\Phi)$ *in the space* $\Sigma \times \Omega$ *then there exists a dominated Cauchy net* $\{\varphi_d\}$ *in* Φ *such that* $\lim_d \tau(\varphi_d) = (\sigma_0, \omega_0)$.

Since $\mathfrak{B} \circ \mu = \mathfrak{H}$ there is a system B of generators for $\mathfrak{B}$ such that $B \circ \mu = H$, where H is the previously chosen system of generators for $\mathfrak{H}$. Denote by $\mathfrak{D}$ the set of all triples (α, β, n), where α is a finite subset of A (the system of generators for $\mathfrak{G}$), β is a finite subset of B and n is a positive integer. For $d = (\alpha, \beta, n)$ and $d' = (\alpha', \beta', n')$, define $d < d'$ iff $d \neq d'$ and $\alpha \subseteq \alpha'$, $\beta \subseteq \beta'$, $n \leq n'$. Then $\mathfrak{D}$ is a directed set under "<". Now, for each $d \in \mathfrak{D}$ consider the neighborhood $T_d = N_{\sigma_0}(\alpha, \frac{1}{n}) \times N_{\omega_0}(\beta, \frac{1}{n})$ of the point (σ_0, ω_0) in $\Sigma \times \Omega$. Since (σ_0, ω_0) is a limit point of the set $\widehat{\tau(K)} \cap \tau(\Phi)$ we may choose $\varphi_d \in \Phi$ such that $\tau(\varphi_d) \in T_d \cap \widehat{\tau(K)}$. By (2), we have $\varphi_d \in \hat{K}$, so $\{\varphi_d\}$ is a net dominated by K. Let T_0 be an arbitrary neighborhood of (σ_0, ω_0) in $\Sigma \times \Omega$. Then there exists $d_0 \in \mathfrak{D}$ such that $T_{d_0} \subseteq T_0$. Since $T_d \subseteq T_{d_0} \subseteq T_0$ for $d_0 \leq d$, we have $\tau(\varphi_d) \in T_0$. In other words, $\lim_d \tau(\varphi_d) = (\sigma_0, \omega_0)$. In particular, $\{\tau(\varphi_d)\}$ is a Cauchy net in $\tau(\Phi)$. Therefore, since $\tau : \Phi \to \tau(\Phi)$ is a homeomorphism $\{\varphi_d\}$ is a dominated Cauchy net in Φ completing the proof of (4).

(5). *Condition* (i) *implies* (iv).

If $[\Phi, \mathfrak{H}]$ is relatively complete then the dominated Cauchy net $\{\varphi_d\}$ constructed in the proof of (4) converges to a point $\varphi_0 \in \Phi$. Also, since $\tau : \Phi \to \tau(\Phi)$ is continuous $\tau(\varphi_0) = \lim_d \tau(\varphi_d) = (\sigma_0, \omega_0)$. Therefore $\widehat{\tau(K)} \cap \tau(\Phi)$ is closed in $\Sigma \times \Omega$. Hence, by (3) the set $\tau(\Phi)$ is $\mathfrak{G} \otimes \mathfrak{H}$-convex in $\Sigma \times \Omega$. Since $[\Sigma \times \Omega$, $\mathfrak{G} \otimes \mathfrak{B}]$ is natural it follows that $[\tau(\Phi), \mathfrak{G} \otimes \mathfrak{B}]$ is natural (Theorem 7.1 (i)) and, as previously noted, this implies that $[\Phi, \mathfrak{H}]$ is natural. ◆

47.4 COROLLARY. *If the projection* $p : \Phi \to \Sigma$ *is a proper mapping (i.e. inverse images of compact sets are compact) and* $p(\Phi)$ *is* $\mathfrak{G}$*-convex then* $[\Phi, \mathfrak{H}]$ *is convex and therefore natural.*

The mapping $\tau : \Phi \to \Sigma \times \Omega$, used in the proof of Theorem 47.3 to transform Φ into a local subvariety of a domain in $\Sigma \times \Omega$, generalizes a technique used by Oka [O1] to represent an analytic polyhedron in $\mathbb{C}^n$ as a subvariety in a higher dimensional space $\mathbb{C}^{m+n}$. It is therefore appropriate to call this a "*generalized Oka mapping*". We have used such mappings previously [R4, §3; R9, §4] for essentially the same purpose but with the arbitrary natural system extension replaced by the special extension into $[\mathbb{C}^\Lambda, \mathfrak{P}]$ given by Proposition 10.2. In the latter form it represents a more or less direct generalization of the Oka technique. The form used here tends to bring out the connection with Quigley's proof of Corollary 40.2 mentioned in §43.

CHAPTER X
HOLOMORPHIC EXTENSIONS OF $[\Sigma, \mathfrak{G}]$-DOMAINS

§48. MORPHISMS AND EXTENSIONS. DOMAINS OF HOLOMORPHY

We are interested in this chapter in the category of all Σ-domains associated with a given fixed system $[\Sigma, \mathfrak{G}]$. As before, we shall assume that Σ is both connected and locally connected. If (Φ, p) and (Ψ, q) are any two Σ-domains then a *morphism* from (Φ, p) to (Ψ, q), or a *domain morphism*

$$\rho : (\Phi, p) \to (\Psi, q)$$

is given by a continuous mapping $\rho : \Phi \to \Psi$ of Φ into Ψ such that $p = q \circ \rho$. Thus we have the commutative diagram as indicated in the figure. If $\rho : \Phi \to \Psi$ is one-to-one onto then the morphism is an *isomorphism*.

$$\begin{array}{ccc} \Phi & \xrightarrow{\rho} & \Psi \\ & {}_{p}\searrow \quad \swarrow_{q} & \\ & \Sigma & \end{array}$$

48.1 LEMMA. *If* $\rho : (\Phi, p) \to (\Psi, q)$ *then* ρ *is automatically an open local homeomorphism that maps* p-*neighborhoods homeomorphically onto* q-*neighborhoods. Thus if* ρ *is one-to-one then it is a homeomorphism.*

Proof. Let φ be an arbitrary point of Φ and set $\psi = \rho(\varphi)$, $\sigma = p(\varphi) = q(\psi)$. Choose an open q-neighborhood V_ψ of ψ in Ψ. Then, since ρ is continuous, there exists an open p-neighborhood U_φ such that $\rho(U_\varphi) \subseteq V_\psi$. Set $N_\sigma = p(U_\varphi)$. Then N_σ is an open neighborhood of σ and p maps U_φ homeomorphically onto N_σ. Also, since

$$N_\sigma = p(U_\varphi) = (q \circ \rho)(U_\varphi) \subseteq q(V_\varphi)$$

the inverse of q on $q(V_\psi)$ maps N_σ homeomorphically onto $\rho(U_\varphi)$. Therefore $\rho(U_\varphi)$ is an open q-neighborhood of ψ. Moreover, when restricted to U_φ, q is equal to the composition of two homeomorphisms q_ψ^{-1} and p. Hence ρ maps U_φ homeomorphically onto the open set $\rho(U_\varphi)$ and so is an open local homeomorphism.

It follows easily from the condition $p = q \circ \rho$ that ρ maps each p-set surjectively onto a q-set. Therefore p-neighborhoods map homeomorphically onto q-neighborhoods. ♦

48.2 LEMMA. *Let* $\rho : (\Phi, p) \to (\Psi, q)$ *and* $\rho' : (\Phi, p) \to (\Psi, q)$ *be morphisms such that* $\rho(\varphi) = \rho(\varphi')$ *for at least one point of* Φ. *If* Φ *is connected then* $\rho = \rho'$.

Proof. Set $\Phi_0 = \{\varphi \in \Phi : \rho(\varphi) = \rho'(\varphi)\}$. Then Φ_0 is nonempty, by hypothesis, and is closed since the mappings are continuous. Let φ_0 be an arbitrary point of Φ_0 and let U_0 be a connected p-neighborhood of φ_0. Then by the previous lemma $\rho(U_0)$ and $\rho'(U_0)$ are both q-neighborhoods of the point $\rho(\varphi_0) = \rho'(\varphi_0)$. Moreover $q(\rho(U_0)) = p(U_0) = q(\rho'(U_0))$. Since $\rho(U_0) \cap \rho'(U_0) \neq \phi$ it follows by Lemma 44.2 that $\rho(U_0) \cup \rho'(U_0)$ is a q-set. This implies that $\rho(\varphi) = \rho'(\varphi)$, for every $\varphi \in U_0$. In other words, $U_0 \subseteq \Phi_0$, which means that Φ_0 is also open in Φ so must exhaust Φ. ♦

48.3 COROLLARY. *If* $\rho : (\Phi, p) \to (\Psi, q)$ *and* $\rho' : (\Psi, q) \to (\Phi, p)$, *where both* Φ *and* Ψ *are connected, then* $(\rho' \circ \rho)(\varphi) = \varphi$ *for at least one point of* Φ *implies that* $\rho : \Phi \to \Psi$ *is a surjective homeomorphism and* $\rho' = \rho^{-1}$.

Consider an arbitrary morphism $\rho : (\Phi, p) \to (\Psi, q)$ of Σ-domains and let g be an $\mathcal{a}$-holomorphic function defined on a subset X of Ψ. Then since ρ is a local homeomorphism the function $g \circ \rho$ is obviously $\mathcal{a}$-holomorphic on the set $\rho^{-1}(X \cap \rho(\Phi))$ in Φ. In particular, we always have the inclusion $\mathcal{O}_\Psi \circ \rho \subseteq \mathcal{O}_\Phi$. In other words, ρ always defines a pair morphism $\rho : [\Phi, \mathcal{O}_\Phi] \to [\Psi, \mathcal{O}_\Psi]$. Now let $\mathcal{H}$ be an arbitrary subset of $\mathcal{O}_\Phi$ and suppose that there exists a set $\mathcal{K} \subseteq \mathcal{O}_\Psi$ such that $\mathcal{K} \circ \rho = \mathcal{H}$. Then we say that ρ defines an *extension of* (Φ, p) *relative to* $\mathcal{H}$, or simply an $\mathcal{H}$-*extension*, and write

$$\rho : (\Phi, p, \mathcal{H}) \Rightarrow (\Psi, q, \mathcal{K}).$$

If the target space Ψ is connected, then the extension is also said to be *connected*. If ρ defines an $\mathcal{O}_\Phi$-extension then we must have $\mathcal{O}_\Psi \circ \rho = \mathcal{O}_\Phi$. An $\mathcal{O}_\Phi$-extension is also called an $\mathcal{a}$-*holomorphic* or $\mathcal{a}$-h *extension* and denoted simply by

$$\rho : (\Phi, p) \Rightarrow (\Psi, q).$$

If $\rho : (\Phi, p) \to (\Psi, q)$ is an isomorphism then obviously $\rho : (\Phi, p) \Rightarrow (\Psi, q)$. Also, if $\mathcal{H}$ separates points and $\rho : (\Phi, p, \mathcal{H}) \Rightarrow (\Psi, q, \mathcal{K})$ then ρ is injective so maps Φ homeomorphically onto an open subset of Ψ.

Two extensions, $\rho' : (\Phi, p, \mathcal{H}) \Rightarrow (\Psi', q', \mathcal{K}')$ and $\rho'' : (\Phi, p, \mathcal{H}) \Rightarrow (\Psi'', q'', \mathcal{K}'')$, are said to be *isomorphic* if there exists a surjective homeomorphism $\mu : \Psi' \to \Psi'$ such that $\rho'' = \mu \circ \rho'$.

An extension $\rho : (\Phi, p, \mathcal{H}) \Rightarrow (\Psi, q, \mathcal{K})$ is said to be *maximal* if any other extension $\rho' : (\Phi, p, \mathcal{H}) \Rightarrow (\Psi', q', \mathcal{K}')$ of (Φ, p) may be "lifted" to the given one. In other words, there exists an extension $\mu : (\Psi', q', \mathcal{K}') \Rightarrow (\Psi, q, \mathcal{K})$ such that $\rho = \mu \circ \rho'$, i.e. the accompanying diagram is commutative. If ρ defines a maximal connected $\mathcal{O}_\Phi$-extension then (Ψ, q) is called an *envelope of $\mathcal{G}$-holomorphy* for (Φ, p).

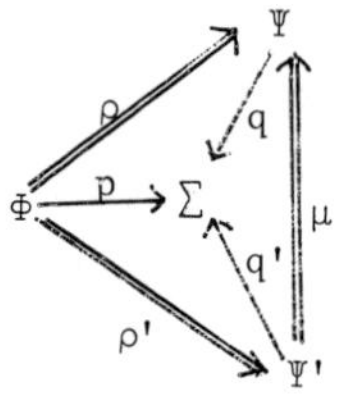

A Σ-domain (Φ, p) is said to be *maximal relative to* $\mathcal{H}$ if every connected extension $\rho : (\Phi, p, \mathcal{H}) \Rightarrow (\Psi, q, \mathcal{K})$ is an isomorphism. If (Φ, p) is connected and maximal relative to $\mathcal{O}_\Phi$ then it is called a *Σ-domain of $\mathcal{G}$-holomorphy*.

A partial indication of the relationship between maximal extensions and maximal domains is provided by the following proposition.

48.4 PROPOSITION. *If* $\rho : (\Phi, p, \mathcal{H}) \Rightarrow (\Psi, q, \mathcal{K})$ *is a maximal connected extension then* (Ψ, q) *is a maximal Σ-domain relative to* $\mathcal{K}$.

Proof. Let $\mu : (\Psi, q, \mathcal{K}) \Rightarrow (\Psi', q', \mathcal{K}')$ be any connected extension of (Ψ, q) relative to $\mathcal{K}$ and observe that the composition mapping $\mu \circ \rho$ defines a connected extension $\mu \circ \rho : (\Phi, p, \mathcal{H}) \Rightarrow (\Psi', q', \mathcal{K}')$. Therefore since the given extension is maximal there exists $\mu' : (\Psi', q', \mathcal{K}') \Rightarrow (\Psi, q, \mathcal{K})$ such that $\rho = \mu' \circ (\mu \circ \rho)$. Note that both Ψ and Ψ' are connected and $(\mu' \circ \mu)(\psi) = \psi$ for each $\psi \in \rho(\Phi)$. Therefore Corollary 48.3 applies and we conclude that μ is a surjective homeomorphism, with $\mu^{-1} = \mu'$, and hence that (Ψ, q) is maximal relative to $\mathcal{K}$. ◊

48.5 COROLLARY. *Any two maximal connected extensions of* (Φ, p) *relative to* $\mathcal{H}$ *are isomorphic.*

A converse to the result in Proposition 48.4 will be proved below (Proposition 50.1) under an additional hypothesis on $[\Sigma, \mathfrak{G}]$. The converse depends on the existence of maximal extensions, which is the next item in our discussion.

§49. EXISTENCE OF MAXIMAL EXTENSIONS

We shall use a standard "sheaf of germs" approach to the construction of maximal Σ-domains and extensions (cf., for example, [N2, Chapter 6]). In order for the sheaf in question to be a Hausdorff space the $\mathfrak{G}$-h functions must satisfy the following uniqueness principle.

49.1 DEFINITION. *Let* (Φ, p) *be a* Σ*-domain. Then the* $\mathfrak{G}$-h *functions in* Φ *are said to satisfy the* ***uniqueness principle*** *if any* $\mathfrak{G}$-h *function, which is defined on an open connected set* G *and vanishes on an open subset of* G, *must vanish on all of* G.

49.2 PROPOSITION. *If the* $\mathfrak{G}$-h *functions in* Σ *satisfy the uniqueness principle then the* $\mathfrak{G}$-h *functions in any* Σ*-domain also satisfy the uniqueness principle.*

Proof. Let G be an open connected subset of Φ and h a holomorphic function defined on G. Suppose that h vanishes on an open set $U \subset G$. Denote by G_0 the union of all p-neighborhoods contained in G on which h vanishes. The set G_0 is obviously open and, since $U \subseteq G_0$, also $G_0 \neq \emptyset$. Let φ_0 be a limit point of G_0 in G and choose an open connected p-neighborhood U_{φ_0} contained in G. Then $h \circ p_{\varphi_0}^{-1}$ is holomorphic on the open set $p(U_{\varphi_0})$ in Σ. Also, $h \circ p_{\varphi_0}^{-1}$ vanishes on the open set $p(G_0 \cap U_{\varphi_0})$ contained in $p(U_{\varphi_0})$. Therefore by the uniqueness principle in Σ, $h \circ p_{\varphi_0}^{-1}$ vanishes on $p(U_{\varphi_0})$. Hence h vanishes on U_{φ_0}, so $\varphi_0 \in G_0$. Thus G_0 is both open and closed in G so must exhaust G. ◆

49.3 COROLLARY. *Under the uniqueness principle, if* $\rho : (\Phi, p, \mathcal{H}) \Rightarrow (\Psi, q, \mathcal{K})$ *then the map* $\rho^* : \mathcal{K} \to \mathcal{H}$, $k \mapsto k \circ \rho$ *is bijective and preserves algebra operations.*

Consider the presheaf $\mathfrak{G}$ of all $\mathfrak{G}$-h functions in Σ and denote by Λ an index set with cardinality equal to the cardinality of the set of all $\mathfrak{G}$-h functions defined on open subsets of Σ. The index set Λ, along with the system $[\Sigma, \mathfrak{G}]$, will be fixed throughout the remainder of this chapter. We shall also assume that the $\mathfrak{G}$-h functions in Σ satisfy the uniqueness principle. Observe that if (Φ, p) is an arbitrary connected Σ-domain then, by the uniqueness principle, card $\mathfrak{O}_\Phi \leq$ card $\mathfrak{O}_G$ for any open set $G \subseteq \Phi$. In particular, if G is a p-set then $\mathfrak{O}_{p(G)} \circ p = \mathfrak{O}_G$, so card $\mathfrak{O}_\Phi \leq$ card Λ. We shall need this fact in Theorem 49.4 below.

Now let

$$\check{f} : X \to \mathbb{C}^\Lambda, \ \sigma \mapsto \check{f}(\sigma) = \{f_\lambda(\sigma)\}$$

denote a function defined on a set $X \subseteq \Sigma$ with values in the product space $\mathbb{C}^\Lambda$. If each of the complex-valued functions f_λ is $\mathfrak{G}$-holomorphic then $\check{f}$ defines a pair morphism $\check{f} : [X, \mathfrak{O}_X] \to [\mathbb{C}^\Lambda, \mathfrak{P}]$ $(\mathfrak{P} \circ \check{f} \subseteq \mathfrak{O}_X)$, so $\check{f}$ is a holomorphic map in the sense of Definition 18.1. The collection of all such holomorphic maps from subsets of Σ to $\mathbb{C}^\Lambda$ will be denoted by $\check{\mathfrak{O}}$. Let $\check{f}$ be an element of $\check{\mathfrak{O}}$ defined on a neighborhood of a point $\sigma \in \Sigma$. Then the *germ* of elements of $\check{\mathfrak{O}}$ determined by $\check{f}$ at the point σ will be denoted by $(\check{f})_\sigma$. Thus

$$(\check{f})_\sigma = \{\check{g} \in \check{\mathfrak{O}} : \check{g} = \check{f} \text{ on some neighborhood of } \sigma\}.$$

The *stalk* of all such germs will be denoted by $(\check{\mathfrak{O}})_\sigma$ and the *sheaf* of germs associated in this way with $\check{\mathfrak{O}}$ will be denoted by $(\check{\mathfrak{O}})$. Recall that a basis for the topology in $(\check{\mathfrak{O}})$ is given by neighborhoods of a point $(\check{f})_\sigma$ in $(\check{\mathfrak{O}})$ of the form

$$N_{(\check{f})_\sigma}(\check{g}, U_\sigma) = \{(\check{g})_{\sigma'} : \sigma' \in U_\sigma\}$$

where U_σ is a neighborhood of σ in Σ and $\check{g}$ is an element of the germ $(\check{f})_\sigma$ defined on U_σ. Since we are assuming the uniqueness principle the topology in $(\check{\mathfrak{O}})$ is Hausdorff. It will be notationally convenient to replace $(\check{\mathfrak{O}})$ by a homeomorphic image in the product space $\Sigma \times (\check{\mathfrak{O}})$, *viz*

$$\Sigma \# (\check{\mathfrak{O}}) = \{(\sigma, (\check{f})_\sigma) : (\check{f})_\sigma \in (\check{\mathfrak{O}})\}.$$

If we define

$$\pi : \Sigma \# (\check{\mathfrak{O}}) \to \Sigma, \ (\sigma, (\check{f})_\sigma) \mapsto \sigma$$

then π is an open local homeomorphism, so $(\Sigma \# (\check{\mathfrak{O}}), \pi)$ is a Σ-domain. Note however that $(\Sigma \# (\check{\mathfrak{O}}), \pi)$ is not connected. We shall continue to refer to the topology in $\Sigma \# (\check{\mathfrak{O}})$ as the *sheaf topology*.

For each $\lambda \in \Lambda$ define

$$F_\lambda(\sigma, (\check{f})_\sigma) = f_\lambda(\sigma), \ (\sigma, (\check{f})_\sigma) \in \Sigma \# (\check{\mathfrak{G}}).$$

Then F_λ is easily seen to be a well-defined $\mathfrak{a}$-holomorphic function on $\Sigma \# (\check{\mathfrak{G}})$. Denote the set $\{F_\lambda : \lambda \in \Lambda\}$ by $\mathfrak{F}$. We are now ready to prove the existence of maximal Σ-domains.

49.4 THEOREM. *Let* Γ_0 *be an arbitrary component of the space* $\Sigma \# (\check{\mathfrak{G}})$. *Then* (Γ_0, π) *is a maximal* Σ*-domain relative to any subset* $\mathfrak{G}$ *of* $\mathfrak{G}_{\Gamma_0}$ *that contains the functions* F_λ. *In particular* (Γ_0, π) *is a domain of* $\mathfrak{a}$*-holomorphy.*

Proof. It is obvious that (Γ_0, π) is a Σ-domain. Therefore let $\rho : (\Gamma_0, \pi, \mathfrak{G}) \Rightarrow (\Phi, p, \mathfrak{H})$ to be an arbitrary connected holomorphic extension of (Γ_0, π) relative to $\mathfrak{G}$. Then by Lemma 48.1, $\rho(\Gamma_0)$ is an open connected subset of Φ. Let φ_0 be a limit point of $\rho(\Gamma_0)$ in Φ and V_0 a connected p-neighborhood of φ_0. Set $\sigma_0 = p(\varphi_0)$ and $U_0 = p(V_0)$, so U_0 is a connected neighborhood of σ_0 in Σ. Since $\mathfrak{H} \circ \rho = \mathfrak{G}$, there exists for each $\lambda \in \Lambda$ a function $G_\lambda \in \mathfrak{G}$ such that $G_\lambda \circ \rho = F_\lambda$. Let $g_\lambda = G_\lambda \circ p_{\varphi_0}^{-1}$, so g_λ is $\mathfrak{a}$-holomorphic on U_0. Hence, if $\check{g} = \{g_\lambda\}$ then $(\sigma_0, (\check{g})_{\sigma_0}) \in \Sigma \# (\check{\mathfrak{G}})$. Also, if $W_0 = \{(\sigma, (\check{g})_\sigma) : \sigma \in U_0\}$ then W_0 is a connected neighborhood of the point $(\sigma_0, (\check{g})_{\sigma_0})$ in the space $\Sigma \# (\check{\mathfrak{G}})$. Observe next that $V_0 \cap \rho(\Gamma_0)$ is a nonempty open subset of Φ, so $W = \rho^{-1}(V_0 \cap \rho(\Gamma_0))$ is a nonempty open subset of Γ_0. Consider any point $(\sigma_1, (\check{f})_{\sigma_1}) = \gamma_1 \in W$. Since W is open there exists a neighborhood U of σ_1 in Σ such that $\check{f}$ is holomorphic on U and $(\sigma, (\check{f})_\sigma) = \gamma \in W$ for all $\sigma \in U$. Moreover, for each $\sigma \in U$

$$\begin{aligned}\check{f}(\sigma) &= \check{F}(\gamma) = (\check{G} \circ \rho)(\gamma) \\ &= (\check{g} \circ p \circ \rho)(\gamma) = \check{g}(\pi(\gamma)) = \check{g}(\sigma).\end{aligned}$$

Therefore $\check{f} = \check{g}$ on a neighborhood of σ_1, so $(\check{f})_{\sigma_1} = (\check{g})_{\sigma_1}$. It follows that $W \subseteq W_0$. Since W_0 is open and connected and intersects the component Γ_0 it must be contained in Γ_0. Now since $\rho : \Gamma_0 \to \Phi$ is a local homeomorphism (Lemma 48.1) we may, after shrinking the neighborhood U_0 and adjusting V_0 and W_0 accordingly, assume that $\rho : W_0 \to \rho(W_0)$ is one-to-one. Then since $\pi = p \circ \rho$ and $\pi : W_0 \to U_0 = \pi(W_0)$ is one-to-one it follows that $p : \rho(W_0) \to U_0 = p(\rho(W_0))$ is one-to-one. Note that $p : V_0 \to U_0 = p(V_0)$ is also one-to-one. Since $V_0 \cap \rho(\Gamma_0) = \rho(W) \subseteq \rho(W_0)$ it follows that $V_0 \cap \rho(W_0) \neq \emptyset$. Moreover, $p(V_0) = p(\rho(W_0)) = U_0$, so we must have $V_0 = \rho(W_0)$ by Lemma 44.2. In particular, $\varphi_0 \in \rho(\Gamma_0)$, proving that $\rho(\Gamma_0)$ is both open and closed in Φ and hence must exhaust Φ. Moreover, since

$$W_0 \subseteq \rho^{-1}(V_0) = \rho^{-1}(V_0 \cap \rho(\Gamma_0)) = W \subseteq W_0$$

it follows that $\rho^{-1}(V_0) = W_0$. In particular, $\rho^{-1}(\varphi_0)$ is a singleton. But, since $\rho(\Gamma_0) = \Phi$ and Φ is connected every point of Φ is a limit point of $\rho(\Gamma_0)$. Therefore the mapping $\rho : \Gamma_0 \to \Phi$ is one-to-one and hence is a homeomorphism. ◆

Now let (Φ, p) be an arbitrary connected Σ-domain and $\mathfrak{H}$ a subset of $\mathfrak{G}_\Phi$. Since card $\mathfrak{G}_\Phi \leq$ card Λ we may choose for each $\lambda \in \Lambda$ an element $h_\lambda \in \mathfrak{H}$ so that $\{h_\lambda : \lambda \in \Lambda\} = \mathfrak{H}$. The elements h_λ, of course, need not be distinct. The set $\{h_\lambda\}$, which we call a *Λ-indexing* of $\mathfrak{H}$, will now be used to construct a maximal extension of (Φ, p) relative to $\mathfrak{H}$. Although the choice of the Λ-indexing $\{h_\lambda\}$ is quite arbitrary, Corollary 48.5 shows that the resulting maximal extension will be independent of $\{h_\lambda\}$, at least up to Σ-domain isomorphism. Recall that we are assuming the uniqueness principle for $\mathfrak{G}$-h functions.

49.5 THEOREM. *Let*

$$\tau : \Phi \to \Sigma \# (\check{\mathfrak{G}}), \ \varphi \to (p(\varphi), (\check{h} \circ p_\varphi^{-1})_{p(\varphi)})$$

where $\check{h}(\varphi) = \{h_\lambda(\varphi)\}$, *and denote by* Γ_Φ *the component of the space* $\Sigma \# (\check{\mathfrak{G}})$ *that contains* $\tau(\Phi)$. *Then* τ *defines a maximal connected extension*

$$\tau : (\Phi, p, \mathfrak{H}) \Rightarrow (\Gamma_\Phi, \pi, \mathfrak{F}).$$

If $\mathfrak{H} = \mathfrak{G}_\Phi$ *then* $\mathfrak{F}|\Gamma_\Phi = \mathfrak{G}_{\Gamma_\Phi}$, *so* τ *defines a maximal* $\mathfrak{G}$*-holomorphic extension.*

Proof. Note that $p = \pi \circ \tau$ and if V_φ is a p-neighborhood of $\varphi \in \Phi$ then $\tau(V_\varphi)$ is a π-neighborhood of $\tau(\varphi)$ in $\Sigma \# (\check{\mathfrak{G}})$. It follows that τ is an open local homeomorphism, so $\tau(\Phi)$ is connected and consequently is contained in a component Γ_Φ of $\Sigma \# (\mathfrak{G})$. Moreover, $h_\lambda = F_\lambda \circ \tau$ for each $\lambda \in \Lambda$, so $\mathfrak{H} = \mathfrak{F} \circ \tau$ and hence $\tau : (\Phi, p, \mathfrak{H}) \Rightarrow (\Gamma_\Phi, \pi, \mathfrak{F})$. Now let $\rho : (\Phi, p, \mathfrak{H}) \Rightarrow (\Psi, q, \mathfrak{K})$ be an arbitrary connected extension of (Φ, p) relative to $\mathfrak{H}$ and recall (Corollary 49.3) that the map $\rho^* : \mathcal{K} \to \mathcal{H}$, $k \mapsto k\circ\rho$ is bijective. Hence for each $\lambda \in \Lambda$ there exists $g_\lambda \in \mathcal{K}$ such that $g_\lambda\circ\rho = h_\lambda$. Then $\{g_\lambda : \lambda \in \Lambda\} = \mathcal{K}$. Now using $\{g_\lambda\}$, we apply the preceding construction to obtain an extension $\tau' : (\Psi, q, \mathcal{K}) \Rightarrow (\Gamma_\Psi, \pi, \mathfrak{F})$, where $\tau' : \psi \mapsto (q(\psi), (\check{g}\circ q_\psi^{-1})_{q(\psi)})$ and Γ_ψ is the component of $\Sigma \# (\check{\mathfrak{G}})$ that contains $\tau'(\Psi)$. Since $p = q\circ\rho$ it follows that $q_{\rho(\varphi)}^{-1} = \rho\circ p_\varphi^{-1}$. Therefore $\check{g}\circ q_{\rho(\varphi)}^{-1} = \check{g}\circ\rho\circ p_\varphi^{-1} = \check{h}\circ p_\varphi^{-1}$, so $\tau'\circ\rho = \tau$. This implies that $\tau(\Phi) \subseteq \Gamma_\Phi$ and hence that $\Gamma_\Phi = \Gamma_\Psi$, so $\tau : (\Phi, p, \mathcal{H}) \Rightarrow (\Gamma_\Phi, \pi, \mathfrak{F})$ is maximal. That $\mathfrak{G}_{\Gamma_\Phi} = \mathfrak{F}|\Gamma_\Phi$ when $\mathcal{H} = \mathfrak{G}_\Phi$, follows by another application of Corollary 49.3. ◆

Note that an extension $\mu : (\Phi, p, \mathcal{H}) \Rightarrow (\Gamma_0, \pi, \mathfrak{F})$, where Γ_0 is any component of $\Sigma \# (\check{\mathfrak{G}})$, is automatically maximal since it is simply the extension constructed in Theorem 49.5 using the Λ-indexing $\mathcal{H} = \{F_\lambda\circ\mu : \lambda \in \Lambda\}$.

The property that $\mathfrak{F}|\Gamma = \mathfrak{G}_\Gamma$ is clearly not enjoyed but every component Γ of $\Sigma \# (\check{\mathfrak{G}})$. For example, let h be any fixed $\mathfrak{a}$-h function defined on Σ and set $f_\lambda = h$ for each $\lambda \in \Lambda$. Then $\check{f} \in \check{\mathfrak{G}}$. If $\Gamma_0 = \{(\sigma, (\check{f})_\sigma) : \sigma \in \Sigma\}$ then π maps Γ_0 homeomorphically onto Σ. It is easy to verify that Γ_0 is a component of $\Sigma \# (\check{\mathfrak{G}})$. However $\{F_\lambda|\Gamma_0 : \lambda \in \Gamma\}$ contains only the function $h\circ\pi$. On the other hand, an application of Theorem 49.5 plus Theorem 49.4 gives the following result.

49.6 COROLLARY. *Let* Γ_0 *be an arbitrary component of* $\Sigma \# (\check{\mathfrak{G}})$. *Then there exists another component* Γ_0' *such that the* Σ*-domains* (Γ_0, π) *and* (Γ_0', π) *are isomorphic and* $\mathfrak{G}_{\Gamma_0'} = \mathfrak{F}|\Gamma_0'$.

§50. PROPERTIES OF MAXIMAL DOMAINS

The following proposition contains the promised converse to Proposition 48.4.

50.1 PROPOSITION. *If the uniqueness principle for* $\mathfrak{a}$-h *functions is satisfied then a necessary and sufficient condition for an arbitrary connected extension* $\rho : (\Phi, p, \mathcal{H}) \Rightarrow (\Psi, q, \mathcal{K})$ *to be maximal is that the* Σ*-domain* (Ψ, q) *be maximal relative to* $\mathcal{K}$.

Proof. The necessity is given by Proposition 48.4 even without the uniqueness principle. Therefore assume that (Ψ, q) is maximal and let

$\rho' : (\Phi, p, \mathcal{H}) \Rightarrow (\Psi', q', \mathcal{K}')$ be any other connected $\mathfrak{G}$-extension of (Φ, p). Let $\tau : (\Psi', q', \mathcal{K}') \Rightarrow (\Gamma_{\Psi'}, \pi, \mathfrak{F})$ be the maximal $\mathcal{K}'$-extension of (Ψ', q') given by Theorem 49.4. Then, by the remark after the proof of Theorem 49.5, the extension $\tau\circ\rho' : (\Phi, p, \mathcal{H}) \Rightarrow (\Gamma_{\Psi'}, \pi, \mathfrak{F})$ is also maximal. Hence there exists $\mu : (\Psi, q, \mathcal{K}) \Rightarrow (\Gamma_{\Psi'}, \pi, \mathfrak{F})$ such that $\mu\circ\rho = \tau\circ\rho'$. But since (Ψ, q) is maximal the mapping $\mu : \Psi \to \Gamma_{\Psi'}$ must be a surjective homeomorphism. Therefore $\mu^{-1}\circ\tau : (\Psi', q', \mathcal{K}') \Rightarrow (\Psi, q, \mathcal{K})$ and $\rho = \mu^{-1}\circ(\tau\circ\rho') = (\mu^{-1}\circ\tau)\circ\rho'$, so the extension $\rho : (\Phi, p, \mathcal{H}) \Rightarrow (\Phi, q, \mathcal{K})$ is maximal. ◊

Two Σ-domains (Φ, p) and (Ψ, q) are defined to be *holomorphically equivalent* if there exists a surjective homeomorphism $\rho : \Phi \to \Psi$ such that $\mathfrak{G}_\Psi\circ\rho = \mathfrak{G}_\Phi$. In other words, ρ defines a pair isomorphism $\rho : [\Phi, \mathfrak{G}_\Phi] \Rightarrow [\Psi, \mathfrak{G}_\Psi]$. In fact, it is not difficult to verify that we even have a presheaf isomorphism $\rho : \{\Phi, \mathfrak{G}\} \Rightarrow \{\Sigma, \mathfrak{G}\}$. Although in general $p \neq q\circ\rho$, so ρ need not define a morphism of the Σ-domains, we nevertheless have the following result.

50.2 PROPOSITION. *Let* (Φ, p) *and* (Ψ, q) *be holomorphically equivalent under a homeomorphism* $\rho : \Phi \to \Psi$ *and assume that the base system* $[\Sigma, \mathfrak{G}]$ *is natural. Also assume that* $[\Phi, \mathfrak{G}_\Phi]$ *and* $[\Psi, \mathfrak{G}_\Psi]$ *are systems. Then there exists an open local homeomorphism* $\bar{q} : \Psi \to \Sigma$ *such that* ρ *defines a* Σ*-domain isomorphism* $\rho : (\Phi, p) \Rightarrow (\Psi, \bar{q})$. *In particular* ${}_{(\Psi, q)}\mathfrak{G} = {}_{(\Psi, \bar{q})}\mathfrak{G}$.

Proof. Note that the projection $p : \Phi \to \Sigma$ defines a pair morphism $p : [\Phi, \mathfrak{G}_\Phi] \to [\Sigma, \mathfrak{G}]$. Also, since ρ is a surjective homeomorphism the extension $\rho : [\Phi, \mathfrak{G}_\Phi] \Rightarrow [\Psi, \mathfrak{G}_\Psi]$ is trivially both minimal (Definition 9.2) and faithful (§2), so Theorem 9.3 applies to give a pair morphism $\bar{q} : [\Psi, \mathfrak{G}_\Psi] \to [\Sigma, \mathfrak{G}]$ such that $p = \bar{q}\circ\rho$. Furthermore, since ρ is a homeomorphism and p is an open local homeomorphism $\bar{q} : \Psi \to \Sigma$ is also an open local homeomorphism, *i.e.* $(\Psi, \bar{q})$ is a Σ-domain and $\rho : (\Phi, p) \Rightarrow (\Psi, \bar{q})$. In particular ${}_{(\Psi, \bar{q})}\mathfrak{G}_\Psi\circ\rho = \mathfrak{G}_\Phi$. Also, by hypothesis ${}_{(\Psi, q)}\mathfrak{G}_\Psi\circ\rho = \mathfrak{G}_\Phi$ and hence ${}_{(\Psi, \bar{q})}\mathfrak{G} = {}_{(\Psi, q)}\mathfrak{G}$. ◊

The next result shows that a morphism of Σ-domains extends to the corresponding maximal domains. The proof involves Theorem 49.5 so depends on our assumption of the uniqueness principle.

50.3 THEOREM. *Let* $\rho: (\Phi, p) \to (\Psi, q)$ *be an arbitrary morphism of connected* Σ*-domains and let*

$$\tau_\Phi : (\Phi, p) \Rightarrow (\bar{\Phi}, \bar{p}), \ \tau_\Psi : (\Psi, q) \Rightarrow (\bar{\Psi}, \bar{q})$$

be maximal $\mathfrak{G}$-h *extensions of these domains. Then there exists a morphism* $\bar{\rho} : (\bar{\Phi}, \bar{p}) \to (\bar{\Psi}, \bar{q})$ *of the corresponding maximal domains that extends* ρ, *i.e.* $\tau_\Psi \circ \rho = \bar{\rho} \circ \tau_\Phi$.

Proof. Since any two maximal extensions of a connected Σ-domain are isomorphic we need only make the proof for special choices of the maximal extensions. Since $\mathfrak{G}_\Phi \circ \rho \subseteq \mathfrak{G}_\Phi$ there exists $\Lambda_0 \subseteq \Lambda$, with cardinality equal to Λ, along with indexings $\{g_\lambda : \lambda \in \Lambda_0\}$ and $\{h_\lambda : \lambda \in \Lambda\}$ of $\mathfrak{G}_\Psi$ and $\mathfrak{G}_\Phi$ respectively such that $g_\lambda \circ \rho = h_\lambda$ for $\lambda \in \Lambda_0$. Let $g_\lambda = 0$ for $\lambda \in \Lambda \setminus \Lambda_0$, so g_λ is defined for all $\lambda \in \Lambda$. As in Theorem 49.5, the functions $\check{h}$ and $\check{g}$ determine maximal extensions

$$\tau_\Phi: (\Phi, p) \Rightarrow (\Gamma_\Phi, \pi), \ \tau_\Psi : (\Psi, q) \Rightarrow (\Gamma_\Psi, \pi)$$

where

$$\tau_\Phi : \varphi \mapsto (p(\varphi), (\check{h} \circ p_\varphi^{-1})_{p(\varphi)}), \ \tau_\Psi : \psi \mapsto (q(\psi), (\check{g} \circ q_\psi^{-1})_{q(\psi)}).$$

Now for arbitrary $\check{f} \in \check{\mathfrak{G}}$ set $\check{f}^0 = \{f_\lambda^0\}$, where $f_\lambda^0 = f_\lambda$ for $\lambda \in \Lambda_0$ and $f_\lambda^0 = 0$ for $\lambda \in \Lambda \setminus \Lambda_0$. Then define

$$\bar{\rho} : \Sigma \ \# \ (\check{\mathfrak{G}}) \to \Sigma \ \# \ (\check{\mathfrak{G}}), \ (\sigma, (\check{f})_\sigma) \mapsto (\sigma, (\check{f}^0)_\sigma).$$

It is obvious that $\bar{\rho}$ is continuous, so maps Γ_Φ into another component of $\Sigma \ \# \ (\check{\mathfrak{G}})$. Moreover $\check{h}^0 = \check{g} \circ \rho$. If $\psi = \rho(\varphi)$ then $p(\varphi) = q(\psi)$ and $q_\psi^{-1} = \rho \circ p_\varphi^{-1}$, so $\check{h}^0 \circ p_\varphi^{-1} = \check{g} \circ \rho \circ p_\varphi^{-1} = \check{g} \circ q_\psi^{-1}$. Hence if $\sigma = p(\varphi) = q(\varphi)$ then

$$\varphi \xmapsto{\tau_\varphi} (\sigma, (\check{h} \circ p_\varphi^{-1})_\sigma) \xmapsto{\bar{\rho}} (\sigma, (\check{h}^0 \circ p_\varphi^{-1})_\sigma)$$

and

$$(\sigma, (\check{h}^0 \circ p_\varphi^{-1})_\sigma) = (\sigma, (\check{g}^0 \circ q_\psi^{-1})_\sigma) = \tau_\Psi(\psi).$$

Therefore $\tau_\Psi \circ \rho = \bar{\rho} \circ \tau_\Phi$. In particular, $\bar{\rho}(\Gamma_\Phi)$ intersects Γ_Ψ so must be contained in Γ_Ψ. Also $\pi = \pi \circ \bar{\rho}$, so $\bar{\rho} : (\Gamma_\Phi, \pi) \to (\Gamma_\Psi, \pi)$ is the desired extension of $\rho : (\Phi, p) \to (\Psi, q)$ to the maximal Σ-domains. ◊

The above details are summarized in the following commutative diagram:

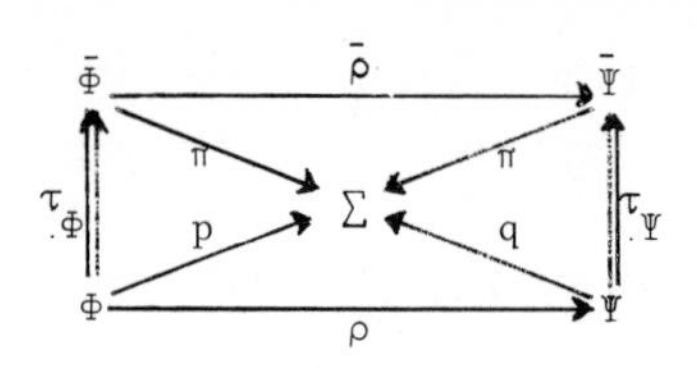

Recall that a connected Σ-domain, which is maximal relative to the $\mathfrak{G}$-holomorphic functions defined on it, is called a *domain of $\mathfrak{G}$-holomorphy* (§48).

50.4 COROLLARY. *Let* (Φ, p) *be a Σ-domain such that the trivial Σ-domain* $(p(\Phi), \iota)$ *is a domain of $\mathfrak{G}$-holomorphy and let* $\tau : (\Phi, p) \Rightarrow (\bar{\Phi}, \bar{p})$ *be a maximal* $\mathfrak{G}$-h *extension of* (Φ, p). *Then* $\bar{p}(\bar{\Phi}) = p(\Phi)$.

Let (Φ, p) be a Σ-domain of $\mathfrak{G}$-holomorphy and let $\rho : (\Phi, p) \to (\Psi, q)$ be an arbitrary connected Σ-domain morphism. If $\rho(\Phi) \neq \Psi$ then since (Φ, p) is maximal the morphism cannot be an $\mathfrak{G}$-h extension of (Φ, p). In other words, $\mathfrak{G}_\Psi \circ \rho$ is a *proper* subset of $\mathfrak{G}_\Phi$. This means that there are elements of $\mathfrak{G}_\Phi$ that do not extend through ρ to Ψ. If there exists a single function h_Φ in $\mathfrak{G}_\Phi$ that cannot be extended through ρ to Ψ for any connected morphism $\rho : (\Phi, p) \to (\Psi, q)$, with $\rho(\Phi) \neq \Psi$, then (Φ, p) is called a *domain of existence* for h_Φ.

Note that if $\mathfrak{G}_\Phi$ separates points then for an arbitrary extension $\rho : (\Phi, p) \Rightarrow (\Psi, q)$ the map $\rho : \Phi \to \rho(\Phi)$ is one-to-one so must be a homeomorphism. Therefore in this case we have the converse property that if $\rho : (\Phi, p) \to (\Psi, q)$ and $\rho(\Phi) \neq \Psi$ always imply $\mathfrak{G}_\Psi \circ \rho \subsetneq \mathfrak{G}_\Phi$ then (Φ, p) is a domain of holomorphy. In particular, a domain of existence is automatically a domain of holomorphy.

By imposing certain countability restrictions on a Σ-domain (Φ, p) we can prove that $\mathfrak{G}$-h convexity of Φ implies that (Φ, p) is a domain of existence and hence a Σ-domain of $\mathfrak{G}$-holomorphy. Two conditions are needed: (1) Φ is σ-compact and (2) there exists a countable sequence $\{B_n\}$ of subsets of Φ having noncompact closures such that every open subset of Φ with noncompact closure contains a set in the sequence. Condition (2) is a kind of countable "boundary condition" on Φ. If (Φ, p) satisfies both (1) and (2) we say that it is *strongly σ-compact*. Observe that a connected Riemann domain is always strongly σ-compact. (See the proof of Proposition 2, p. 44, in [G7].)

The main idea in the proof of the following theorem (*viz.* the construction of the function h_Φ) is well known. (Cf. [B6, p. 110].)

50.5 THEOREM. *If a connected* Σ*-domain* (Φ, p) *is strongly* σ*-compact and* $\mathcal{G}$-h *convex and if* $\mathcal{O}_\Phi$ *separates points then it is a domain of existence for an* $\mathcal{G}$*-holomorphic function so is a domain of* $\mathcal{G}$*-holomorphy.*

Proof. Since Φ is σ-compact there exists a sequence $\{G_n\}$ of open subsets of Φ, with union equal to Φ, such that $\{\bar{G}_n\}$ is compact and $\bar{G}_n \subset G_{n+1}$ for each n. Observe that if K is an arbitrary compact subset of Φ then $K \subset G_n$ for large n.

We shall now construct by induction, a subsequence $\{H_n\}$ of $\{G_n\}$ and a sequence $\{\delta_n\}$ of points of Φ such that for each n, $\delta_n \in (H_{n+1} \setminus \hat{\bar{H}}_n) \cap B_n$, where $\{B_n\}$ is the sequence of sets specified in condition (2) for strong σ-compactness and $\hat{\bar{H}}_n$ is the $\mathcal{G}$-h convex hull of the set $\bar{H}_n$ in Φ. Since Φ is assumed to be $\mathcal{G}$-h convex the hulls $\hat{\bar{H}}_n$ are compact. Start with $H_1 = G_1$. Then since $\bar{B}_1$ is not compact there exists a point $\delta_1 \in B_1 \setminus \hat{\bar{H}}_1$. Next let H_2 denote the first element of $\{G_n\}$ that contains the point δ_1, so $\delta_1 \in H_2 \setminus \hat{\bar{H}}_1$. Now assume that $H_1, \ldots, H_m$ and $\delta_1, \ldots, \delta_{m-1}$ are already defined with the desired property. Again, since $\bar{B}_m$ is not compact there exists $\delta_m \in B_m \setminus \hat{\bar{H}}_m$. Choosing H_{m+1} as the first element of $\{G_n\}$ that contains δ_m, we have $\delta_m \in H_{m+1} \setminus \hat{\bar{H}}_m$, so the desired sequences $\{H_n\}$ and $\{\delta_n\}$ exist by induction. Note also that Φ is still equal to the union of the sets H_n.

Since $\delta_n \notin \hat{\bar{H}}_n$, there exists $h_n \in \mathcal{O}_\Phi$ such that $|h_n|_{\bar{H}_n} < 1 < |h_n(\delta_n)|$. Again by an induction we may define an increasing sequence $\{m_n\}$ of positive integers such that $|h_n^{m_n}|_{\bar{H}_n} < 2^{-n}$ and

$$\left|\sum_{k=1}^{n-1} h_k^{m_k}\right|_{\bar{H}_{n+1}} < |h_n^{m_n}(\delta_n)| - n$$

for each n. Now define

$$h_\Phi(\varphi) = \sum_{n=1}^{\infty} h_n^{m_n}(\varphi), \quad \varphi \in \Phi.$$

Since $\bar{H}_k \subset H_{k+1}$ we have for all $k \geq n$

$$|h_k^{m_k}|_{\bar{H}_n} \leq |h_k^{m_k}|_{\bar{H}_k} < 2^{-k}.$$

Therefore the series for h_Φ converges uniformly on each of the sets H_n. In particular the series converges locally uniformly in G, so $h_\Phi \in \mathcal{O}_\Phi$.

Next let ℓ denote an arbitrary positive integer and consider $n > \ell$. Then since $\delta_n \in H_{n+1} \setminus \hat{\bar{H}}_n$,

$$\left|\sum_{k=1}^{n-1} h_k^{m_k}(\delta_n)\right| \le \left|\sum_{k=1}^{n-1} h_k^{m_k}\right|_{\bar{H}_{n+1}}.$$

Also, $\delta_n \in H_k$ for $k > n$, so

$$|h_k^{m_k}(\delta_n)| \le |h_k^{m_k}|_{\bar{H}_k} < 2^{-k}$$

and hence

$$\sum_{k=n+1}^{\infty} |h_k^{m_k}(\delta_n)| \le \sum_{k=n+1}^{\infty} 2^{-k} = 2^{-n}$$

Therefore

$$|h_\Phi(\delta_n)| \ge |h_n^{m_n}(\delta_n)| - \left|\sum_{k=1}^{n-1} h_k^{m_k}\right|_{\bar{H}_{n+1}} - 2^{-k}$$

$$> n - 2^{-n} \ge \ell + 1 - 2^{-n} > \ell$$

so $|h_\Phi(\delta_n)| > \ell$ for all $n > \ell$.

Finally let $\rho : (\Phi, p) \to (\Psi, q)$ be a morphism of connected Σ-domains with $\rho(\Phi) \ne \Psi$ and suppose that there were an element $g \in \mathcal{O}_\Psi$ such that $g \circ \rho = h_\Phi$. Since Ψ is connected $\mathrm{bd}_\Psi \rho(\Phi)$ contains at least one point ψ_0. Let V_0 be a neighborhood of the point ψ_0 in the space Ψ on which g is bounded. Then since ρ is continuous $\rho^{-1}(V_0 \cap \rho(\Phi))$ is a nonempty open subset of Φ. Furthermore by Lemma 48.1 the set $\rho(\Phi)$ is open in Ψ, so ψ_0 does not belong to $\rho(\Phi)$. Therefore $\rho^{-1}(V_0 \cap \rho(\Phi))$ does not have a compact closure in Φ. Now for an arbitrary positive integer ℓ set $U_\ell = \rho^{-1}(V_0 \cap \rho(\Phi)) \setminus \hat{\bar{H}}_{\ell+1}$. Then U_ℓ is an open subset of Φ with noncompact closure. Hence there exists $B_m \subset U_\ell$. In particular $\delta_m \in U_\ell$. Furthermore $\delta_m \notin \hat{\bar{H}}_{\ell+1}$, so we must have $m > \ell$. Therefore $|h_\Phi(\delta_m)| > \ell$. Since $\rho(\delta_m) \in V_0$ and $g(\rho(\delta_m)) = h_\Phi(\delta_m)$ it follows that $|g|_{V_0} > \ell$. But ℓ is arbitrary, so $|g|_{V_0} = \infty$, contradicting the condition that g be bounded on V_0. In other words h_Φ cannot be extended through p to Ψ. ◆

A fundamental result in the theory of Several Complex Variables is a converse to Theorem 50.5 for Riemann domains. It is the necessity half of the following theorem, the proof of which will be found in standard works on SCV. [G7, Theorem 18, p. 55; N2, Corollary 3, p. 115].

50.6 THEOREM. *Let* (Φ, p) *be a connected Riemann domain such that* $\mathfrak{G}_\Phi$ *separates points. Then a necessary and sufficient condition for* (Φ, p) *to be a domain of holomorphy is that it be holomorphically convex.*

The sufficiency portion of this theorem is obviously covered by Theorem 50.5. The necessity is a much deeper result even in finite dimensions, so the problem of generalizing it is very difficult. For the general case, it does not seem to help a great deal even to assume that (Φ, p) is a domain of existence. Although we shall return to this problem in Chapter XIV, for the present we settle for a rather weak converse to Theorem 50.5. For convenience, we shall call a net $\{\varphi_d : d \in \mathcal{D}\}$ in Φ *ultimately compact* if there exists $d \in \mathcal{D}$ such that the set $\overline{\{\varphi_{d'} : d' \geq d\}}$ is compact.

50.7 LEMMA. *If* $[\Phi, \mathfrak{G}_\Phi]$ *is a system then every dominated Cauchy net in* Φ *which is ultimately compact converges in* Φ.

Proof. Let $\{\varphi_d : d \in \mathcal{D}\}$ be an ultimately compact dominated Cauchy net in Φ. Then there exists a point

$$\varphi_0 \in \bigcap_{d \in \mathcal{D}} \overline{\{\varphi_{d'} : d' \geq d\}}.$$

Let $h_1, \ldots, h_n$ be an arbitrary finite subset of $\mathfrak{G}_\Phi$ and r an arbitrary positive real number. Since $\{\varphi_d\}$ is a Cauchy net there exists $d_0 \in \mathcal{D}$ such that $d', d'' \geq d_0$ implies

$$|h_i(\varphi_{d'}) - h_i(\varphi_{d''})| < \frac{r}{2}, \quad i=1, \ldots, n.$$

Since $\varphi_0 \in \overline{\{\varphi_{d'} : d' \geq d_0\}}$, there also exists $d_1 \geq d_0$ such that

$$|h_i(\varphi_{d_1}) - h_i(\varphi_0)| < \frac{r}{2}, \quad i=1, \ldots, n.$$

Together, these inequalities imply that $|h_i(\varphi_d) - h_i(\varphi_0)| < r$ for $d \geq d_0$ and each i. Therefore, since $[\Phi, \mathfrak{G}_\Phi]$ is a system, $\lim_d \varphi_d = \varphi_0$. ♦

50.8 THEOREM. *Assume that* $[\Phi, \mathfrak{G}_\Phi]$ *is a system and that* $[\Sigma, \mathfrak{G}]$ *is natural. Assume also that for any net* $\{\varphi_d : d \in \mathcal{D}\}$ *in* Φ *which is not ultimately compact there exists a function* $h \in \mathfrak{G}_\Phi$ *such that* $\limsup_d |h(\varphi_d)| = \infty$. *Then* (Φ, p) *is an* $\mathfrak{G}$-h *convex domain of* $\mathfrak{G}$-*holomorphy.*

Proof. Observe first that elements of $\mathfrak{G}_\Phi$ are bounded on dominated Cauchy nets. This implies, in the present case, that every dominated Cauchy net is ultimately compact and hence, by Lemma 50.7, convergent. Therefore (Φ, p) is $\mathfrak{G}$-h convex by Theorem 47.3.

Next let $\rho : (\Phi, p) \to (\Psi, q)$ be a morphism, where Ψ is connected and different from $\rho(\Phi)$, and choose a point $\psi_0 \in bd_\Psi \rho(\Phi)$. Note that since $\rho(\Phi)$ is open $\psi_0 \notin \rho(\Phi)$. Denote by $\mathcal{V}$ the set of all neighborhoods of the point ψ_0 in Ψ and partial order $\mathcal{V}$ by the relation "$V_1 \leq V_2$ iff $V_2 \subseteq V_1$". Then $\mathcal{V}$ is a directed set. For each $V \in \mathcal{V}$, choose a point $\varphi_V \in \Phi$ such that $\rho(\varphi_V) \in V$. Then

$$\lim_V \rho(\varphi_V) = \psi_0.$$

Now suppose that the net $\{\varphi_V\}$ were ultimately compact. Then there exists a point

$$\varphi_1 \in \bigcap_V \overline{\{\varphi_{V'} : V' \subseteq V\}}.$$

Set $\rho(\varphi_1) = \psi_1$. Then $\psi_1 \neq \psi_0$, so there exist disjoint neighborhoods V_1 of ψ_0 and W_1 of ψ_1. Choose a neighborhood U_1 of φ_1 such that $\rho(U_1) \subseteq W_1$. Then there exists $V' \subseteq V_1$ such that $\varphi_{V'} \in U_1$, so $\rho(\varphi_{V'}) \in W_1$. But, on the other hand, $\rho(\varphi_{V'}) \in V' \subseteq V_1$, which is impossible since $V_1 \cap W_1 = \phi$. Therefore the net $\{\varphi_V\}$ is not ultimately compact. Hence there exists $h \in \mathfrak{G}_\Phi$ such that $\lim_V \sup|h(\varphi_V)| = \infty$. In particular, $\lim_V h(\varphi_V)$ does not exist. On the other hand, if $g \in \mathfrak{G}_\Psi$ exists such that $h = g \circ \rho$, then

$$g(\psi_0) = \lim_V g(\rho(\varphi_V)) = \lim_V h(\varphi_V).$$

Therefore such g does not exist, proving that $h \notin \mathfrak{G}_\Psi \circ \rho$ and that (Φ, p) is a domain of holomorphy. ♦

§51. REMARKS

Theorems 49.4 and 49.5 give the existence of Σ-domains of $\mathfrak{G}$-holomorphy and envelopes of $\mathfrak{G}$-holomorphy for Σ-domains based on a system $[\Sigma, \mathfrak{G}]$ that satisfies the uniqueness principle for $\mathfrak{G}$-h functions. These results are direct generalizations of standard results for Riemann domains. In the latter case, domains of holomorphy are characterized by certain intrinsic properties, the study of which constitutes directly or indirectly a substantial portion of the theory of Several Complex Variables. Although it is easy to formulate some of these properties for the general case, the problem of obtaining interesting conditions under which the classical results may be generalized turns out to be very difficult. Theorem 47.3, in which the equivalence of several of the expected conditions for a domain of holomorphy is proved, suggests the type of results that one would like to obtain. On the other hand, an obstruction to progress in this direction appears to lie in the generality of the concept of $\mathfrak{G}$-holomorphic functions, in spite of the many nice properties exhibited by these functions. In fact, it appears to be necessary to specialize in the direction of linear vector spaces in order to obtain satisfactory generalizations of the more subtle properties of Riemann domains of holomorphy. The next two chapters, which constitute an introduction to a holomorphy theory for dual pairs of vector spaces, provide a basis for a more or less satisfactory treatment in Chapters XIII and XIV of Σ-domains of holomorphy, but with respect to a restricted class of $\mathfrak{G}$-holomorphic functions. It turns out that these special $\mathfrak{G}$-holomorphic functions are closely related to the holomorphic functions introduced by R. Arens [A2] in studying uniform algebras.

The problem of generalizing the classical holomorphy theory has also been studied extensively by workers in the field of "Infinite Dimensional Holomorphy" (IDH), where a variety of results for domains spread over certain linear topological spaces have been obtained. (See, e.g., [C1, M1, N3, S2].) However, most of these results involve conditions and techniques rather alien to the function algebra approach so will not be discussed here. Although our consideration of holomorphy in dual pairs of vector spaces obviously overlaps IDH, the actual involvement with the main body of that subject is minimal in both content and approach. On the other hand, the potential exists for a far more extensive interaction between these approaches. This is an area in which much remains to be done.

CHAPTER XI
HOLOMORPHY THEORY FOR DUAL PAIRS OF VECTOR SPACES

§52. GENERALIZED POLYNOMIALS AND HOLOMORPHIC FUNCTIONS IN A CLTS

Let E be a complex convex linear topological space (CLTS) and denote by E^m the cartesian product $E \times \cdots \times E$ (m factors). A function $P : E \to \mathbb{C}$ is called a *homogeneous polynomial of degree* m on E if there exists a nonzero, symmetric, m-linear functional

$$\beta : E^m \to \mathbb{C},\ (x_1,\ldots,x_m) \mapsto \beta(x_1,\ldots,x_m)$$

such that

$$P(x) = \beta(x,\ldots,x),\ x \in E.$$

A constant function is a homogeneous polynomial of degree zero. A function which is a finite sum of homogeneous polynomials is called a *generalized polynomial*. The collection of all generalized polynomials is obviously a point separating algebra of functions on E. Each of its elements P has a unique representation of the form $P = P_0 + \cdots + P_m$, where P_k is either zero or a homogeneous polynomial of degree k, for $0 \leq k \leq m$, and $P_m \neq 0$. The integer m is called the *degree* of P and the polynomials P_k are called its *homogeneous components*. If x_0 is a fixed element of E then $Q(x) = P(x - x_0)$, $x \in E$, defines a generalized polynomial with degree equal to that of P. Observe that a linear functional on E is a polynomial of degree one. Elements of the algebra generated by the linear functionals are called *finite polynomials*.

Now let E be a (locally) convex linear topological space and denote its topology by τ. Then the collection of all those generalized polynomials on E that are continuous with respect to the topology τ constitutes a subalgebra $\mathcal{P}_\tau$ of the algebra of all generalized polynomials on E. It is not difficult to prove that a homogeneous polynomial of degree m will be continuous iff its associated m-linear functional is continuous at the origin of E^m, and an arbitrary generalized polynomial will be continuous iff each of its homogeneous components is continuous. Also, a polynomial will be continuous iff it is bounded on a τ-neighborhood of some point of E. Since E is assumed to be a CLTS its continuous linear functionals separate the points of E. Therefore $\mathcal{P}_\tau$ also separates points. It follows that the $\mathcal{P}_\tau$ topology in E is Hausdorff, although it is in general coarser than τ. Therefore $[E, \mathcal{P}_\tau]$ need not be a system under the given topology τ on E.

Because of the linear space structure of E, it is possible to introduce a notion of derivative for functions in E and thus to develop a holomorphy theory that parallels, more-or-less, the classical development in finite dimensions. The resulting theory, along with its many ramifications, has come to be known as "Infinite Dimensional Holomorphy" (IDH). This subject, which involves a great deal of material having little to do with the questions that interest us, is much too extensive to be dealt with in any detail here. Therefore we mention only the fact that a function h defined on an open set U in E is holomorphic according to this approach iff it admits at each point $x_0 \in U$ a "Taylor expansion" of the form

$$h(x) = \sum_{k=0}^{\infty} \frac{1}{k!} P_k(x - x_0)$$

where P_k is either zero or a continuous homogeneous polynomial of degree k and the series converges uniformly on a neighborhood of x_0 in U. Let us call these functions *standard holomorphic functions*. It is immediate from the above characteristization, that standard holomorphic functions are $\mathcal{P}_\tau$-holomorphic in the sense of Chapter IV. Moreover, since a uniform limit of standard holomorphic functions is standard holomorphic it follows easily that a function defined on an open set in E will be standard holomorphic iff it is $\mathcal{P}_\tau$-holomorphic. It also follows that $\mathcal{P}_\tau$-holomorphic functions defined on open subsets of E are of order at most 1. (Definition 17.3). These remarks show that, with respect to the τ-topology in E, the notions of standard holomorphic and $\mathcal{P}_\tau$-holomorphic coincide for functions defined on open subsets of E.

We have the following useful property of $\mathcal{P}_\tau$-holomorphic functions in E. For $r > 0$, denote by $D^m(0,r)$ the open polydisc of radius r centered at the origin in $\mathbb{C}^m$.

52.1 PROPOSITION. *Let* h *be* $\wp_\tau$*-holomorphic on a neighborhood of the point* x_0 *in* E. *Then, for arbitrary* $u_1,\ldots,u_m \in E$ *there exists* $r > 0$ *such that the function*

$$f(\xi_1,\ldots,\xi_m) = h(x_0 + \sum_{i=1}^m \xi_i u_i)$$

is holomorphic on the polydisc $D^m(0, r)$ *in* $\mathbb{C}^m$.

Proof. Let U_{x_0} be a neighborhood of x_0 on which h is a uniform limit of a sequence $\{P_n\}$ of generalized polynomials. Choose $r > 0$ so that

$$x_0 + \sum_{i=1}^m \xi_i u_i \in U_{x_0}, \check{\xi} = (\xi_1,\ldots,\xi_m) \in D^m(0, r).$$

Since P_n is a generalized polynomial, the function

$$P_n(x_0 + \sum_{i=1}^m \xi_i u_i), \check{\xi} \in \mathbb{C}^m$$

is an ordinary polynomial in the variables $\xi_1,\ldots,\xi_m$. Therefore the uniform limit

$$f(\xi_1,\ldots,\xi_m) = \lim_{n\to\infty} P_n(x_0 + \sum_{i=1}^m \xi_i u_i), \check{\xi} \in D^m(0, r)$$

is holomorphic in the ordinary sense on $D^m(0, r)$. ♦

Observe that, since the $\wp_\tau$-topology in E is in general coarser than the τ-topology, $\wp_\tau$-neighborhoods tend to be larger than τ-neighborhoods. Therefore, even on a $\wp_\tau$-open subset of E, it is conceivable that a standard holomorphic function might fail to be $\wp_\tau$-holomorphic with respect to the $\wp_\tau$-topology. The precise relationship between the $\wp_\tau$-topology and the τ-topology appears to be difficult to determine in general. This problem, as well as the general question of when the pair $[E, \wp_\tau]$ is natural, has been completely answered only for the special class of spaces that we shall now discuss.

§53. DUAL PAIRS $< E, F >$

A pair $< E, F >$ of complex linear vector spaces is called a *dual pair* if there is given a complex-valued bilinear form

$$< > : E \times F \to \mathbb{C}, (x, y) \mapsto < x, y >$$

defined on $E \times F$, which satisfies the following two conditions:

1) $< x, y > = 0$, for all $y \in F$, implies $x = 0$.

2) $< x, y > = 0$, for all $x \in E$, implies $y = 0$.

In other words, the bilinear form is *nondegenerate*.

An important consequence of the nondegeneracy of the bilinear form in a dual pair is given in the following property which we state without proof. (See [S1, p. 124]): *If* $v_1,\ldots,v_n$ *is any finite set of linearly independent elements of* F *then there exist elements* $u_1,\ldots,u_n$ *of* E *such that* $< u_i, v_j > = \delta_{ij}$ *for all* i,j. (δ_{ij} is the Kronecker delta.) Elements $u \in E$ and $v \in F$ such that $< u, v > = 0$ are said to be *orthogonal*. Arbitrary sets of elements $\{u_i\} \subset E$ and $\{v_i\} \subset F$ such that $< u_i, v_j > = \delta_{ij}$ for all i,j, are said to be *biorthogonal*. Obviously, biorthogonality implies that each of the sets $\{u_i\}$ and $\{v_i\}$ consists of linearly independent elements. If L is the subspace of E spanned by $\{u_i\}$ and M the subspace of F spanned by $\{v_i\}$, then L and M constitute a dual pair $< L, M >$ with respect to the given bilinear form restricted to $L \times M$. The collection $\{(u_i, v_i)\}$ of pairs of the basis elements is called a *dual basis* for $< L, M >$. If L and M are arbitrary subspaces of E and F respectively that form a dual pair $< L, M >$ with respect to the given bilinear form in $< E, F >$ then we say that $< L, M >$ is contained in $< E, F >$ and write $< L, M > \subseteq < E, F >$. In general, $< L, M >$ may or may not admit a dual basis. However, if L (and hence M) is finite dimensional, then a dual basis for $< L, M >$ does exist by the property stated above.

Each element $y \in F$ determines, *via* the bilinear form, a linear functional

$$f_y : E \to \mathbb{C},\ x \mapsto < x, y >$$

on the space E. The map, $y \mapsto f_y$, obviously defines a linear isomorphism of F with a linear subspace of E', the full dual of E consisting of *all* linear functionals on E. Therefore F may be regarded as a subspace of E'. Then $< E, F >$ is contained in $< E, E' >$, where $< x, f > = f(x)$ for $x \in E$ and $f \in E'$. Similarly, $< E, F >$ may be regarded as contained in $< F', F >$. The topology determined on E by F (the F-topology) is denoted by $\sigma(E, F)$. Similarly, $\sigma(F, E)$ denotes the E-topology on F. Unless otherwise indicated, these are the topologies on E and F alluded to in the following discussion.

If A is an arbitrary subset of E the *orthogonal complement* of A in F is the set

$$A^{\perp} = \{y \in F : \langle A, y \rangle = (0)\}.$$

It is obvious that $A^{\perp}$ is a linear subspace of F. Similarly, if $B \subseteq F$ then the orthogonal complement $B^{\perp}$ is a linear subspace of E.

53.1 LEMMA. *Let* $\langle L, M \rangle \subseteq \langle E, F \rangle$, *where* L *and* M *are finite dimensional. Then*

$$E = L \oplus M^{\perp}, \ F = M \oplus L^{\perp}, \ \langle M^{\perp}, L^{\perp} \rangle \subseteq \langle E, F \rangle$$

and the associated projections are both open and continuous.

Proof. Choose a biorthogonal basis $\{(u_i, v_i)\}$ for $\langle L, M \rangle$. Let $x \in E$ and set

$$x_L = \sum_i \langle x, v_i \rangle u_i, \ x_L' = x - x_L.$$

Then $\langle x_L', v_i \rangle = 0$ for each i, so $x_L' \in M^{\perp}$. Hence $E = L + M^{\perp}$. Furthermore, since $\langle L, M \rangle$ is a dual pair $L \cap M^{\perp} = (0)$, so the sum is a direct sum $E = L \oplus M^{\perp}$. Similarly, $F = M \oplus L^{\perp}$. As in the case of E, we define for each $y \in F$

$$y_M = \sum_i \langle u_i, y \rangle v_i, \ y_M' = y - y_M.$$

Moreover, for arbitrary $x \in E$ we have $\langle x, y_M \rangle = \langle x_L, y_M \rangle$ and $\langle x, y_M' \rangle = \langle x_L', y_M' \rangle$, so $\langle x, y \rangle = \langle x_L, y_M \rangle + \langle x_L', y_M' \rangle$. Therefore it follows that $\langle M^{\perp}, L^{\perp} \rangle \subseteq \langle E, F \rangle$. It also follows that sets of the form $V + V'$, where V is a basic $\sigma(L, M)$-neighborhood of the origin in L and V' is a basic neighborhood of the origin in $M^{\perp}$, constitute a subbase at the origin for the $\sigma(E, F)$-topology in E. This implies that the projections $\varepsilon_L : E \to L$ and $\varepsilon_L' : E \to M^{\perp}$ are both open and continuous. ♦

A well-known result for dual pairs is that a linear functional $f : E \to \mathbb{C}$ will be continuous in the $\sigma(E, F)$-topology iff it belongs to F. (See [S1, p. 124].) In other words, $f = f_y$ for some $y \in F$. We prove next a generalization of this result for the algebra consisting of all generalized polynomials continuous in the $\sigma(E, F)$-topology.

53.2 THEOREM. *A generalized polynomial on* E *will be continuous in the* $\sigma(E, F)$*-topology iff it belongs to the algebra of polynomials generated by elements of* F.

Proof. Note that each element of the algebra generated by F is equal to a finite sum of finite products of linear functionals on E that belong to F. It is obvious that all such polynomials are $\sigma(E, F)$-continuous. Therefore let P be an arbitrary $\sigma(E, F)$-continuous generalized polynomial on E and let $P = \sum_{k=1}^{n} P_k$ be its representation in terms of homogeneous polynomials P_k. Thus, P_0 is a constant and $P_k(x) = \beta_k(x,\dots,x)$, where β_k is either zero or a symmetric k-linear form on E for $1 \le k \le n$. Since P is continuous there exists a basic $\sigma(E, F)$-neighborhood $N_0(v_1,\dots,v_m ; \varepsilon)$ of $0 \in E$ on which P is bounded. We may assume that the elements $v_1,\dots,v_m$ are linearly independent in F and choose elements $u_1,\dots,u_m$ in E such that $< u_i, v_j > = \delta_{ij}$. Let L and M be the subspaces of E and F spanned by the elements $\{u_i\}$ and $\{v_i\}$ respectively. Observe that, if $x \in N_0$, then also $x_L + \xi x_L' \in N_0$ for all $\xi \in \mathbb{C}$. Moreover, since the forms β_k are symmetric and linear in each variable we have

$$\begin{aligned} P(x_L + \xi x_L') &= \sum_{k=0}^{n} \beta_k(x_L+\xi x_L', \dots ,x_L+\xi x_L') \\ &= \sum_{k=0}^{n} \sum_{\ell=0}^{k} \binom{k}{\ell}\beta_k(\overbrace{x_L,\dots,x_L}^{k-\ell}, \overbrace{x_L',\dots,x_L'}^{\ell})\xi^\ell \\ &= \sum_{\ell=0}^{n} \xi^\ell \sum_{k=\ell}^{n} \binom{k}{\ell}\beta_k(\overbrace{x_L,\dots,x_L}^{k-\ell}, \overbrace{x_L',\dots,x_L'}^{\ell}). \end{aligned}$$

In order for $P(x_L+\xi x_L')$ to be a bounded function of $\xi \in \mathbb{C}$, the polynomial on the right must reduce to a constant. Therefore

$$P(x_L+\xi x_L') = P(x_L),\ \xi \in \mathbb{C}$$

and hence $P(x) = P(x_L)$ for all $x \in N_0$. Now let x be an arbitrary element of E and choose $r > 0$ such that $\xi x \in N_0$ for $|\xi| \le r$. Then $P(\xi x) = P(\xi x_L)$, so

$$\sum_{k=0}^{n} \xi^k P_k(x) = \sum_{k=0}^{n} \xi^k P_k(x_L),\quad |\xi| \le r.$$

This implies that $P_k(x) = P_k(x_L)$ for each k, and hence that $P(x) = P(x_L)$ for all $x \in E$. Since $x_L = \sum < x, v_i > u_i$ it follows that P is generated by the functionals corresponding to $v_1,\dots,v_m$. ♦

53.3 COROLLARY. *If a generalized polynomial is bounded on a basic* $\sigma(E, F)$-*neighborhood* $N(v_1,\ldots,v_m;\ \varepsilon)$ *of any point in* E *then it is generated by the elements* $v_1,\ldots,v_m$.

53.4 COROLLARY. *Let* $\mathcal{P}_{<E,\ F>}$ *denote the algebra of all* $\sigma(E, F)$-*continuous generalized polynomials on* E. *Then the* $\sigma(E, F)$ *and the* $\mathcal{P}_{<E,\ F>}$-*topologies are equivalent in* E *and* $[E, \mathcal{P}_{<E,\ F>}]$ *is therefore a system.*

The system $[\mathbb{C}^{\Lambda}, \mathcal{P}]$, discussed in §5 (Example 5.2) and §9 (Example 19.2), is included in the dual pair setting. In fact, let $\mathbb{C}^{(\Lambda)}$ be the subspace of $\mathbb{C}^{\Lambda}$ consisting of elements $\check{\xi} = \{\xi_\lambda\}$ such that $\xi_\lambda = 0$ for all but a finite set of the indices λ. Then $< \mathbb{C}^{\Lambda}, \mathbb{C}^{(\Lambda)} >$ is a dual pair under the bilinear form

$$< \check{\zeta}, \check{\xi} > = \sum_{\lambda} \zeta_\lambda \xi_\lambda, \ \check{\zeta} \in \mathbb{C}^{\Lambda}, \ \check{\xi} \in \mathbb{C}^{(\Lambda)}.$$

The $\sigma(\mathbb{C}^{\Lambda}, \mathbb{C}^{(\Lambda)})$-topology in $\mathbb{C}^{\Lambda}$ is obviously the ordinary product space topology and $\mathcal{P}_{<\mathbb{C}^{\Lambda},\ \mathbb{C}^{(\Lambda)}>}$ is the algebra $\mathcal{P}$ of all polynomials in the variables ζ_λ. It turns out that many properties of $[\mathbb{C}^{\Lambda}, \mathcal{P}]$ extend nicely to $[E, \mathcal{P}_{<E,\ F>}]$ for an arbitrary dual pair $< E, F >$. We extend first the notion of "finite determination" for functions defined in E.

53.5 DEFINITION. *Let* f *be a function defined on a set* $U \subset E$ *and* M *a linear subspace of* F. *Then* f *is said to be* ***determined in*** U *by* M *if* $f(x_1) = f(x_2)$ *whenever* $x_1, x_2 \in U$ *and* $x_1 - x_2 \in M^{\perp}$. *If* M *is finite dimensional then* f *is said to be* ***finitely determined.***

A function f defined on a set $G \subseteq E$ is said to be *determined at a point* x in G by the space $M \subseteq F$ if the point x admits a neighborhood U such that f is determined by M on $U \cap G$. If f is determined by M at each point of G then it is said to be *determined locally* in G by M. Observe that if f is determined by M and $M \subseteq M'$ then f is also determined by M'. If $P \in \mathcal{P}_{<E,\ F>}$ then since P is generated by a finite number of elements of F it is obviously determined by the subspace of F spanned by the generators. In particular, elements of $\mathcal{P}_{<E,\ F>}$ are finitely determined. If $\check{v} = (v_1,\ldots,v_n)$ is a set of linearly independent elements of F such that f is determined by the subspace of F spanned by $\check{v}$ then we shall also say that f is *determined by* $\check{v}$.

§54. HOLOMORPHIC FUNCTIONS IN A DUAL PAIR

We introduce some notations for the proof of the next theorem. Let $< L, M > \subseteq < E, F >$, where L and M have dimension m, and let $\{(u_i, v_i)\}$ be a biorthogonal basis for $< L, M >$. Consider an open set $U \subseteq E$. Then, for a fixed $u \in U$ there exists $\delta > 0$ such that

$$u + \sum_{i=1}^{m} \xi_i u_i \in U, \ \check{\xi} \in D^m(0, \delta)$$

where $D^m(0, \delta)$ is the open polydisc with center 0 and radius equal to δ in $\mathbb{C}^m$. Set

$$U_\delta = \{x \in U : |< x-u, v_i > | < \delta, \ i=1,\dots,m >\}$$

Now let f be a function determined in the set U by the space M. For $x \in U_\delta$ set $x' = u + \sum_{i=1}^{m} < x-u, v_i > u_i$. Then both $x, x' \in U_\delta$ and $x - x' \in M^\perp$. Therefore

$$f(x) = f(u + \sum_{i=1}^{m} < x-u, v_i > u_i), \ x \in U_\delta.$$

For the case of $\mathbb{C}^\Lambda$ the following result was proved by Hirschowitz [H3]. (Cf. [N1].)

54.1 THEOREM. *Let* h *be a* $\mathcal{P}_{<E, F>}$*-holomorphic function defined on an open connected set* $G \subseteq E$. *Then there exists a minimal finite dimensional subspace* M_h *of* F *that determines* h *locally in* G. *The space* M_h *is uniquely minimal in the strong sense that* M_h *will be contained in any subspace of* F *that determines* h *at a single point of* G.

Proof. Since $\mathcal{P}_{<E, F>}$-holomorphic functions are of order most equal to 1 (on open sets), there exists for each point of G a basic $\sigma(E, F)$-neighborhood $N = N(v_1,\dots,v_m; \varepsilon)$ contained in G, on which h is bounded, and a sequence $\{P_n\} \subset \mathcal{P}_{<E, F>}$ such that $|h(x) - P_n(x)| < \frac{1}{n}$, $x \in N$. Note that each of the polynomials P_n is also bounded on N and hence is determined by the linear space M spanned in F by the elements $v_1,\dots,v_m$. Let $z \in M^\perp$ and $x \in N$. Then also $x + z \in N$, so $P_n(x + z) = P_n(x)$. Therefore

$$|h(x + z) - h(x)| \le |h(x + z) - P_n(x + z)| + |P_n(x) - h(x)| < \frac{2}{n}$$

for all n. It follows that $h(x + z) = h(x)$, so h is also determined on N by M. This means that each point of G admits a neighborhood on which h is determined by a finite dimensional subspace of F. Let M_0 be one such subspace with *minimal* dimension and denote by G_0 the union of all those open subsets of G on which h is determined by M_0. Then G_0 is not empty and h is determined locally in G_0 by M_0. We prove that $G_0 = G$.

Let s be a limit point of G_0 in G. Then there exists a basic (E, F)-neighborhood $N_s(v_1,\ldots,v_m; \varepsilon)$ contained in G on which h is determined by the subspace M of F spanned by the elements $\{v_i\}$. We may also assume that $M_0 \subseteq M$ and that the elements $\{v_1,\ldots,v_m\}$ constitute a basis for M with $\{v_1,\ldots,v_{m_0}\}$ a basis for M_0 $(m_0 \leq m)$. Now choose elements $\{u_1,\ldots,u_m\} \subset E$ such that $\langle u_i, v_j \rangle = \delta_{ij}$ for $i,j=1,\ldots,m$. Since s is a limit point of G_0 there exists an open set $U_0 \subseteq N_s \cap G_0$ on which h is determined by M_0. Note that x will belong to N_s iff it has the (unique) form

$$x = s + \sum_{i=1}^{m} \xi_i u_i + z, \ \check{\xi} \in D^m(0, \varepsilon), \ z \in M^{\perp}.$$

Therefore by the remarks preceding the theorem

$$h(x) = h(s + \sum_{i=1}^{m} \xi_i u_i), \ x \in N_s.$$

Next define

$$\tilde{h}(\xi_1,\ldots,\xi_m) = h(s + \sum_{i=1}^{m} \xi_i u_i), \ \check{\xi} \in D^m(0, \varepsilon)$$

then $\tilde{h}(\xi_1,\ldots,\xi_m)$ is holomorphic on $D^m(0, \varepsilon) \subset \mathbb{C}^m$. Choose $\check{\xi}^0 \in D^m(0, \varepsilon)$ and $z_0 \in M^{\perp}$ so that

$$s + \sum_{i=1}^{m} \xi_i^0 u_i + z_0 \in U_0.$$

Then there exists $\delta > 0$ such that $\check{\xi} - \check{\xi}^0 \in D^m(0, \varepsilon)$ implies

$$t = s + \sum_{i=1}^{m} \xi_i u_i + z_0 \in U_0.$$

Note also that

$$t' = s + \sum_{i=1}^{m_0} \xi_i u_i + \sum_{i=m_0+1}^{m} \xi_i^0 u_i + z_0 \in U_0$$

and

$$t - t' = \sum_{i=m_0+1}^{m} (\xi_i - \xi_i^0) u_i \in M_0^{\perp}.$$

Therefore $\tilde{h}(\xi_1,\ldots,\xi_{m_0},\xi_{m_0+1}^0,\ldots,\xi_m^0) = h(t') = h(t) = \tilde{h}(\xi_1,\ldots,\xi_m)$ for $\check{\xi} - \check{\xi}^0 \in D^m(0, \delta)$

Thus h is independent of the variables $\xi_{m_0+1},\dots,\xi_m$ near $\check{\xi}^0$ so must be independent of these variables throughout $D^m(0, \varepsilon)$. Hence

$$\tilde{h}(\xi_1,\dots,\xi_m) = \tilde{h}(\xi_1,\dots,\xi_{m_0},0,\dots,0),\ \check{\xi} \in D^m(0, \varepsilon).$$

Now let $x \in N_s$ and $x + z_0 \in N_s$, where $z_0 \in M_0^{\perp}$. Then

$$x = s + \sum_{i=1}^{m} \xi_i u_i + z,\ z \in M^{\perp},\ \check{\xi} \in D^m(0, \varepsilon)$$

and

$$x + z_0 = s + \sum_{i=1}^{m} \xi_i' u_i + z',\ z' \in M^{\perp},\ \check{\xi}' \in D^m(0, \varepsilon).$$

Moreover, if $1 \le i \le m_0$ then $\xi_i' = \xi_i$, so

$$x + z_0 = s + \sum_{i=1}^{m_0} \xi_i u_i + \sum_{i=m_0+1}^{m} \xi_i' u_i + z'.$$

Therefore, $h(x + z_0) = \tilde{h}(\xi_1,\dots,\xi_{m_0},0,\dots,0) = h(x)$. In other words, h is determined in N_s by M_0. In particular $s \in G_0$, which implies $G_0 = G$. Thus, if $M_h = M_0$ then h is determined locally in G by M_h.

It remains to prove that M_h is unique as indicated. Therefore let M be an arbitrary finite dimensional subspace of F that determines h at some point x_0 in G. We must prove that $M_h \subseteq M$. Choose a basic $\sigma(E, F)$-neighborhood N_0 of x_0 contained in G on which both M_h and M determine h. Let x and $x + z$ be arbitrary elements of N_0 such that $z \in (M_h \cap M)^{\perp}$ and define

$$U = \{\zeta \in \mathbb{C} : x + \zeta z \in N_0\}.$$

Then U is an open subset of $\mathbb{C}$ that contains both of the points $\zeta = 0$ and $\zeta = 1$. Furthermore, since N_0 is a basic neighborhood it is easy to verify that U is linearly convex and hence connected. Now define

$$\tilde{h}(\zeta) = h(x + \zeta z),\ \zeta \in U.$$

Then $\tilde{h}$ is holomorphic in U. Since $z \in (M_h \cap M)^{\perp}$ it may be decomposed in the form $z = z_0 + z_1$, where $z_0 \in M_h^{\perp}$ and $z_1 \in M^{\perp}$. This may be done using biorthogonal bases. It follows that $x + \zeta z_0 \in N_0$ for all ζ in a neighborhood of $0 \in U$. Therefore, using the fact that M_h and M determine h on N_0, we have

$$\tilde{h}(\zeta) = h(x + \zeta z_0 + \zeta z_1) = h(x + \zeta x_0) = h(x) = \tilde{h}(0)$$

for all $\zeta \in U$ such that $x + \zeta z_0 \in N_0$. In other words, $\tilde{h}$ is constant on a neighborhood of $0 \in U$, and hence, being holomorphic, must be constant on U. In particular, $\tilde{h}(1) = \tilde{h}(0)$ so $h(x + z) = h(x)$. In other words, $M_h \cap M$ determines h locally in G. Therefore $\dim(M_h \cap M) = \dim(M_h)$, which implies that $M_h \subseteq M$. ♦

Let $x, x + z \in G$, where $z \in M_h^{\perp}$, and define $U' = \{\zeta \in \mathbb{C} : x + \zeta z \in G\}$. Then U' is an open subset of $\mathbb{C}$ that contains 0 and 1, and the function $\tilde{h}(\zeta) = h(x + \zeta z)$, $\zeta \in U'$, is holomorphic on U'. Since M_h determines h locally in G, the function $\tilde{h}$ must be constant on a neighborhood of $0 \in U'$. Therefore $\tilde{h}$ is constant on the component of U' that contains 0. Thus, if U' is connected we have $h(x + z) = h(x)$. This gives the following result.

54.2 COROLLARY. *If* G *is linearly convex (in particular, if* G *is a basic* $\sigma(E, F)$*-neighborhood in* E*) then* M_h *determines* h *on all of* G.

Note that, by the example discussed in §19 for the special system $[\mathbb{C}^{\Lambda}, \mathcal{P}]$, there will generally not exist a finite dimensional subspace of F that determines h on all of its domain of definition.

Next is the uniqueness principle for $\mathcal{P}_{<E, F>}$-holomorphic functions.

54.3 PROPOSITION. *Let* h *be a* $\mathcal{P}_{<E, F>}$*-holomorphic function defined on an open connected set* $G \subseteq E$. *Then if* h *vanishes on any open subset of* G *it must vanish on all of* G.

Proof. Let Z_0 be the interior of the set of zeros of the function h in G. Assume that $Z_0 \neq \emptyset$ and let x_0 be a limit point of Z_0 in G. By Theorem 54.1, there exists a basic $\sigma(E, F)$-neighborhood $N_0(v_1,\ldots,v_n : \varepsilon)$ of the point x_0 which is contained in G and on which h is determined by the subspace M of F spanned by $v_1,\ldots,v_n$. Choose $u_1,\ldots,u_n \in E$ such that $< u_i, v_j > = \delta_{ij}$ for all i, j. Then an element $x \in E$ will belong to N_0 iff it has the form

$$x = x_0 + \sum_{i=1}^{n} \xi_i u_i + z, \ \check{\xi} \in D^n(0, \varepsilon), \ z \in M^{\perp}.$$

Set

$$U = \{\check{\xi} \in D^n(0, \varepsilon) : x_0 + \sum_{i=1}^{n} \xi_i u_i + z \in Z_0, \text{ for some } z \in M^{\perp}\}.$$

Then U is a nonempty open subset of $D^n(0, \varepsilon)$. Now define

$$\tilde{h}(\xi_1,\ldots,\xi_n) = h(x_0 + \sum_{i=1}^{n} \xi_i u_i), \ \check{\xi} \in D^n(0, \varepsilon).$$

Then $\tilde{h}$ is holomorphic on $D^n(0, \varepsilon)$ and vanishes on U so must vanish on all of $D^n(0, \varepsilon)$. Since $h(x) = \tilde{h}(\xi_1,\ldots,\xi_n)$ for $x = x_0 + \sum_{i=1}^{n} \xi_i u_i + z \in N_0$ it follows that h vanishes on N_0. In particular, $x_0 \in Z_0$ and Z_0 is both open and closed in G so must equal G. ♦

We close this section with the following theorem which is a generalization of a familiar result for $\mathbb{C}^n$ and has a similar proof. (Cf. [N2, Proposition 1, p. 50].)

54.4 THEOREM. *Let* Θ *be a* $\mathcal{P}_{\langle E, F\rangle}$*-subvariety of an open connected set* $G \subseteq E$. *If* $\Theta \neq G$ *then* $G \setminus \Theta$ *is a dense connected subset of* G.

Proof. Denote by Θ° the interior of the set Θ. Then $G\setminus\Theta$ will be dense in G iff $\Theta^\circ = \emptyset$. Let x_0 be a limit point of Θ° in G. Since Θ is relatively closed in G, $x_0 \in \Theta$. Hence there exists a connected open neighborhood N_0 of x_0, contained in G, such that $N_0 \cap \Theta$ is the set of common zeros of $\mathcal{P}_{\langle E, F\rangle}$-holomorphic functions defined on N_0. But since $N_0 \cap \Theta^\circ$ is a nonempty open subset of N_0, it follows that $N_0 \subseteq \Theta^\circ$ by the uniqueness principle. Therefore Θ° is both open and closed in G so must equal G. This proves that if $\Theta \neq G$ then $\Theta^\circ = \emptyset$, so $G\setminus\Theta$ is dense in G.

Now suppose that $G\setminus\Theta = G_1 \cup G_2$, where G_1 and G_2 are not empty and $G_1 \cap \bar{G}_2 = \bar{G}_1 \cap G_2 = \emptyset$. Since $G\setminus\Theta$ is dense in G, it follows that $G \subseteq \bar{G}_1 \cup \bar{G}_2$. Therefore since G is connected $G \cap \bar{G}_1 \cap \bar{G}_2 \neq \emptyset$. Let x_0 be a point of $G \cap \bar{G}_1 \cap \bar{G}_2$ and note that $x_0 \in \Theta$. Choose a basic $\sigma(E, F)$-neighborhood N_0 of x_0, contained in G, such that $N_0 \cap \Theta$ is the set of common zero of $\mathcal{P}_{\langle E, F\rangle}$-holomorphic functions defined on N_0. Next choose $x_1 \in G_1 \cap N_0$, $x_2 \in G_2 \cap N_0$ and set

$$U = \{\zeta \in \mathbb{C} : \zeta x_1 + (1 - \zeta)x_2 \in N_0\}.$$

Also set

$$N_0' = \{\zeta x_1 + (1 - \zeta)x_2 : \zeta \in U\}.$$

Then U and N_0' are linearly convex subsets of $\mathbb{C}$ and N_0 respectively. In particular, both sets are connected. Note further that the map $\zeta \mapsto \zeta x_1 + (1 - \zeta)x_2$

establishes a homeomorphism of U onto N_0'. Since $x_1, x_2 \in N_0'$ the set $N_0' \setminus \Theta$ is nonempty, so there exists a $\mathcal{P}_{<E, F>}$-holomorphic function h defined on N_0 which vanishes on N_0 but is not identically zero on N_0'. Therefore the function $\tilde{h}(\zeta) = h(\zeta x_1 + (1 - \zeta)x_2)$, $\zeta \in U$, is holomorphic and not identically zero on U. It follows that the set of zeros of $\tilde{h}$ in U is discrete. Therefore the set of zeros of h in N_0' is also discrete. In particular, $N_0' \cap \Theta$ is discrete and $N_0' \setminus \Theta$ is therefore connected. But this contradicts the fact that $x_1 \in N_0 \cap G_1$ and $x_2 \in N_0 \cap G_2$ and completes the proof that $G \setminus \Theta$ must be connected. ♦

§55. ARENS HOLOMORPHIC FUNCTIONS

As was indicated earlier (19.1), R. Arens [A2] has given a definition of holomorphic functions defined in a general linear topological space E that we shall now examine. The Arens definition asserts that a function h defined on a subset of E is *holomorphic* if it is of the form $h = f \circ \tau$, where τ is a linear continuous mapping of E into $\mathbb{C}^n$ (for some n depending on h) and f is an ordinary holomorphic function defined on some open set $W \subseteq \mathbb{C}^n$. Thus, the domain of definition of h is the open set $\tau^{-1}(W) \subseteq E$. We shall call these functions *Arens holomorphic* functions.

In the above definition, the linear continuous mapping $\tau : E \to \mathbb{C}^n$ obviously has the form $\tau(x) = (\tau_1(x), \ldots, \tau_n(x))$, $x \in E$, where each of the mappings $\tau_i : E \to \mathbb{C}$ $(i=1,\ldots,n)$ is a linear continuous functional on E. Denote by E' the linear space of all linear continuous functionals on E. For an arbitrary linear topological space E the space E' may be trivial, in which case the Arens holomorphic functions will also be trivial. Therefore, in order to avoid awkward qualifying statements in the comments that follow, we shall assume throughout that E is a *locally convex* linear topological space and thereby assure the existence of an abundance of linear continuous functionals. In fact E and E' then form a dual pair $< E, E'>$ under the bilinear form $< x, y >$ whose value for $(x, y) \in E \times E'$ is the value of the functional y on x. In this notation, if $\tau : E \to \mathbb{C}^n$ is linear and continuous then there exists a finite set $v = \{v_1, \ldots, v_n\} \subset E'$ such that

$$\tau(x) = < x, \check{v} > = (< x, v_1 >, \ldots, < x, v_n >), \; x \in E.$$

We shall accordingly denote the mapping by $\check{v}$ instead of τ. Observe also that the domain of definition of an Arens holomorphic function, being of the form $\check{v}^{-1}(W)$ where W is an open set in $\mathbb{C}^n$, is always a $\sigma(E, E')$-open subset of E. Therefore we may as well start with an arbitrary dual pair $< E, F >$ with the $\sigma(E, F)$-topology on E. The next theorem answers completely the question of the relationship between Arens holomorphic and $\mathcal{P}_{<E, F>}$-holomorphic functions in E.

55.1 THEOREM. *A function* h *defined on an open subset of* E *will be* $\mathcal{P}_{<E, F>}$-*holomorphic iff it is locally Arens holomorphic.*

Proof. First let h be Arens holomorphic with $h = f \circ \check{v}$, where $\check{v} = (v_1, \ldots, v_n)$ and f is holomorphic on an open set $W \subseteq \mathbb{C}^n$. Then

$$h(x) = f(< x, v_1 >, \ldots, < x, v_n >), \; x \in \check{v}^{-1}(W).$$

Since each of the functionals v_i defines an element of the algebra $\mathcal{P}_{<E, F>}$ it follows immediately from Theorem 17.4 that h is $\mathcal{P}_{<E, F>}$-holomorphic. Therefore a locally Arens holomorphic function is locally $\mathcal{P}_{<E, F>}$-holomorphic and hence is $\mathcal{P}_{<E, F>}$-holomorphic.

Now assume that h is $\mathcal{P}_{<E, F>}$-holomorphic on a $\sigma(E, F)$-open set $G \subseteq E$ and let $x_0 \in G$. Then by Theorem 54.1 there exists a basic $\sigma(E, F)$-neighborhood $N_0(v_1, \ldots, v_n; \varepsilon)$ of x_0, contained in G, such that h is determined on N_0 by the subspace M of F spanned by the elements $v_1, \ldots, v_n$. We may assume as usual that the elements $v_1, \ldots, v_n$ are linearly independent and choose $u_1, \ldots, u_n \in E$ such that $< u_i, v_j > = \delta_{ij}$ for all i,j. Let $\check{\xi}^0 = (< x_0, v_1 >, \ldots, < x_0, v_n >)$. Since elements of N_0 are of the form

$$x = \sum_{i=1}^{n} \xi_i u_i + z, \; \check{\xi} \in D^n(\check{\xi}^0, \varepsilon), \; z \in M^{\perp}$$

it follows that $N_0 = \check{v}^{-1}(D^n(\check{\xi}^0, \varepsilon))$. Now define

$$f(\xi_1, \ldots, \xi_n) = h(\sum_{i=1}^{n} \xi_i u_i), \; \check{\xi} \in D^n(\check{\xi}^0, \varepsilon).$$

Then f is holomorphic on $D^n(\check{\xi}^0, \varepsilon)$ and, for arbitrary $x \in N_0$,

$$x = \sum_{i=1}^{n} < x, v_i > u_i + z, \; z \in M^{\perp}.$$

Also $\check{v}(x) = (< x, v_1 >, \ldots, < x, v_n >) \in D^n(\check{\xi}^0, \varepsilon)$, so

$$h(x) = h(\sum_{i=1}^{n} < x, v_i > u_i) = f(< x, v_1 >, \ldots, < x, v_n >) = f\circ\check{v}(x).$$

In other words, h is Arens holomorphic on N_0. ◆

§56. CANONICAL REPRESENTATION OF DUAL PAIRS

We examine next a "canonical" embedding of an arbitrary dual pair $< E, F >$ in the special dual pair $< \mathbb{C}^\Lambda, \mathbb{C}^{(\Lambda)} >$ discussed in §53. Let $\{v_\lambda : \lambda \in \Lambda\}$ denote a basis for the linear space F, *i.e.* each element $y \in F$ has a unique representation of the form $y = \sum_\lambda \eta_\lambda v_\lambda$, where $\{\eta_\lambda\} \in \mathbb{C}^{(\Lambda)}$. Obviously the correspondence $y \mapsto \check{\eta}$ defines a linear isomorphism (in general not topological) between the two spaces F and $\mathbb{C}^{(\Lambda)}$. On the other hand, it is easy to verify that the topology determined in E by the set $\{v_\lambda\}$ of basis elements is equivalent to the $\sigma(E, F)$-topology. Therefore the mapping

$$\check{v} : E \to \mathbb{C}^\Lambda,\ x \mapsto < x, \check{v} > = \{< x, v_\lambda >\}$$

is a linear homeomorphism with respect to the $\sigma(E, F)$-topology in E and the $\sigma(\mathbb{C}^\Lambda, \mathbb{C}^{(\Lambda)})$-topology (product space topology) in $\mathbb{C}^\Lambda$. We accordingly write $\check{v} : < E, F > \Rightarrow < \mathbb{C}^\Lambda, \mathbb{C}^{(\Lambda)} >$ since, as already noted, $\mathbb{C}^{(\Lambda)}\circ\check{v} = F$. In fact, we even have $\wp\circ\check{v} = \wp_{<E, F>}$, so v defines a pair extension

$$\check{v} : [E, \wp_{<E, F>}] \Rightarrow [\mathbb{C}^\Lambda, \wp].$$

It follows that if ${}_\Lambda\mathfrak{G}$ denotes all $\wp$-holomorphic functions defined on subsets of $\mathbb{C}^\Lambda$ and ${}_{<E, F>}\mathfrak{G}$ all $\wp_{<E, F>}$-holomorphic functions defined on subsets of E then ${}_\Lambda\mathfrak{G}\circ\check{v} = {}_{<E, F>}\mathfrak{G}$, so $\check{v}$ defines a *presheaf extension*

$$\check{v} : \{E, {}_{<E, F>}\mathfrak{G}\} \Rightarrow \{\mathbb{C}^\Lambda, {}_\Lambda\mathfrak{G}\}.$$

On the other hand, a $\wp$-holomorphic function defined on a relatively open subset of $\check{v}(E)$ need not admit an extension to an open subset of $\mathbb{C}^\Lambda$, so we generally cannot replace ${}_\Lambda\mathfrak{G}$ and ${}_{<E, F>}\mathfrak{G}$ by ${}_\Lambda\mathfrak{G}^o$ and ${}_{<E, F>}\mathfrak{G}^o$, the holomorphic functions defined on open sets. It is still true, of course, that ${}_\Lambda\mathfrak{G}^o\circ\check{v} \subseteq {}_{<E, F>}\mathfrak{G}^o$, so we have the presheaf morphism

$$\check{v} : \{E, {}_{<E,\,F>}\mathfrak{G}^{\mathcal{O}}\} \to \{\mathbb{C}^{\Lambda}, {}_{\Lambda}\mathfrak{G}^{\mathcal{O}}\}.$$

The following proposition shows, however, that this morphism is, so-to-speak, a "local" presheaf extension.

56.1 PROPOSITION.

(i) Each element of ${}_{<E,\,F>}\mathfrak{G}^{\mathcal{O}}$ *belongs locally to* ${}_{\Lambda}\mathfrak{G}^{\mathcal{O}}\circ\check{v}$.

(ii) Let H *be a connected open subset of* $\mathbb{C}^{\Lambda}$ *such that* $H \cap \check{v}(E) \neq \emptyset$ *and let* g, h *be* $\wp$*-holomorphic functions defined on* H. *Then* $g|\check{v}(E) = h|\check{v}(E)$ *implies that* $g = h$ *on all of* H.

Proof. Let h be an element of ${}_{<E,\,F>}\mathfrak{G}^{\mathcal{O}}$ defined on an open set $G \subseteq E$ and let $x_0 \in G$. Then, by Theorem 54.1 there exists a finite set $\pi \subseteq \Lambda$ and $\varepsilon > 0$ such that the basic $\sigma(E, F)$-neighborhood of x_0 determined by $\{v_\lambda : \lambda \in \pi\}$ and ε is contained in G and the subspace M_π of F spanned by the elements $\{v_\lambda : \lambda \in \pi\}$ determines h on N_0. Choose elements $\{u_\lambda : \lambda \in \pi\} \subset E$ such that $< u_{\lambda'}, v_{\lambda''} > = \delta_{\lambda'\lambda''}$ for all $\lambda', \lambda'' \in \pi$ and set $\check{\xi}^0 = \{< x_0, v_\lambda > : \lambda \in \pi\} \in \mathbb{C}^{\pi}$. Also denote by D_0^π the open polydisc in $\mathbb{C}^\pi$ with center $\check{\xi}^0$ and radius ε, so $x \in N_0$ iff it is of the form $x = \sum_{\lambda\in\pi} \xi_\lambda u_\lambda + z$, $\check{\xi} \in D_0^\pi$, $z \in M_\pi^\perp$, where $\check{\xi} = \{< x, u_\lambda > : \lambda \in \pi\}$. Also $h(x) = h(\sum_{\lambda\in\pi} \xi_\lambda u_\lambda) = \tilde{h}(\check{\xi})$, where $\tilde{h}$ is holomorphic on D_0^π. Set

$$\tilde{\pi} : \mathbb{C}^{\Lambda} \to \mathbb{C}^{\pi},\ \{\zeta_\lambda : \lambda \in \Lambda\} \mapsto \{\zeta_\lambda : \lambda \in \pi\}.$$

Then $\tilde{h}\circ\tilde{\pi}$ is $\wp$-holomorphic on the neighborhood

$$N_{\check{v}(x_0)}(\pi, \varepsilon) = D_0^\pi \times \mathbb{C}^{\Lambda\setminus\pi} = \tilde{\pi}^{-1}(D_0^\pi)$$

of the point $\check{v}(x_0)$ in $\mathbb{C}^\Lambda$. Moreover, for $x \in N_0$

$$((\tilde{h}\circ\tilde{\pi})\circ\check{v})(x) = \tilde{h}(\{< x, v_\lambda >: \lambda \in \pi\}) = h(\sum_{\lambda\in\pi} < x, v_\lambda > u\}) = h(x)$$

i.e. h belongs locally to ${}_{\Lambda}\mathfrak{G}^{\mathcal{O}}\circ\check{v}$, completing the proof of (i).

For the proof of (ii), it will be sufficient to show that if g is $\wp$-holomorphic on the open set $H \subseteq \mathbb{C}^\Lambda$ and $g|\check{v}(E) = 0$ then $g = 0$ on all of H. Let $\check{\delta} \in H \cap \check{v}(E)$ and choose a basic neighborhood $V_{\check{\delta}}(\pi, \varepsilon)$ of $\check{\delta}$ in $\mathbb{C}^\Lambda$ on which g is defined and determined by π. Since $\check{\delta} \in H \cap \check{v}(E)$, there exists $x_0 \in E$ such that $\check{v}(x_0) = \check{\delta}$. Consider the neighborhood N_0 of x_0 in E determined by $\{v_\lambda : \lambda \in \pi\}$ and ε. Then, by the characterization of N_0 in the proof of (i), it follows that $\tilde{\pi}(V_{\check{\delta}} \cap \check{v}(E)) = \tilde{\pi}(V_{\check{\delta}})$. Hence, for arbitrary $\check{\zeta} \in V_{\check{\delta}}$ there exists $\check{\xi} \in V_{\check{\delta}} \cap \check{v}(E)$ such that $\tilde{\pi}(\check{\zeta}) = \tilde{\pi}(\check{\xi})$. Since g is determined on $V_{\check{\delta}}$ by π, it follows that $g(\check{\zeta}) = g(\check{\xi}) = 0$, $\check{\xi} \in V_{\check{\delta}}$. Therefore $g = 0$ on all of H by the uniqueness principle. ♦

56.2 COROLLARY. *The mapping* $\check{v} : E \to \mathbb{C}^{\Lambda}$ *induces a homeomorphism of the sheaf of germs of* $\wp$*-holomorphic functions in* $\check{v}(E)$ *with the sheaf of germs of* $\wp_{<E, F>}$*-holomorphic functions in* E.

§57. DERIVATIVES

We turn next to the problem of defining derivatives for $\wp_{<E, F>}$-holomorphic functions. These will be needed in the chapters that follow. The problem could be approached through the general theory of differentiation in vector spaces, but we prefer to exploit at the outset the special properties of dual pairs. The difference is, of course, primarily one of emphasis rather than substance.

Let h be an arbitrary $\wp_{<E, F>}$-holomorphic function defined on an open connected set G in E. The derivatives of h at a point $x_0 \in G$ will be taken "with respect to" a set $\check{v} = (v_1,\ldots,v_n)$ of linearly independent elements of F that determines h at the point x_0. Since G is connected, it follows by Theorem 54.1 that h will be determined by $\check{v}$ at every point of G. Therefore the derivatives of h with respect to $\check{v}$ will be defined everywhere in G. We denote by M the subspace of F spanned by $\check{v}$. Now, as usual, choose elements $u_1,\ldots,u_n$ in E such that $< u_i, v_i > = \delta_{ij}$ for all i, j. Let x be an arbitrary element of G. Then, since G is open there exists $\delta > 0$ such that

$$x + \sum_{i=1}^{n} \xi_i u_i \in G, \check{\xi} \in D^n(0, \delta).$$

Define

$$\tilde{h}(x, \check{\xi}) = h(x + \sum_{i=1}^{n} \xi_i u_i), \check{\xi} \in D^n(0, \delta).$$

Then, as a function of $\check{\xi}$, $\tilde{h}(x, \check{\xi})$ is holomorphic on the polydisc $D^n(0, \delta)$. Next let $\varkappa = (k_1,\ldots,k_n)$ be a multiple index and define

$$\mathscr{D}_{\check{\xi}}^{\varkappa}\tilde{h}(x, \check{\xi}) = \frac{\partial^{|\varkappa|}}{(\partial\check{\xi})^{\varkappa}}\tilde{h}(x, \check{\xi})$$

where $|\varkappa| = k_1 + \cdots + k_m$ and $(\partial\check{\xi})^{\varkappa} = \partial\xi_1^{k_1} \ldots \partial\xi_n^{k_n}$. We make the following definition.

57.1 DEFINITION. *The* $\varkappa$*-derivative of* h *with respect to* $\check{v}$ *at the point* $x \in G$ *is given by*

$$\mathscr{D}_{\check{v}}^{\varkappa}h(x) = \mathscr{D}_{\check{\xi}}^{\varkappa}\tilde{h}(x, 0).$$

By Theorem 54.1, the determining space M_h for h in G must be contained in the space M spanned by $\check{v}$. If $\dim M_h = m$ and $\check{v}$ is chosen so that $v_1,\ldots,v_m$ is a basis for M_h then the associated elements $u_i \in E$ belong to $M_h^{\perp}$ for $i > m$. Hence $h(x + \sum_{i=1}^{n} \xi_i u_i) = h(x + \sum_{i=1}^{m} \xi_i u_i)$ so $\mathscr{D}_{\check{v}}^{\varkappa}h(x) = 0$ for any multiple index $\varkappa = (k_1,\ldots,k_n)$ with $k_{m+1} + \cdots + k_n \neq 0$.

Observe that the derivatives $\mathscr{D}_{\check{v}}^{\varkappa}h(x)$ are quite independent of the choice of the elements $u_1,\ldots,u_n$. In fact, if $u_1',\ldots,u_n'$ are any other elements of E such that $< u_i', v_j > = \delta_{ij}$ then $u_i' - u_i \in M^{\perp}$ for each i, so $h(x + \sum_{i=1}^{n} \xi_i u_i) = h(x + \sum_{i=1}^{n} \xi_i u_i')$. On the other hand, the derivatives obviously must depend on the choice of the basis $\check{v}$ for M. Thus, if $\check{v}'$ is another basis for M, so $v_i = \sum_{j=1}^{n} \alpha_{ji} v_j'$ for $i=1,\ldots,n$, and if $< u_i', v_j' > = \delta_{ij}$ then $u_i' = \sum_{j=1}^{n} \alpha_{ij} u_j + z_i$, $z_i \in M^{\perp}$. Hence $h(x + \sum_{i=1}^{n} \xi_i' u_i') = h(x + \sum_{j=1}^{n} \xi_j u_j)$ where

$$\xi_j = \sum_{i=1}^{n} \xi_i' \alpha_{ij}, \; j=1,\ldots,n.$$

Therefore, if we denote by δ_k the multiple index $(\delta_{k1},\ldots,\delta_{kn})$ with a "1" in the kth place and "0's" elsewhere then

$$\mathscr{D}_{\check{v}'}^{\varkappa} = \prod_{i=1}^{n} \left(\sum_{j=1}^{n} \alpha_{ij} \mathscr{D}_{\check{v}}^{\delta_j} \right)^{k_i}$$

for $\varkappa = (k_1,\ldots,k_n)$. Thus $\mathscr{D}_{\check{v}'}^{\varkappa}$ is a linear combination of derivatives with respect to $\check{v}$. Therefore, in view of the preceding remark, every derivative of h is a linear combination of derivatives with respect to a basis for the determining space M_n.

We prove next that the derivatives of holomorphic functions are also holomorphic.

57.2 THEOREM. *The derivative function* $\mathscr{D}_{\check{v}}^{\varkappa}h(x)$ *is* $\mathcal{P}_{<E,\ F>}$*-holomorphic and determined locally in* G *by the space* M_h.

Proof. Let $x_0 \in G$ and choose a neighborhood $U_0 \subseteq G$ of the point x_0 on which h is determined by M_h. Also choose $\varepsilon_0 > 0$ such that

$$x_0 + \sum_{i=1}^{n} \xi_i u_i \in U_0, \ \check{\xi} \in D^n(0, \varepsilon_0).$$

For $x \in N_{x_0}(\check{v}, \varepsilon_0) \cap U_0$ set $\eta_i = < x - x_0, v_i >$ for $i=1,\ldots,n$. Then $\check{\eta} \in D^n(0, \varepsilon_0)$ and $x = x_0 + \sum_{i=1}^{n} \eta_i u_i + z$, where $z \in M^{\perp}$. Now choose $\varepsilon > 0$ such that

$$x + \sum_{i=1}^{n} \xi_i u_i \in N_{x_0}(\check{v}, \varepsilon_0) \cap U_0, \ \check{\xi} \in D^n(0, \varepsilon).$$

Then

$$x + \sum_{i=1}^{n} \xi_i u_i = x_0 + \sum_{i=1}^{n} (\eta_i + \xi_i) u_i + z$$

and $\tilde{h}(x, \check{\xi}) = \tilde{h}(x_0, \check{\eta} + \check{\xi})$ for $\check{\xi} \in D^n(0, \varepsilon)$. Hence

$$\mathcal{D}_{\check{v}}^{\varkappa} h(x) = \mathcal{D}_{\check{\xi}}^{\varkappa} \tilde{h}(x, 0) = \mathcal{D}_{\check{\xi}}^{\varkappa} \tilde{h}(x_0, \check{\eta}).$$

Since the function $\mathcal{D}_{\check{\xi}}^{\varkappa} \tilde{h}(x_0, \check{\eta})$ is a holomorphic function of $\check{\eta} \in D^n(0, \varepsilon_0)$, it admits a series expansion in $\eta_1,\ldots,\eta_n$ uniformly convergent for $\check{\eta} \in D^n(0, \varepsilon_0/2)$. Recall that $\eta_i = < x - x_0, v_i >$. Hence it follows that $\mathcal{D}_{\check{v}}^{\varkappa} h(x)$ is a uniform limit on $N_{x_0}(\check{v}, \varepsilon_0/2) \cap U_0$ of polynomials in the linear fucntionals $\{< x - x_0, v_i >\}$. Therefore $\mathcal{D}_{\check{v}}^{\varkappa} h(x)$ is $\mathcal{P}_{<E, F>}$-holomorphic at each point $x_0 \in G$. ◆

Observe that, for $x \in N_{x_0}(\check{v}, \varepsilon_0) \cap U_0$, $h(x) = \tilde{h}(x_0, \check{\eta})$, $\check{\eta} = \{< x - x_0, v_i >\}$. Substituting the series expansion of $\tilde{h}(x_0, \check{\eta})$ in terms of $\check{\eta}$, we obtain the following result.

57.3 COROLLARY. *The Taylor's series expansion*

$$h(x) = \sum_{\varkappa} \frac{\mathcal{D}_{\check{v}}^{\varkappa} h(x_0)}{\varkappa!} < x - x_0, \check{v} >^{\varkappa}$$

where $\varkappa! = k_1!k_2!\cdots k_m!$ *and* $< x - x_0, \check{v} >^{\varkappa} = < x - x_0, v_1 >^{k_1} \cdots < x - x_0, v_n >^{k_n}$, *is valid for all* $x \in N_{x_0}(\check{v}, \varepsilon_0) \cap U_0$ *and convergence is uniform on* $N_{x_0}(\check{v}, \varepsilon) \cap U_0$ *for* $\varepsilon < \varepsilon_0$.

57.4 PROPOSITION. *Let* g *and* h *be* $\mathcal{P}_{<E, F>}$*-holomorphic functions defined on an open connected set* $G \subseteq E$ *and let* M *be a finite dimensional subspace of* F *that determines both* g *and* h *on* G. *Let* $\check{v}$ *be a basis for* M *and assume that* $\mathcal{D}_{\check{v}}^{\varkappa} g(x_0) = \mathcal{D}_{\check{v}}^{\varkappa} h(x_0)$ *for some point* $x_0 \in G$ *and all* $\varkappa$. *Then* $g(x) = h(x)$ *for all* $x \in G$.

Proof. By Corollary 57.3, the functions g and h coincide on a neighborhood of x_0 and therefore coincide on all of G by the uniqueness principle (Proposition 54.3).

§58. NATURALITY

We close this chapter with a brief examination of the question of naturality of the system $[E, \mathcal{P}_{<E, F>}]$. Let $\varphi : P \mapsto \hat{P}(\varphi)$ be a continuous homomorphism of $\mathcal{P}_{<E, F>}$ onto $\mathbb{C}$, where, as usual, the topology in $\mathcal{P}_{<E, F>}$ is uniform convergence on $\sigma(E, F)$-compact subsets of E. Since the space F is contained in $\mathcal{P}_{<E, F>}$ the homomorphism φ induces on F a linear functional f_φ. Also, since $\mathcal{P}_{<E, F>}$ is generated by F the homomorphism $\varphi : \mathcal{P}_{<E, F>} \to \mathbb{C}$ will be a point evaluation in E iff the functional f_φ is continuous for the $\sigma(F, E)$-topology in F. In the space F, the topology of uniform convergence on $\sigma(E, F)$-compact convex circled subsets of E is equivalent to the "Mackey topology". The latter is the finest topology with respect to which E is the space of all continuous linear functionals on F [S1, p. 131]. Now, by the continuity of φ there exists a $\sigma(E, F)$-compact set $K_\varphi \subset E$ such that $|\hat{P}(\varphi)| \le |P|_{K_\varphi}$, $P \in \mathcal{P}_{<E, F>}$. In particular, $|f_\varphi(y)| \le \max_{x \in K_\varphi} |< x, y >|$, $y \in F$. The last inequality also holds if K_φ is replaced by its closed convex circled hull. Hence, if this hull is $\sigma(E, F)$-compact then f_φ is continuous with respect to the Mackey topology so there exists $x_\varphi \in E$ such that $f_\varphi(y) = < x_\varphi, y >$. Therefore $\varphi : \mathcal{P}_{<E, F>} \to \mathbb{C}$ is a point evaluation and we have the following result.

58.1 PROPOSITION. *If for the dual pair* $<E, F>$ *the closed convex circled hull of every* $\sigma(E, F)$ *-compact set is also compact then* $[E, \mathcal{P}_{<E, F>}]$ *is a natural system.*

We have the following special cases covered by Proposition 58.1 [S1, pp. 131,148].

58.2 COROLLARY. *If* E *is quasi-complete in the* $\sigma(E, F)$*-topology then* $[E, \mathcal{P}_{<E, F>}]$ *is natural.*

58.3 COROLLARY. *If* F *is a Banach space and* E *is the space of all bounded linear functionals on* F *then* $[E, \mathcal{P}_{<E, F>}]$ *is natural.*

CHAPTER XII
<E, F> -DOMAINS OF HOLOMORPHY

§59. HOLOMORPHIC FUNCTIONS IN <E, F>-DOMAINS

This chapter, which is concerned with domains spread over the complex linear vector space E of a dual pair $<E, F>$, consists mainly of extensions of results for the special case $\mathbb{C}^{\Lambda}$ due to M. Matos [M1, M2] (countable Λ) and V. Aurich [A4] (arbitrary Λ). (See also [H3, R6].) The dual pair $<E, F>$ is fixed throughout and the topology in E is always assumed to be the $\sigma(E, F)$-topology. Also, instead of using the rather cumbersome expression, "$[E, \mathcal{P}_{<E, F>}]$-domain", for a domain spread over E, we shall use the simpler term, "*<E, F>-domain*". Similarly, an <E, F>-domain of $\mathcal{P}_{<E, F>}$-holomorphy will be called simply an *<E, F>-domain of holomorphy*.

Now let (Φ, p) be an arbitrary <E, F>-domain and h a $\mathcal{P}_{<E, F>}$-holomorphic function defined on an open set $G \subseteq \Phi$. A finite dimensional space $M \subseteq F$ is said to *determine* h *locally* in G iff for each $\varphi \in G$ there exists a p-neighborhood U_φ contained in G such that the function $h \circ p_\varphi^{-1}$ is determined by M on $p(U_\varphi)$ in the sense of Definition 53.5. Theorem 54.1 extends without change to the present situation, giving the following proposition. The proof is a straightforward adaptation of the proof of Theorem 54.1 so will be omitted.

59.1 PROPOSITION. *Let* h *be a* $\mathcal{P}_{<E, F>}$*-holomorphic function defined on an open connected set* $G \subseteq \Phi$. *Then there exists a uniquely determined minimal finite dimensional subspace* M_h *of* F *that determines* h *locally in* G.

We may also use the local homeomorphism $p : \Phi \to E$ to define derivatives for $\mathcal{P}_{<E, F>}$-holomorphic functions in Φ in terms of derivatives of functions in E. Thus if h is determined locally in G by $M \subseteq F$ and if U_φ is an arbitrary p-neighborhood of a point $\varphi \in \Phi$ then the function $h \circ p_\varphi^{-1}$ is obviously determined locally in $p(U_\varphi)$ by M. Therefore if $\check{v} = (v_1, \ldots, v_n)$ is a basis for M and $\varkappa = (k_1, \ldots, k_n)$

is a multiple index then the derivative $\mathcal{D}_{\check{v}}^{\varkappa}(h \circ p_{\varphi}^{-1})(x)$ is defined for $x \in p(U_{\varphi})$. We accordingly make the following definition.

59.2 DEFINITION. *Let* h *be* $\mathcal{P}_{<E, F>}$*-holomorphic on an open connected set* $G \subseteq \Phi$ *and let* M *be a finite dimensional subspace of* F *that determines* h *locally in* G. *If* $\check{v} = (v_1, \ldots, v_n)$ *is a basis for* M *and* $\varkappa = (k_1, \ldots, k_n)$ *is an arbitrary multiple index, then the* $\varkappa$-*derivative of* h *with respect to* $\check{v}$ *at the point* $\varphi \in G$ *is given by*

$$\mathcal{D}_{\check{v}}^{\varkappa} h(\varphi) = \mathcal{D}_{\check{v}}^{\varkappa}(h \circ p_{\varphi}^{-1})(p(\varphi)).$$

As in the case of functions in E, derivatives of h with respect to an arbitrary basis for M are linear combinations of derivatives with respect to any given basis for the minimal space M_h.

§60. SUBDOMAINS DETERMINED BY A SUBSPACE OF F

Recall (§45) that a basic p-neighborhood of a point in a general $[\Sigma, \mathfrak{G}]$-domain is a p-neighborhood that projects homeomorphically onto a basic neighborhood in Σ determined by elements from a given linearly independent system of generators for $\mathfrak{G}$. We apply the general definition here, recognizing that the algebra $\mathcal{P}_{<E, F>}$ is generated by the space F. (See also §65 below.) More precisely, if (Φ, p) is an arbitrary $<E, F>$-domain then a p-neighborhood of a point $\varphi \in \Phi$ is a *basic* p-*neighborhood* provided its image in E under p is of the form $N_{p(\varphi)}(\{v_i\};\varepsilon)$, where $\{v_i\}$ is an arbitrary finite set of linearly independent elements of F and $\varepsilon > 0$. We accordingly denote such a neighborhood by

$$W_{\varphi}(\{v_i\}; \varepsilon) = p_{\varphi}^{-1}(N_{p(\varphi)}(\{v_i\}; \varepsilon))$$

and say that it is *defined by* $\{v_i\}$. Note that for a given set $\{v_i\} \subset F$ the neighborhood $W_{\varphi}(\{v_i\}; \varepsilon)$ need not be defined for any $\varepsilon > 0$. On the other hand, the basic p-neighborhoods do generate the topology of Φ and, in the present case, each is connected. Observe that if $W_{\varphi}(\{v_i\}; \varepsilon)$ is defined and M is the subspace of F spanned by the elements $\{v_i\}$, and $\{v_i'\}$ is any other basis for M, then there exists $\varepsilon' > 0$ such that $W_{\varphi}(\{v_i'\}; \varepsilon')$ is defined. We accordingly make the following definition.

60.1 DEFINITION. *Let* M *be a finite dimensional subspace of* F *and* $\{v_i\}$ *a basis for* M. *If* $\varepsilon > 0$ *exists such that* $W_\varphi(\{v_i\}; \varepsilon)$ *is defined then we say that* M *determines* (Φ, p) *at the point* φ. *The subset of* Φ *consisting of all points at which* M *determines* (Φ, p) *is denoted by* Φ^M. *If* $\Phi^M = \Phi$ *then we say that* M *determines* (Φ, p).

Note that there always exists M such that $\Phi^M \neq \emptyset$. Also, if $M \subseteq M'$ then $\Phi^M \subseteq \Phi^{M'}$. If $W_\varphi(\{v_i\}; \varepsilon)$ is defined and $\varphi' \in W_\varphi(\{v_i\}; \varepsilon/2)$, then $W_{\varphi'}(\{v_i\}; \varepsilon/2)$ is also defined and contained in $W_\varphi(\{v_i\}; \varepsilon)$. Therefore Φ^M is an open subset of Φ, so (Φ^M, p) is an <E, F>-domain.

Let M be a finite dimensional subspace of F with basis $\{v_i\}$ and let φ be an arbitrary point of Φ^M, so $\varepsilon > 0$ exists such that $W_\varphi(\{v_i\}; \varepsilon)$ is defined. Now, if $z \in M^\perp$ then $p(\varphi) + z \in N_{p(\varphi)}(\{v_i\}; \varepsilon)$, so there exists $\varphi' \in W_\varphi(\{v_i\}; \varepsilon)$ such that $p(\varphi') = p(\varphi) + z$. This suggests defining φ' to be the "*sum*" of φ and z. We write

$$\varphi + z = p_\varphi^{-1}(p(\varphi) + z).$$

Observe that the sum $\varphi + z$ defined in this way depends only on the space M and not on the basis $\{v_i\}$ or ε. We note in passing that the local homeomorphism p could be used to lift more of the local linear structure of E to the domain Φ. However, the additional structure is not needed for our purposes, so we shall limit attention to addition of elements of Φ^M to elements of $M^\perp$ as defined above. The following lemma contains some elementary properties of the operation.

60.2 LEMMA.

(i) $\Phi^M + M^\perp = \Phi^M$.

(ii) If $\varphi \in \Phi^M$, *then* $\varphi + 0 = \varphi$. *Also, addition is associative, i.e.* $\varphi + (z + z') = (\varphi + z) + z'$ *for arbitrary* $z, z' \in M^\perp$.

(iii) If $\varphi \in \Phi^M$ *and* $p(W_\varphi(\{v_i\}, \varepsilon)) = N_{p(\varphi)}(\{v_i\}, \varepsilon)$, *where* $\{v_i\}$ *is a basis for* M, *then for each* $z \in M^\perp$

$$W_{\varphi+z}(\{v_i\}; \varepsilon) = W_\varphi(\{v_i\}; \varepsilon) = W_\varphi(\{v_i\}; \varepsilon) + M^\perp$$

(iv) Let h *be* $\mathcal{P}_{<E, F>}$*-holomorphic on an open set* $G \subseteq \Phi^M$ *such that* $G + M^\perp = G$. *If* h *is determined locally in* G *by a space* $M_h \supseteq M$, *then* h *is determined "globally" in* G *by* M_h.

Proof. Properties (i)-(iii) are obvious consequences of the definition. Property (iv) means that $h(\varphi + z) = h(\varphi)$ for $\varphi \in G$ and $z \in M_h^{\perp}$. Note that since $M \subseteq M_h$ we have $M_h^{\perp} \subseteq M^{\perp}$, so $\varphi + z$ is defined. Let $\{v_i\}$ be a basis for M and choose $\varepsilon > 0$ so that $p(W_\varphi(\{v_i\}; \varepsilon)) = N_{p(\varphi)}(\{v_i\}; \varepsilon)$. By definition, M_h determines the function $h \circ p_\varphi^{-1}$ locally in $N_{p(\varphi)}(\{v_i\}; \varepsilon)$. Since basic neighborhoods in E are linearly convex, it follows from Corollary 11.8 that M_h determines $h \circ p_\varphi^{-1}$ on $N_{p(\varphi)}(\{v_i\}; \varepsilon)$. Therefore, if $z \in M_h^{\perp}$, then $p(\varphi) + z \in N_{p(\varphi)}(\{v_i\}; \varepsilon)$, so

$$h(\varphi + z) = h(p_\varphi^{-1}(p(\varphi) + z)) = (h \circ p_\varphi^{-1})(p(\varphi) + z) = (h \circ p_\varphi^{-1})(p(\varphi)) = h(\varphi)$$

completing the proof of (iv). ♦

§61. ENVELOPES OF HOLOMORPHY

By Propositions 49.2 and 54.3 the $\mathcal{P}_{\langle E, F\rangle}$-holomorphic functions in an $\langle E, F\rangle$-domain (Φ, p) satisfy the uniqueness principle. Therefore by Theorem 49.5 the "sheaf of germs" construction may be used to obtain the envelope of $\mathcal{P}_{\langle E, F\rangle}$-holomorphy for an arbitrary connected $\langle E, F\rangle$-domain as a component of the "universal" $\langle E, F\rangle$-domain $(E \# (\check{\mathcal{O}}), \pi)$. We shall follow the notations developed in §49 for general $[\Sigma, \mathfrak{a}]$-domains. Recall that $(\check{\mathcal{O}})$ is the sheaf of germs $(\check{f})_x$ of holomorphic maps $\check{f}$ from open subsets of E to $\mathbb{C}^\Lambda$, $E \# (\check{\mathcal{O}})$ is the subspace of the product $E \times (\check{\mathcal{O}})$ consisting of the pairs $(x, (\check{f})_x)$, and $\pi(x, (\check{f})_x) = x$. Also, F_λ is the $\mathcal{P}_{\langle E, F\rangle}$-holomorphic function

$$F_\lambda(\gamma) = f_\lambda(x),\ \gamma = (x, (\check{f})_x) \in E \# (\check{\mathcal{O}}).$$

61.1 PROPOSITION. *Let $\mathfrak{F}$ denote the algebra of all functions on $E \# (\check{\mathcal{O}})$ generated by $\mathcal{P}_{\langle E, F\rangle} \circ \pi$ plus the functions $\{F_\lambda : \lambda \in \Lambda\}$ and their derivatives. Then $\mathfrak{F}$ separates the points of $E \# (\check{\mathcal{O}})$.*

Proof. Let $\gamma_1 = (x_1, (\check{f})_x)$ and $\gamma_2 = (x_2, (\check{g})_{x_2})$ be two points of $E \# (\check{\mathcal{O}})$ and suppose that $h(\gamma_1) = h(\gamma_2)$ for every $h \in \mathfrak{F}$. Since $\mathcal{P}_{\langle E, F\rangle} \circ \pi \subseteq \mathfrak{F}$ and $\mathcal{P}_{\langle E, F\rangle}$ separates the points of E we must have $x_1 = x_2$, so $\gamma_1 = (x_1, (\check{f})_{x_1})$ and $\gamma_2 = (x_1, (\check{g})_{x_1})$. Choose a basic $\sigma(E, F)$-neighborhood N_{x_1} on which germ representatives $\check{f}$ and $\check{g}$ are defined. Then by Corollary 54.2 there exists for each $\lambda \in \Lambda$

a finite dimensional subspace $M_\lambda \subseteq F$ which determines f_λ and g_λ on N_{x_1}. Let $\check{v}$ be a basis for M_λ. Then for an arbitrary derivative $\mathcal{D}^{\varkappa}_{\check{v}}$ with respect to $\check{v}$ we have $\mathcal{D}^{\varkappa}_{\check{v}}F_\lambda \in \mathfrak{F}$, so

$$\mathcal{D}^{\varkappa}_{\check{v}}f_\lambda(x_1) = \mathcal{D}^{\varkappa}_{\check{v}}F_\lambda(\lambda_1) \doteq \mathcal{D}^{\varkappa}_{\check{v}}F_\lambda(\lambda_2) = \mathcal{D}^{\varkappa}_{\check{v}}g_\lambda(x_1).$$

Therefore by Proposition 57.4 the functions f_λ and g_λ are equal on N_{x_1}. Since this is true for each $\lambda \in \Lambda$ it follows that $\check{f} = \check{g}$ on N_{x_1}. Therefore $(\check{f})_{x_1} = (\check{g})_{x_1}$, so $\gamma_1 = \gamma_2$. ♦

The next theorem is of central importance in the theory of <E, F>-domains.

61.2 THEOREM. *Let* Γ *be an arbitrary component of the space* $E \# (\check{\mathfrak{G}})$ *and let* M *be any finite dimensional subspace of* F. *Then either* $\Gamma^M = \emptyset$ *or* $\Gamma^M = \Gamma$.

Proof. Assume that $\Gamma^M \neq \emptyset$. Then since Γ^M is open and Γ is connected the proof reduces to showing that Γ^M is also closed in Γ. Therefore let γ_1 be a limit point of Γ^M in Γ and choose a finite dimensional subspace M' in F, with $M \subseteq M'$, such that (Γ, π) is determined at γ_1 by M'. Let $\{v_i\}$ be a basis for the space M and $\{v_i'\}$ a basis for M' that contains $\{v_i\}$. Then there exists $\varepsilon' > 0$ such that $W_{\gamma_1}(\{v_i'\}; 2\varepsilon')$ is defined. Choose $\gamma_0 \in \Gamma^M \cap W_{\gamma_1}(\{v_i'\}; \varepsilon')$. Since

$$N_{\pi(\gamma_0)}(\{v_i'\}; \varepsilon') \subseteq N_{\pi(\gamma_1)}(\{v_i'\}; 2\varepsilon')$$

it follows that $W_{\gamma_0}(\{v_i'\}; \varepsilon')$ is defined and contained in $W_{\gamma_1}(\{v_i'\}; 2\varepsilon')$. Moreover, $\gamma_1 \in W_{\gamma_0}(\{v_i'\}; \varepsilon')$. Now, since $\gamma_0 \in \Gamma^M$ there exists $\varepsilon > 0$ such that $W_{\gamma_0}(\{v_i\}; \varepsilon)$ is defined. If $\varepsilon' \leq \varepsilon$ then $W_{\gamma_0}(\{v_i'\}; \varepsilon') \subseteq W_{\gamma_0}(\{v_i\}; \varepsilon)$, so $\gamma_1 \in W_{\gamma_0}(\{v_i\}; \varepsilon)$ and hence $\gamma_1 \in \Gamma^M$. Therefore we assume that $\varepsilon' > \varepsilon$ and set

$$W = W_{\gamma_0}(\{v_i\}; \varepsilon) \cup W_{\gamma_0}(\{v_i'\}; \varepsilon').$$

Observe that the set

$$\pi(W_{\gamma_0}(\{v_i\}; \varepsilon)) \cap \pi(W_{\gamma_0}(\{v_i'\}; \varepsilon')) = N_{\pi(\gamma_0)}(\{v_i\}; \varepsilon) \cap N_{\pi(\gamma_0)}(\{v_i'\}; \varepsilon')$$

being the intersection of two linearly convex sets, is itself linearly convex and hence connected. Therefore, by Lemma 44.2, W is a π-set. Let $V = \pi(W)$ and $x_0 = \pi(\gamma_0)$. Then

$$V = N_{x_0}(\{v_i\}; \varepsilon) \cup N_{x_0}(\{v_i'\}; \varepsilon')$$

and $\pi^{-1}_{\gamma_0}$ is defined on V.

For fixed $\lambda \in \Lambda$ consider the function F_λ on Γ and set $f_\lambda = F_\lambda \circ \pi_{\gamma_0}^{-1}$ on V. Choose a subspace $M_\lambda \subseteq F$, with $M' \subseteq M_\lambda$, such that M_λ determines F_λ locally in Γ. Observe that $W \subseteq \Gamma^{M'} \subseteq \Gamma^{M_\lambda}$ and $W + M_\lambda^\perp = W$. Therefore by Lemma 60.2 the space M_λ determines F_λ globally in W. Now let $\{v_i''\}$ be a basis for M_λ that contains the basis $\{v_i'\}$. Write $\check{v}'' = \{v_1'', \ldots, v_n'', \ldots, v_{n'}'', \ldots, v_{n_\lambda}''\}$ where $\{v_1'', \ldots, v_n''\} = \{v_i\}$, the basis for M, and $\{v_1'', \ldots, v_{n'}''\} = \{v_i'\}$, the basis for M'. Denote by U the image of the set V in the space $\mathbb{C}^{n_\lambda}$ under the mapping

$$\check{v}'' : V \to \mathbb{C}^{n_\lambda}, \; x \mapsto \{< x - x_0, v_i'' >\}.$$

Then

$$U = (D^n(0, \varepsilon) \times \mathbb{C}^{n_\lambda - n}) \cup (D^{n'}(0, \varepsilon') \times \mathbb{C}^{n_\lambda - n'}).$$

Observe that if x_1 and x_2 are elements of V such that $\check{v}''(x_1) = \check{v}''(x_2)$ then $x_1 - x_2 \in M_\lambda^\perp$. Hence $f_\lambda(x_1) = f_\lambda(x_2)$. Therefore if we set

$$h_\lambda(\check{\xi}) = f_\lambda(x), \; \check{v}''(x) = \check{\xi}$$

then h_λ is a well defined function on U which, by Proposition 52.1, is holomorphic in the ordinary sense. Observe also that, since $\varepsilon' > \varepsilon$, it follows from well-known results in SCV theory (see [B2, Theorem 6, p. 19]) that the holomorphic hull of the set U in $\mathbb{C}^{n_\lambda}$ is equal to

$$\tilde{U} = D^n(0, \varepsilon') \times \mathbb{C}^{n_\lambda - n}.$$

Therefore h_λ extends to a holomorphic function $\tilde{h}_\lambda$ on the domain $\tilde{U}$. Observe that $\check{v}''$ maps the neighborhood $N_{x_0}(\{v_i\}; \varepsilon')$ onto $\tilde{U}$. If we define

$$g_\lambda(x) = \tilde{h}_\lambda(\check{v}''(x)), \; x \in N_{x_0}(\{v_i\}; \varepsilon')$$

then g_λ is a $\mathcal{P}_{<E, F>}$-holomorphic extension of f_λ to the neighborhood $N_{x_0}(\{v_i\}; \varepsilon')$. Furthermore, the neighborhood $N_{x_0}(\{v_i\}; \varepsilon')$ is independent of λ. Therefore the function $\check{f}$ extends to an element $\check{g} \in \check{\mathfrak{G}}$ defined on $N_{x_0}(\{v_i\}; \varepsilon')$. Hence the set

$$\tilde{W} = \{(x, (\check{g})_x) : x \in N_{x_0}(\{v_i\}; \varepsilon')\} = W_{\gamma_0}(\{v_i\}; \varepsilon')$$

is a π-neighborhood of γ_0 in $E \# (\check{\mathfrak{G}})$. Moreover since $\tilde{W}$ is connected and intersects Γ it follows that $\tilde{W} \subseteq \Gamma$. In particular $\gamma_1 \in \Gamma^M$, so Γ^M is both open and closed in Γ and consequently must exhaust Γ. ◆

By Theorem 49.5 and Corollary 48.5 we have the following corollary.

61.3 COROLLARY. *Let* (Φ, p) *be a connected* $\langle E, F\rangle$*-domain and choose* $M \subseteq F$ *such that* $\Phi^M \neq \emptyset$. *Then for any maximal extension* $\rho : (\Phi, p) = (\Psi, q)$ *we have* $\Psi^M = \Psi$. *If* (Φ, p) *is an* $\langle E, F\rangle$*-domain of holomorphy then already* $\Phi^M = \Phi$.

61.4 PROPOSITION. *Let* $\langle L, M\rangle \subseteq \langle E, F\rangle$, *where* M *is a finite dimensional subspace of* F, *and let* (Φ, p) *be an* $\langle E, F\rangle$*-domain such that* $\Phi^M = \Phi$. *Define*

$$\Phi_L = \{\varphi \in \Phi : p(\varphi) \in L\}.$$

Then (Φ_L, p) *is an* $\langle L, M\rangle$*-domain and* $\Phi = \Phi_L \oplus M^{\perp}$.

Proof. Since $\langle L, M\rangle \subseteq \langle E, F\rangle$ the $\sigma(L, M)$-topology in L is equivalent to that induced on L by the $\sigma(E, F)$-topology in E. Therefore the assertion that (Φ_L, p) is an $\langle L, M\rangle$-domain follows immediately from the fact that (Φ, p) is an $\langle E, F\rangle$-domain.

In order to prove the "direct sum" $\Phi = \Phi_L \oplus M^{\perp}$, we must show that each element of $\varphi \in \Phi$ has a *unique* representation in the form $\varphi = \varphi_L + z_\varphi$, where $\varphi_L \in \Phi_L$ and $z_\varphi \in M^{\perp}$. Recall that $E = L \oplus M^{\perp}$ (Lemma 53.1), so we have the decomposition

$$p(\varphi) = x_L + z_\varphi, \; x_L \in L, \; z_\varphi \in M^{\perp}.$$

Now choose $\varepsilon > 0$ so that $W_\varphi(\{v_i\}; \varepsilon)$ is defined. Then

$$p(W_\varphi(\{v_i\}; \varepsilon)) = N_{p(\varphi)}(\{v_i\}; \varepsilon) = N_{x_L}(\{v_i\}; \varepsilon).$$

Since $x_L \in N_{p(\varphi)}(\{v_i\}; \varepsilon)$ there exists $\varphi_L \in W_\varphi(\{v_i\}; \varepsilon)$ such that $p(\varphi_L) = x_L$. Therefore $\varphi_L \in \Phi_L$ and

$$\varphi_L + z_\varphi = p_{\varphi_L}^{-1}(p(\varphi_L) + z_\varphi) = p_\varphi^{-1}(x_L + z_\varphi) = \varphi.$$

It remains to prove the uniqueness. Thus suppose that also $\varphi = \varphi'_L + z'_\varphi$, where $\varphi'_L \in \Phi_L$ and $z'_\varphi \in M^{\perp}$. Then $p(\varphi'_L) + z'_\varphi = p(\varphi) = p(\varphi_L) + z_\varphi$, so $p(\varphi'_L) - p(\varphi_L) = z_\varphi - z'_\varphi$. But $p(\varphi'_L) - p(\varphi_L) \in L$, $z_\varphi - z'_\varphi \in M^{\perp}$, and $L \cap M^{\perp} = (0)$. Therefore $z_\varphi = z'_\varphi$, which gives $\varphi'_L = \varphi - z'_\varphi = \varphi - z_\varphi = \varphi_L$, proving the uniqueness. ♦

Observe that if $N'_{x_L} = N_{x_L}(\{v_i\}; \varepsilon) \cap L$ and $W'_{\varphi_L} = W_{\varphi_L}(\{v_i\}; \varepsilon) \cap \Phi_L$ then W'_{φ_L} is a p-neighborhood of φ_L in (Φ_L, p) and $p(W'_{\varphi_L}) = N'_{x_L}$. Also, the mapping $\varepsilon_L : \Phi \to \Phi_L$, $\varphi \mapsto \varphi_L$, is continuous. Thus we have the following corollary.

61.5 COROLLARY. *Let* $\Phi^M = \Phi$ *and define*

$$\tilde{p} : \Phi_L \times M^\perp \to E, \ (\varphi, z) \mapsto p(\varphi) + z$$

Then $(\Phi_L \times M^\perp, \tilde{p})$ *is an* $\langle E, F\rangle$*-domain. Also, if*

$$\mu : \Phi \to \Phi_L \times M^\perp, \ \varphi \mapsto (\varphi_L, z_\varphi)$$

then $\mu : (\Phi, p) \to (\Phi_L \times M^\perp, \tilde{p})$ *is an isomorphism of* $\langle E, F\rangle$*-domains.*

By the above corollary, if $\Phi^M = \Phi$ then we may, whenever it is convenient, replace the $\langle E, F\rangle$-domain (Φ, p) by $(\Phi_L \times M^\perp, \tilde{p})$. For the next proposition, recall that the Kronecker product of the algebras $\mathfrak{G}_{(\Phi_L, p)}$ and $\mathfrak{G}_{M^\perp}$, denoted by $\mathfrak{G}_{(\Phi_L, p)} \otimes \mathfrak{G}_{M^\perp}$, consists of all finite sums of products of an element from $\mathfrak{G}_{(\Phi_L, p)}$ with an element of $\mathfrak{G}_{M^\perp}$ regarded as an algebra of functions on $\Phi_L \times M^\perp$.

61.6 PROPOSITION.

(i) $\mathcal{P}_{\langle L, M\rangle} \otimes \mathcal{P}_{\langle M^\perp, L^\perp\rangle} = \mathcal{P}_{\langle E, F\rangle}$

(ii) $\mathfrak{G}_{(\Phi_L, p)} \otimes \mathfrak{G}_{M^\perp} \subseteq \mathfrak{G}_{(\Phi_L \times M^\perp, \tilde{p})}$

(iii) Fix a point $\varphi_0 \in \Phi_L$ *and for* $h \in \mathfrak{G}_{(\Phi_L \times M^\perp, \tilde{p})}$ *define* $h^{(1)}(\varphi) = h(\varphi, 0)$, $h^{(2)}(z) = h(\varphi_0, z)$. *Then* $h \mapsto h^{(1)}$ *and* $h \mapsto h^{(2)}$ *define homomorphisms of* $\mathfrak{G}_{(\Phi_L \times M^\perp, \tilde{p})}$ *onto* $\mathfrak{G}_{(\Phi_L, p)}$ *and* $\mathfrak{G}_{M^\perp}$ *respectively. Thus in particular* $\mathfrak{G}_{(\Phi_L, p)} = \mathfrak{G}_{(\Phi, p)}|\Phi_L$.

Proof. Since the proof is straightforward and easy it is omitted. ◆

If $\langle L, M\rangle \subseteq \langle E, F\rangle$, where L and M are finite dimensional as in the above discussion, then each $\langle L, M\rangle$-domain may be identified with a Riemann domain spread over $\mathbb{C}^n$, where n is the dimension of the space L. More precisely, let $\{(u_i, v_i)\}$ be a dual basis for $\langle L, M\rangle$. Then each $x \in L$ is of the form $x = \sum_{i=1}^{n} \langle x, v_i\rangle u_i$. Define

$$\check{v} : L \to \mathbb{C}^n, \ x \mapsto (\langle x, v_1\rangle, \ldots, \langle x, v_n\rangle) = \langle x, \check{v}\rangle.$$

Then $\check{v}$ is a linear homeomorphism which defines a pair isomorphism

$$\check{v} : [L, \mathcal{P}_{\langle L, M\rangle}] \to [\mathbb{C}^n, \mathcal{P}].$$

Therefore ${}_n\mathcal{O}\circ\check{v} = {}_{<L,\ M>}\mathcal{O}$. Thus we may identify an arbitrary $<L, M>$-domain (Φ, p) with the Riemann domain $(\Phi, \check{v}\circ p)$ spread over $\mathbb{C}^n$. Note that every Riemann domain over $\mathbb{C}^n$ is obtained in this way from an $<L, M>$-domain. We shall accordingly treat an $<L, M>$-domain directly as a Riemann domain without bothering in general to mention the homeomorphism $\check{v} : L \to \mathbb{C}^n$.

§62. SERIES EXPANSIONS

In order to simplify the statement and proof of the next lemma, we need to introduce a few definitions. Let (Φ, p) be an $<E, F>$-domain such that $\Phi^M = \Phi$ for a finite dimensional subspace M in F, and let h be a $\mathcal{P}_{<E,\ F>}$-holomorphic function defined on a connected open set $G \subseteq \Phi$ such that $G + M^\perp = G$. Then there exists a finite dimensional subspace M_h' in F, with $M \subseteq M_h'$, such that M_h' determines h on G (Lemma 60.2 (iv)). Choose a subspace $M' \subseteq M_h'$, such that $M \oplus M' = M_h'$, and subspaces L, L' and L_h' of E such that $L \oplus L' = L_h'$, $<L_h', M_h'> \subseteq <E, F>$, $<L, M> \subseteq <L_h', M_h'>$, and $<L', M'> \subseteq <L_h', M_h'>$. Also choose biorthogonal bases $\{(u_i, v_i)\}$ and $\{(u_j', v_j')\}$ for $<L, M>$ and $<L', M'>$ respectively. Then the union of these bases is a biorthogonal basis for $<L_h', M_h'>$. Since $G + M^\perp = G$ there exists for each $\varphi \in G$ a p-neighborhood $W_\varphi(\{v_i\}; \varepsilon) \subseteq G$. Define

$$G_L = G \cap \Phi_L = \{\varphi \in G : p(\varphi) \in L\}$$

and for $\varphi \in G_L$ set

$$W_\varphi^L(\varepsilon) = W_\varphi(\{v_i\}; \varepsilon) \cap \Phi_L$$

so W_φ^L is a p-neighborhood in (G_L, p). Also, for arbitrary $\rho > 0$ set

$$M^\perp(\rho) = \{z \in M^\perp;\ |<z, v_j'>| \le \rho, \text{ for each } j\}.$$

62.1 LEMMA. *Let* h *be a* $\mathcal{P}_{<E,\ F>}$*-holomorphic function defined on a connected open set* $G \subseteq \Phi$ *with* $G + M^\perp = G$. *Then for each multiple index* $\varkappa$ *associated with the basis* $\check{v}'$ *there exists* $h_\varkappa \in \mathcal{O}_{(G_L,\ p)}$ *such that*

$$h(\varphi + z) = \sum_\varkappa h_\varkappa(\varphi)<z, v'>^\varkappa$$

where the series converges for all $(\varphi, z) \in G_L \times M^\perp$. *Moreover, for arbitrary* $\rho > 0$ *and compact* $K_L \subset G_L$ *the convergence is uniform on* $K_L \times M^\perp(\rho)$.

Proof. In order to avoid complicated notations in dealing with the ambiguity in the derivative notations, we define temporarily

$$h(\varphi, z) = h(\varphi + z), \ (\varphi, z) \in G_L \times M^{\perp}.$$

Also, let $\varphi_0 \in G_L$ and choose $\varepsilon_0 > 0$ so that $\overline{W_{\varphi_0}(\{v_i\}; 2\varepsilon_0)}$ is a p-neighborhood contained in G and define $W^L_{\varphi_0}(2\varepsilon_0)$ as before. Then $\overline{W^L_{\varphi_0}(\varepsilon_0)} \subset W^L_{\varphi_0}(2\varepsilon_0)$ and $p(\overline{W^L_{\varphi_0}(\varepsilon_0)})$ is a bounded closed neighborhood in L. Now, by Corollary 57.3

$$h(\varphi, z) = \sum_{\varkappa^0, \varkappa} \frac{\mathcal{D}^{\varkappa^0}_{\check{v}} \mathcal{D}^{\varkappa}_{\check{v}'} h(\varphi_0, 0)}{\varkappa^0! \varkappa!} < p(\varphi) - p(\varphi_0), \check{v} >^{\varkappa^0} < z, \check{v}'>^{\varkappa}$$

where the series converges for all $(\varphi, z) \in W^L_{\varphi_0}(2\varepsilon_0) \times M^{\perp}$. Moreover, for arbitrary $\rho > 0$ the convergence is uniform for $(\varphi, z) \in W^L_{\varphi_0}(\varepsilon_0) \times M^{\perp}(\rho)$. We may accordingly sum the series in the following manner:

$$h(\varphi, z) = \sum_{\varkappa} \left(\frac{1}{\varkappa!} \sum_{\varkappa^0} \frac{\mathcal{D}^{\varkappa^0}_{\check{v}} \mathcal{D}^{\varkappa}_{\check{v}'} h(\varphi_0, 0)}{\varkappa^0!} < p(\varphi) - p(\varphi_0), \check{v} >^{\varkappa^0} \right) < z, \check{v}'>^{\varkappa}.$$

Observe that

$$\sum_{\varkappa^0} \frac{\mathcal{D}^{\varkappa^0}_{\check{v}} \mathcal{D}^{\varkappa}_{\check{v}'} h(\varphi_0, 0)}{\varkappa^0!} < p(\varphi) - p(\varphi_0), \check{v} >^{\varkappa^0} = \mathcal{D}^{\varkappa}_{\check{v}'} h(\varphi, 0).$$

Therefore, defining $h_{\varkappa}(\varphi) = \mathcal{D}^{\varkappa}_{\check{v}'} h(\varphi, 0)$ we have $h_{\varkappa} \in \mathcal{O}_{(G_L, p)}$. The argument applies for arbitrary $\varphi_0 \in G_L$ and $h_{\varkappa}$ is independent of φ_0, so

$$h(\varphi + z) = \sum_{\varkappa} h_{\varkappa}(\varphi) < z, \check{v}'>^{\varkappa}, \ (\varphi, z) \in G_L \times M^{\perp}.$$

Since for each $\varphi_0 \in G_L$ the convergence is uniform on $W^L_{\varphi_0}(\varepsilon_0) \times M^{\perp}(\rho)$ and an arbitrary compact set $K_L \subset G_L$ is covered by a finite number of the neighborhoods $W^L_{\varphi_0}(\varepsilon_0)$ the convergence is uniform on $K_L \times M^{\perp}(\rho)$. ♦

62.2 COROLLARY. *The algebra* $\mathcal{O}_{(\Phi_L, p)} \otimes \mathcal{P}_{<M^{\perp}, L^{\perp}>}$ *is dense in* $\mathcal{O}_{(\Phi_L \times M^{\perp}, p)}$ *with respect to the compact-open topology.*

62.3 LEMMA. *Assume that each of the function* $h_{\varkappa}$ *in the preceding lemma extends to an element* $\tilde{h}_{\varkappa} \in \mathcal{O}_{(\Phi_L, p)}$. *Also assume that the* $\mathcal{O}_{(\Phi_L, p)}$*-hull of* G_L *is equal to* Φ_L. *Then the function* h *may be extended to an element* $\tilde{h} \in \mathcal{O}_{(\Phi, p)}$.

Proof. Since (Φ_L, p) is a Riemann domain the $\mathcal{O}_{(\Phi_L, p)}$-hull of G_L is equal to $G_L^{\#}$, *i.e.* each point of Φ_L belongs to the $\mathcal{O}_{(\Phi_L, p)}$-hull $\hat{K}$ of some compact set $K \subseteq G_L$. Therefore let K be a compact set in G_L. As in the proof of the above lemma, for each $\varphi_0 \in K$ there exists $\varepsilon_0 > 0$ such that $W^L_{\varphi_0}(2\varepsilon_0)$ is a p-neighborhood contained in G_L and for arbitrary $\rho > 0$ the series

$$h(\varphi, z) = \sum_{\varkappa^0, \varkappa} \frac{\mathcal{D}_{\check{v}}^{\varkappa^0} \mathcal{D}_{\check{v}'}^{\varkappa} h(\varphi_0, 0)}{\varkappa^0! \varkappa!} < p(\varphi) - p(\varphi_0), \check{v} >^{\varkappa^0} < z, \check{v}' >^{\varkappa}$$

converges uniformly for $(\varphi, z) \in W^L_{\varphi_0}(2\varepsilon_0) \times M^{\perp}(\rho)$. Since K is compact, there exists a finite number of the neighborhoods $W^L_{\varphi_0}(\varepsilon_0)$, say $W^L_{\varphi_1}(\varepsilon_1), \ldots, W^L_{\varphi_n}(\varepsilon_n)$, that cover K. Set $\varepsilon = \min(\varepsilon_1, \ldots, \varepsilon_n)$. If $\varphi \in W^L_{\varphi_k}(\varepsilon)$ then $\overline{W^L_{\varphi}(\varepsilon)} \subset W^L_{\varphi_k}(2\varepsilon)$. Define

$$K(\varepsilon) = \overline{\bigcup_{\varphi \in K} W^L_{\varphi}(\varepsilon)} \subseteq \bigcup_{k=1}^{n} \overline{W^L_{\varphi_k}(2\varepsilon)}.$$

Since L is finite dimensional each $\overline{W^L_{\varphi_k}(2\varepsilon)}$ is compact, so it follows that $K(\varepsilon)$ is also compact. Now, for arbitrary $\rho > 0$ define

$$K(\varepsilon, \rho) = K(\varepsilon) + M^{\perp}(\rho) \cap L'.$$

Since the sets $K(\varepsilon)$ and $M^{\perp}(\rho) \cap L'$ are compact the set $K(\varepsilon, \rho)$ is also compact. Therefore $\beta(\varepsilon, \rho) = |h|_{K(\varepsilon, \rho)} < \infty$. Hence by the Cauchy inequalities

$$\left| \frac{\mathcal{D}_{\check{v}}^{\varkappa^0} \mathcal{D}_{\check{v}'}^{\varkappa} h(\varphi_0, 0)}{\varkappa^0! \varkappa!} \right| \leq \frac{\beta(\varepsilon, \rho)}{\varepsilon^{|\varkappa^0|} \rho^{|\varkappa|}}, \quad \varphi_0 \in K.$$

Observe that

$$\frac{\mathcal{D}_{\check{v}}^{\varkappa^0} \mathcal{D}_{\check{v}'}^{\varkappa} h(\varphi_0, 0)}{\varkappa^0!} = \frac{\mathcal{D}_{\check{v}}^{\varkappa^0} h_{\varkappa}(\varphi_0)}{\varkappa^0!}$$

so

$$\left| \frac{\mathcal{D}_{\check{v}}^{\varkappa^0} h_{\varkappa}}{\varkappa^0!} \right|_K = \left| \frac{\mathcal{D}_{\check{v}}^{\varkappa^0} h_{\varkappa}}{\varkappa^0!} \right|_K \leq \frac{\beta(\varepsilon, \rho)}{\varepsilon^{|\varkappa^0|} \rho^{|\varkappa|}}.$$

Therefore, since $\mathcal{D}_{\check{v}}^{\varkappa^0} \tilde{h}_{\varkappa} \in \mathcal{O}_{(\Phi_L, \rho)}$ it follows that $\varphi_0 \in \hat{K} \subset \Phi_L$ implies

$$\left| \frac{\mathcal{D}_{\check{v}}^{\varkappa^0} \tilde{h}_{\varkappa}(\varphi_0)}{\varkappa^0!} \right| \leq \frac{\beta(\varepsilon, \rho)}{\varepsilon^{|\varkappa^0|} \rho^{|\varkappa|}}$$

so the series

$$\sum_{\varkappa^0,\ \varkappa} \frac{\mathcal{D}^{\varkappa^0}_{\check{v}} \tilde{h}_{\varkappa}(\varphi_0)}{\varkappa^0!} < p(\varphi) - p(\varphi_0),\ \check{v} >^{\varkappa^0} < z,\ \check{v}' >^{\varkappa}$$

converges uniformly for $(\varphi,\ z) \in W^L_{\varphi_0}(\varepsilon') \times M^{\perp}(\rho)$, where $\rho > 0$, ε is chosen so that $W^L_{\varphi_0}(\varepsilon)$ is a p-neighborhood in Φ_L, and $0 < \varepsilon' < \varepsilon$. Therefore, summing first with respect to $\varkappa^0$ we obtain the series $\sum_{\varkappa} \tilde{h}_{\varkappa}(\varphi) < z,\ \check{v}' >^{\varkappa}$ which converges uniformly for $(\varphi,\ z) \in W^L_{\varphi_0}(\varepsilon') \times M^{\perp}(\rho)$ and hence defines a holomorphic function. Moreover this is true for all $\rho > 0$ and $\varphi_0 \in \hat{K} \subset \Phi_L$. Therefore, since the $\mathfrak{G}_{(\Phi_L,\ p)}$-hull of G_L is equal to Φ_L, the function

$$\tilde{h}(\varphi + z) = \sum_{\varkappa} \tilde{h}_{\varkappa}(\varphi) < z,\ \check{v}' >^{\varkappa}$$

is defined and holomorphic for all $(\varphi,\ z) \in \Phi_L \times M^{\perp}$. Since $\Phi_L + M^{\perp} = \Phi$ the function $\tilde{h}$ is an extension of h to all of Φ. ◆

§63. THE FINITE DIMENSIONAL COMPONENT OF A DOMAIN OF HOLOMORPHY

We prove next the following important result for <E, F>-domains of holomorphy.

63.1 THEOREM. *Let* $(\Phi,\ p)$ *be an* <E, F>*-domain of holomorphy for which* $\mathfrak{G}_{(\Phi,\ p)}$ *separates points and choose* $\langle L, M\rangle \subseteq \langle E, F\rangle$ *such that* $\Phi^M = \Phi$. *Then* $(\Phi_L,\ p)$ *is an* <L, M>*-domain of holomorphy.*

Proof. By Proposition 61.4, $(\Phi_L,\ p)$ is an <L, M>-domain and is thus essentially a Riemann domain spread over $\mathbb{C}^n$, where $n = \dim L$. Let $\rho : (\Phi_L,\ p) \Rightarrow (\Omega,\ q)$ be a maximal $\mathfrak{P}_{\langle L,\ M\rangle}$-holomorphic extension of $(\Phi_L,\ p)$. Then $(\Omega,\ q)$ is essentially a Riemann domain of holomorphy. Also, since $\mathfrak{G}_{(\Phi,\ p)}$ separates points in Φ the mapping $\rho : \Phi_L \to \Omega$ is a homeomorphism of Φ_L onto an open subset of Ω (Lemma 48.1). We must prove that $\rho(\Phi_L) = \Omega$.

Consider the space $\Omega \times M^{\perp}$ and the projection

$$\tilde{q} : \Omega \times M^{\perp} \to E,\ (\omega,\ z) \mapsto q(\omega) + z.$$

Then $(\Omega \times M^{\perp},\ \tilde{q})$ is obviously an <E, F>-domain. Consider also the <E, F>-domain $(\Phi_L \times M^{\perp},\ \tilde{p})$ of Corollary 61.5 and define

$$\tilde{\rho} : \Phi_L \times M^\perp \to \Omega \times M^\perp, \ (\varphi, z) \mapsto (\rho(\varphi), z)$$

Then $\tilde{q}\circ\tilde{\rho}(\varphi, z) = \tilde{q}(\rho(\varphi), z) = (q\circ\rho)(\varphi) + z = p(\varphi) + z = \tilde{p}(\varphi, z)$, so $\tilde{q}\circ\tilde{\rho} = \tilde{p}$, proving that $\tilde{\rho}$ defines a morphism

$$\tilde{\rho} : (\Phi_L \times M^\perp, \tilde{p}) \to (\Omega \times M^\perp, \tilde{q})$$

of <E, F>-domains. Note that $\tilde{\rho}$ maps $\Phi_L \times M^\perp$ homeomorphically onto the open set $\rho(\Phi_L) \times M^\perp$ in $\Omega \times M^\perp$. We prove next that this morphism is actually a holomorphic extension, i.e.

$$\mathcal{O}_{(\Omega \times M^\perp, \tilde{q})}\circ\tilde{\rho} = \mathcal{O}_{(\Phi_L \times M^\perp, \tilde{p})}.$$

Let $h \in \mathcal{O}_{(\Phi_L \times M^\perp, \tilde{p})}$ and apply Lemma 62.1 to obtain h in the form

$$h(\varphi, z) = \sum_{\varkappa} h_\varkappa(\varphi)< z, \check{v}'>^\varkappa$$

where $h_\varkappa \in \mathcal{O}_{(\Phi_L, p)}$ and the series is locally uniformly convergent in $\Phi_L \times M^\perp$. Since $\mathcal{O}_{(\Omega, q)}\circ\rho = \mathcal{O}_{(\Phi_L, p)}$ there exists for each $\varkappa$ a function $\tilde{g}_\varkappa \in \mathcal{O}_{(\Omega, q)}$ such that $\tilde{g}_\varkappa\circ\rho = h_\varkappa$. Consider the series

$$g(\omega, z) = \sum_{\varkappa} \tilde{g}_\varkappa(\omega)< z, \check{v}'>^\varkappa .$$

Since $\rho : \Phi_L \to \Omega$ is a homeomorphism the series converges locally uniformly in $\rho(\Phi_L) \times M^\perp$ to an element of $\mathcal{O}_{(\rho(\Phi_L) \times M^\perp, \tilde{q})}$. Now (Ω, q) is the envelope of holomorphy of the Riemann domain $(\rho(\Phi_L), q)$, so the holomorphic hull of $\rho(\omega_L)$ in Ω is equal to Ω. We may therefore apply Lemma 62.3 to obtain an extension of $g(\omega, z)$ to an element $\tilde{g} \in \mathcal{O}_{(\Omega \times M^\perp, \tilde{q})}$. Moreover $g\circ\tilde{\rho} = g\circ\tilde{\rho} = h$ which proves that $\tilde{\rho}$ defines an extention $\tilde{\rho} : (\Phi_L \times M^\perp, \tilde{p}) \Rightarrow (\Omega \times M^\perp, \tilde{q})$. Since $(\Phi_L \times M^\perp, \tilde{p})$ is isomorphic with (Φ, p) it is a maximal <E, F>-domain. Therefore $\tilde{\rho}$ must be a surjective homeomorphism and this implies that $\rho : \Phi_L \to \Omega$ is also a surjective homeomorphism. ♦

§64. THE ALGEBRA OF HOLOMORPHIC FUNCTIONS

64.1 THEOREM. *Let* (Φ, p) *be an arbitrary connected* <E, F>*-domain for which* $\mathcal{O}_{(\Phi, p)}$ *separates points. Then* $[\Phi, \mathcal{O}_{(\Phi, p)}]$ *is a system.*

Proof. The problem is to prove that the $\mathcal{O}_{(\Phi, p)}$-topology is equivalent to the given topology in Φ. Observe first that since $\mathcal{O}_{(\Phi, p)}$ separates points (Φ, p) is contained (*via* an isomorphism) in its envelope of $\mathcal{P}_{<E, F>}$-holomorphy. Hence there is no loss in assuming that (Φ, p) is itself an <E, F>-domain of holomorphy. Now by Corollary 61.5 and Theorem 63.1 the <E, F>-domain (Φ, p) may be replaced by $(\Phi_L \times M^\perp, \tilde{p})$, where (Φ_L, p) is an <L, M>-domain of holomorphy. Also, since the Kronecker product $\mathcal{O}_{(\Phi_L, p)} \otimes \mathcal{O}_{M^\perp}$ is contained in $\mathcal{O}_{(\Phi_L \times M^\perp, \tilde{p})}$ it will be sufficient to prove that $[\Phi_L \times M^\perp, \mathcal{O}_{(\Phi_L, p)} \otimes \mathcal{O}_{M^\perp}]$ is a system. Observe that this pair is simply the direct product (§4) of the two pairs $[\Phi_L \times \mathcal{O}_{(\Phi_L, p)}]$ and $[M^\perp, \mathcal{O}_{M^\perp}]$, so the problem is reduced to showing that they are systems. In the case of $[M^\perp, \mathcal{O}_{M^\perp}]$, we note that the topology in $M^\perp$ is the $\sigma(E, F)$-topology which is obviously equivalent to the $\mathcal{P}_{<E, F>}$-topology and hence to the $\mathcal{O}_{M^\perp}$-topology. Thus $[M^\perp, \mathcal{O}_{M^\perp}]$ is a system. Finally, since (Φ_L, p) is essentially a Riemann domain of holomorphy the fact that $[\Phi_L, \mathcal{O}_{(\Phi_L, p)}]$ is a system is a consequence of well-known results from SCV. (See, for example, [N2, p. 130; G7, Theorem 10, p. 224].) ♦

64.1 THEOREM. *Let* (Φ, p) *be an arbitrary connected* <E, F>*-domain for which* $\mathcal{O}_{(\Phi, p)}$ *separates points. Then* (Φ, p) *will be a domain of holomorphy iff there exists a finite dimensional* <L, M> $\subseteq$ <E, F> *such that* $\Phi^M = \Phi$ *and* (Φ_L, p) *is an* <L, M>*-domain of holomorphy.*

Proof. That the condition is satisfied for a domain of holomorphy is given by Theorem 63.1. Therefore assume that the desired <L, M> exists and let $\rho : (\Phi, p) \Rightarrow (\Psi, q)$ be a connected maximal extension of (Φ, p). Since $\mathcal{O}_{(\Phi, p)}$ separates points the mapping $\rho : \Phi \mapsto \Psi$ is a homeomorphism. The problem is to prove that it is surjective.

Since ρ maps basic p-neighborhoods in Φ homeomorphically onto basic q-neighborhoods in Ψ it follows that $\Psi^M \neq \emptyset$ and hence, by Corollary 61.3, that $\Psi^M = \Psi$. Therefore $\Psi = \Psi_L \oplus M^\perp$ by Proposition 61.4. Also, since $p = q \circ \rho$ it follows that $\rho(\Phi_L) \subseteq \Psi_L$. Now, by Proposition 61.6 (iii), if $\rho_L = \rho | \Phi_L$ then

$$\mathcal{O}_{(\Psi_L, q)} \circ \rho_L = (\mathcal{O}_{(\Psi, q)} \circ \rho) | \Phi_L = \mathcal{O}_{(\Phi, p)} | \Phi_L = \mathcal{O}_{(\Phi_L, p)}.$$

Thus ρ induces on (Φ_L, p) an extension $\rho_L : (\Phi_L, p) \Rightarrow (\Psi_L, q)$. Since (Φ_L, p)

is a domain of holomorphy the mapping $\rho_L : \Phi_L \to \Psi_L$ must be a surjective homeomorphism. Next define

$$\rho' : \Psi \to \Phi, \ \psi \mapsto \rho_L^{-1}(\psi_L) + z_\psi$$

where $\psi = \psi_L + z_\psi$ is the $\Psi_L \oplus M^\perp$ decomposition of the element ψ. By definition of the sum $\rho_L^{-1}(\psi_L) + z_\psi$ we have

$$p(\rho_L^{-1}(\psi_L) + z_\psi) = p(\rho_L^{-1}(\psi_L)) + z_\psi$$

and similarly

$$q(\psi) = q(\psi_L + z_\psi) = q(\psi_L) + z_\psi.$$

Also, since $p = q \circ \rho$, it follows that $p(\rho_L^{-1}(\psi)) = q(\psi)$ for $\psi \in \Psi_L$. Therefore

$$(p \circ \rho')(\psi) = p(\rho_L^{-1}(\psi_L) + z_\psi) = p(\rho_L^{-1}(\psi_L)) + z_\psi$$

$$= q(\psi_L) + z_\psi = q(\psi)$$

for arbitrary $\psi \in \Psi$. In other words, ρ' defines a morphism $\rho' : (\Psi, q) \to (\Phi, p)$. Finally, if $\varphi \in \Phi_L$ then $\rho(\varphi) \in \Psi_L$, so $\rho'(\rho(\varphi)) = \rho_L^{-1}(\rho_L(\varphi)) = \varphi$. Since Φ and Ψ are connected, it follows by Corollary 48.3 that the mapping $\rho : \Phi \to \Psi$ must be a surjective homeomorphism. ◆

§65. HOLOMORPHIC CONVEXITY AND NATURALITY

65.1 THEOREM. *Let* (Φ, p) *be an arbitrary* $\langle E, F\rangle$*-domain of holomorphy. Then* (Φ, p) *will be* $\mathcal{P}_{\langle E, F\rangle}$*-holomorphically convex iff* E *is* $\mathcal{P}_{\langle E, F\rangle}$*-holomorphically convex.*

Proof. Choose finite dimensional $\langle L, M\rangle \subseteq \langle E, F\rangle$ such that $\Phi^M = \Phi$. Then again by Corollary 61.5 we may replace (Φ, p) by $(\Phi_L \times M^\perp, \tilde{p})$. Assume first that E is holomorphically convex and let K be an arbitrary compact subset of $\Phi_L \times M^\perp$. Let K_L and Z_K be the projections of K into Φ_L and $M^\perp$ respectively. Then K_L and Z_K are compact and $\hat{K} \subseteq \widehat{K_L \times Z_K} \subseteq \hat{K}_L \times \hat{Z}_K$, where the indicated hulls are with respect to holomorphic functions on the corresponding spaces $\Phi_L \times M^\perp$, Φ_L and $M^\perp$ respectively. Since by Theorem 64.2 the $\langle L, M\rangle$-domain (Φ_L, p) is essentially a Riemann domain of holomorphy, the hull $\hat{K}_L$ is compact. Observe also that since

$M^{\perp} = \{x \in E : \langle x, v\rangle = 0, \text{ for } v \in M\}$ it follows that $M^{\perp}$ is a $\mathcal{P}_{\langle E, F\rangle}$-convex subset of E. Therefore the holomorphic hull of Z_K in E is contained in $M^{\perp}$. Since this is compact and contains $\hat{Z}_K$ it follows that $\hat{Z}_K$ is compact. Therefore $\hat{K}$ must be compact, proving that (Φ, p) is $\mathcal{P}_{\langle E, F\rangle}$-holomorphically convex.

Now assume that (Φ, p) is $\mathcal{P}_{\langle E, F\rangle}$-holomorphically convex. Since L is finite dimensional the hull $\hat{K}_L$ is automatically compact. Now fix a point $\varphi_0 \in \Phi_L$. Then $\{\varphi_0\} \times Z_K$ is a compact subset of $\Phi_L \times M^{\perp}$, so the hull $\widehat{\{\varphi_0\} \times Z_K}$ is compact. By Proposition 61.6 (iii) $\widehat{\{\varphi_0\} \times Z_K} = \{\varphi_0\} \times \hat{Z}_K$. Therefore $\hat{Z}_K$ is compact and hence $\hat{K}_L \times \hat{Z}_K$ is compact. This implies that $\hat{K}$ is compact, so E is holomorphically convex. ◆

65.2 THEOREM. *Let* (Φ, p) *be a connected* $\langle E, F\rangle$*-domain for which* $\mathcal{O}_{(\Phi, p)}$ *separates points and there exists* $M \subseteq F$ *such that* $\Phi^M = \Phi$. *If* (Φ, p) *is* $\mathcal{P}_{\langle E, F\rangle}$*-holomorphically convex then* (Φ, p) *will be an* $\langle E, F\rangle$*-domain of holomorphy.*

Proof. Choose $L \subseteq E$ so that $\langle L, M\rangle \subseteq \langle E, F\rangle$ and identify (Φ, p) with $(\Phi_L \times M^{\perp}, \tilde{p})$. By Theorem 64.2, we have only to prove that $\mathcal{P}_{\langle E, F\rangle}$-holomorphic convexity of (Φ, p) implies that (Φ_L, p) is an $\langle L, M\rangle$-domain of holomorphy. Since (Φ_L, p) is essentially a Riemann domain it will suffice to prove that (Φ_L, p) is $\mathcal{P}_{\langle L, M\rangle}$-holomorphically convex. Therefore let $K \subset\subset \Phi_L$ and denote by $\hat{K}$ its $\mathcal{O}_{(\Phi_L, p)}$-convex hull in Φ_L. Consider an arbitrary point $\varphi_L \in \hat{K}$ and let $h \in \mathcal{O}_{(\Phi_L \times M^{\perp}, \tilde{p})}$. As in Proposition 61.6, set $h^{(1)}(\varphi) = h(\varphi, 0)$ for $\varphi \in \Phi_L$. Then $h^{(1)} \in \mathcal{O}_{(\Phi_L, p)}$, so

$$|h(\varphi_L, 0)| = |h^{(1)}(\varphi_L)| \le |h^{(1)}|_K = |h|_{K \times \{0\}}.$$

Therefore $(\varphi_L, 0)$ belongs to the hull $\widehat{K \times \{0\}}$ in $\Phi_L \times M^{\perp}$. Thus $\hat{K} \times \{0\} \subseteq \widehat{K \times \{0\}}$, which implies that $\hat{K}$ is compact. In other words, (Φ_L, p) is $\mathcal{P}_{\langle E, F\rangle}$-holomorphically convex. ◆

Using Theorem 65.1, we obtain the following corollary.

65.3 COROLLARY. *If* (Φ, p) *is* $\mathcal{P}_{\langle E, F\rangle}$*-holomorphically convex, then* E *is* $\mathcal{P}_{\langle E, F\rangle}$*-holomorphically convex.*

65.4 THEOREM. *Let* (Φ, p) *be an arbitrary* $\langle E, F\rangle$*-domain of holomorphy and choose* $\langle L, M\rangle \subseteq \langle E, F\rangle$ *such that* $\Phi^M = M$. *Then the following are true:*

(i) A continuous homomorphism $\chi : \mathcal{O}_{(\Phi, p)} \to \mathbb{C}$, $h \mapsto \hat{h}(\chi)$, *will be a point evaluation in* Φ *iff the induced homomorphism* $\chi_0 : \mathcal{P}_{\langle E, F\rangle} \to \mathbb{C}$, $P \mapsto \widehat{P\circ p}(\chi)$, *is a point evaluation in* E.

(ii) The system $[\Phi, \mathcal{O}_{(\Phi, p)}]$ *will be natural iff* $[E, \mathcal{O}_E]$ *is natural.*

Proof. It is obvious that if χ is a point evaluation in Φ then χ_0 is a point evaluation in E. Therefore assume that χ_0 is a point evaluation. Replace (Φ, p) by $(\Phi_L \times M^\perp, \tilde{p})$ and consider the two induced homomorphisms

$$\chi_L : \mathcal{O}_{(\Phi_L, p)} \to \mathbb{C}, \ \chi_{M^\perp} : \mathcal{P}_{\langle M^\perp, L^\perp\rangle} \to \mathbb{C}$$

obtained by restricting χ to the isomorphic images of $\mathcal{O}_{(\Phi_L, p)}$ and $\mathcal{P}_{\langle M^\perp, L^\perp\rangle}$ in $\mathcal{O}_{(\Phi_L \times M^\perp, \tilde{p})}$ given by Proposition 61.6 (ii). Note that these homomorphisms are continuous. In fact, if χ is dominated by a compact set $K \subseteq \Phi$ and K_L, K_L' are compact sets contained respectively in Φ_L, $M^\perp$ such that $K \subseteq K_L \times K_L'$ then χ_L, $\chi_{M^\perp}$ are dominated respectively by K_L, K_L'. Since (Φ_L, p) is essentially a Riemann domain of holomorphy χ_L is automatically a point evaluation in Φ_L. Furthermore, since $\mathcal{P}_{\langle E, F\rangle} = \mathcal{P}_{\langle L, M\rangle} \otimes \mathcal{P}_{\langle M^\perp, L^\perp\rangle}$ and every homomorphism of $\mathcal{P}_{\langle L, M\rangle}$ is a point evaluation in L the condition that χ_0 be a point evaluation in E is equivalent to the condition that $\chi_{M^\perp}$ be a point evaluation in $M^\perp$. Hence the restriction of χ to $\mathcal{O}_{(\Phi_L, p)} \otimes \mathcal{P}_{\langle M^\perp, L^\perp\rangle}$ is a point evaluation in $\Phi_L \times M^\perp$. Therefore, by continuity and Corollary 62.2, it follows that χ is itself a point evaluation in $\Phi_L \times M^\perp$, completing the proof of (i).

In the case of (ii), we observe first that each continuous homomorphism $\chi : \mathcal{O}_{(\Phi, p)} \to \mathbb{C}$ induces a continuous homomorphism

$$f \to f\circ p \to \widehat{f\circ p}(\varphi), \ f \in \mathcal{O}_E$$

of $\mathcal{O}_E$ onto $\mathbb{C}$. Hence by property (i) naturality of $[E, \mathcal{O}_E]$ implies naturality of $[\Phi, \mathcal{O}_{(\Phi, p)}]$. For the opposite implication, assume that $[\Phi, \mathcal{O}_{(\Phi, p)}]$, and hence $[\Phi_L \times M^\perp, \mathcal{O}_{(\Phi_L \times M^\perp, \tilde{p})}]$, is natural. Since L is finite dimensional the system $[L, \mathcal{O}_L]$ is always natural. Therefore if we show that $[M^\perp, \mathcal{O}_{M^\perp}]$ is natural then the product system $[L \times M^{\perp}, \mathcal{O}_L \otimes \mathcal{O}_M]$ will be natural. Hence by Theorem 40.3 the system $[L \times M^{\perp}, \mathcal{O}_{L \times M^{\perp}}]$ must be natural. But the latter system is isomorphic with $[E, \mathcal{O}_E]$, so the problem is reduced to proving that $[M^{\perp}, \mathcal{O}_{M^{\perp}}]$ is natural.

Let $\chi : \mathcal{O}_{M^{\perp}} \to \mathbb{C}$ be a continuous homomorphism of $\mathcal{O}_{M^{\perp}}$ and, as in Proposition 61.6 (iii), for fixed $\varphi_0 \in \Phi_L$ and $h \in \mathcal{O}_{(\Phi_L \times M^{\perp}, \tilde{p})}$ define $h^{(2)}(z) = h(\varphi_0, z)$, $z \in M^{\perp}$. Then $h^{(2)} \in \mathcal{O}_{M^{\perp}}$ and the map

$$\chi' : \mathcal{O}_{(\Phi_L \times M^{\perp}, \tilde{p})} \to \mathbb{C}, \; h \mapsto \widehat{h^{(2)}}(\chi)$$

defines a continuous homomorphism of $\mathcal{O}_{(\Phi_L \times M^{\perp}, \tilde{p})}$. Hence there exists $(\varphi_\chi, z_\chi) \in \Phi_L \times M^{\perp}$ such that $\widehat{h^{(2)}}(\chi) = h(\varphi_\chi, z_\chi)$, $h \in \mathcal{O}_{(\Phi_L \times M^{\perp}, \tilde{p})}$. Finally, let $g \in \mathcal{O}_{M^{\perp}}$ and define $\tilde{g}(\varphi, z) = g(z)$, $(\varphi, z) \in \Phi_L \times M^{\perp}$. Then $(\tilde{g})^{(2)} = g$. Therefore

$$\hat{g}(\chi) = \widehat{(\tilde{g})^{(2)}}(\chi) = \tilde{g}(\varphi_\chi, z_\chi) = g(z_\chi)$$

so $[M^{\perp}, \mathcal{O}_{M^{\perp}}]$ is natural. ◆

65.5 THEOREM. *Let* (Φ, p) *be a connected* $\langle E, F\rangle$*-domain for which* $\mathcal{O}_{(\Phi, p)}$ *separates points and there exists* $M \subseteq F$ *such that* $\Phi^M = \Phi$. *Also assume that* $[E, \mathcal{O}_E]$ *is natural. Then the following conditions are equivalent:*

(i) (Φ, p) *is an* $\langle E, F\rangle$*-domain of holomorphy.*

(ii) $[\Phi, \mathcal{O}_{(\Phi, p)}]$ *is natural.*

(iii) (Φ, p) *is* $\mathcal{P}_{\langle E, F\rangle}$*-holomorphically convex.*

(iv) (Φ, p) *is* $\mathcal{P}_{\langle E, F\rangle}$*-subharmonically convex.*

Proof. The implication (i) ⇒ (ii) is given by Theorem 65.4 (ii). Since $[\Phi, \mathcal{O}_{(\Phi, p)}]$ is a system (Theorem 64.1) the implication (ii) ⇒ (iii) is given by the general result in Theorem 7.3. Also the implication (iii) ⇒ (iv) is trivial. It remains to prove that (iv) ⇒ (i).

Choose $L \subseteq E$ with $\langle L, M\rangle \subseteq \langle E, F\rangle$ and identify (Φ, p) with $(\Phi_L \times M^{\perp}, \ddot{p})$ so (Φ_L, p) is identified with $(\Phi_L \times \{0\}, \tilde{p})$. As in Proposition 61.6 (iii), we associate with each function f defined on $\Phi_L \times M^{\perp}$ the function $f^{(1)}(\varphi) = f(\varphi, 0)$ on Φ_L. (This amounts to restricting a function defined on Φ to the subset Φ_L.) Then the mapping $f \to f^{(1)}$ carries the set $\mathcal{CS}_\Phi$ of continuous $\mathcal{P}_{\langle E, F\rangle}$-sh functions defined on Φ onto the set $\mathcal{CS}_{\Phi_L}$ of continuous $\mathcal{P}_{<L, M>}$-sh functions defined on Φ_L. Therefore condition (iv) implies that (Φ_L, p) is $\mathcal{P}_{<L, M>}$-sh convex. But, since (Φ_L, p) is essentially a Riemann domain $\mathcal{P}_{<L, M>}$-sh convexity is equivalent to p-convexity which in turn implies that (Φ_L, p) is an <L, M>-domain of holomorphy [G7, Theorem 4, p. 283]. Hence (Φ, p) is an <E, F>-domain of holomorphy, by Theorem 64.2.♦

65.6 COROLLARY. *Let* (Φ, p) *be a connected* $[C^\Lambda, \mathcal{P}]$*-domain of holomorphy. Then* $[\Phi, \mathcal{O}_{(\Phi, p)}]$ *is natural, so* (Φ, p) *is* $\mathcal{P}$*-holomorphically convex.*

§66. A CARTAN-THÜLLEN THEOREM

We close this chapter with a result for <E, F>-domains involving the distance functions d_α introduced in Definition 45.1. In the case of <E, F>-domains, these functions are direct generalizations of the usual distance function for Riemann domains [N2, p. 106], so are accordingly quite well-behaved. Thus, with the machinery developed above, results for Riemann domains extend without much difficulty to <E, F>-domains. As indicated in the introduction, we hope to give a more extensive treatment of this special subject at another time.

Observe first that if A is a basis for the vector space F then the linear functionals on E corresponding to elements of A constitute a linearly independent system of generators for the algebra $\mathcal{P}_{<E, F>}$ of continuous generalized polynomials on E. Thus, if (Φ, p) is an arbitrary <E, F>-domain and α is a finite subset of A then the distance function $d_\alpha(\varphi)$ is defined and $0 \leq d_\alpha(\varphi) \leq \infty$ for each $\varphi \in \Phi$. Recall that $d_\alpha(\varphi)$ is either equal to zero or to the supremum of all $r > 0$ such that the basic neighborhood $W_\varphi(\alpha, r)$ is defined. Moreover, if $d_\alpha(\varphi) > 0$ then the basic neighborhood $W_\varphi(\alpha, d_\alpha(\varphi))$ is also defined. Recall also that for any set $S \subseteq \Phi$

$$\delta_\alpha(S) = \inf\{d_\alpha(\varphi) : \varphi \in S\}.$$

If M_α denotes the subspace of F spanned by the elements of α then

$$\Phi^{M_\alpha} = \{\varphi \in \Phi : d_\alpha(\varphi) > 0\}$$

so $\Phi^{M_\alpha} = \Phi$ iff $d_\alpha(\varphi) > 0$ for each $\varphi \in \Phi$. If $\rho: (\Phi, p) \to (\Psi, q)$ is any

$\langle E, F\rangle$-domain morphism then clearly $d_\alpha(\varphi) \leq d_\alpha(\rho(\varphi))$ for each $\varphi \in \Phi$ and arbitrary α. Now assume that $0 < d_\alpha(\varphi)$ for each $\varphi \in \Phi$. Set $M = M_\alpha$ and choose $L \subseteq E$ such that $\langle L, M\rangle \subseteq \langle E, F\rangle$. If $\varphi = \varphi_L \oplus z_\varphi$, where $\varphi_L \in \Phi_L$ and $z_\varphi \in M^\perp$, then $W_\varphi(\alpha, r) = W_{\varphi_L}(\alpha, r)$ provided either one of the basic neighborhoods is defined. Therefore $d_\alpha(\varphi) = d_\alpha(\varphi_L)$ for each $\varphi \in \Phi$, and if S is any subset of Φ then $\delta_\alpha(S) = \delta_\alpha(S_L)$, where S_L denotes the projection of S into L. Note that the restriction of d_α to Φ_L is essentially the usual distance function for a Riemann domain. This fact enables us to obtain generalizations of the classical Cartan-Thüllen results for Riemann domains.

66.1 LEMMA. *If* (Φ, p) *is an* $\langle E, F\rangle$*-domain of holomorphy then there exists* α *such that* $0 < d_\alpha(\varphi)$ *for each* $\varphi \in \Phi$.

Proof. By Corollary 61.3, there exists a finite dimensional space $M \subseteq F$ such that $\Phi^M = \Phi$. Since A is a basis for F, there exists a finite set $\alpha \subseteq A$ such that $M \subseteq M_\alpha$, so also $\Phi^{M_\alpha} = \Phi$ and hence $0 < d_\alpha(\varphi)$ for each $\varphi \in \Phi$. ♦

66.2 THEOREM. *Let* (Φ, p) *be a connected* $\langle E, F\rangle$*-domain for which* $\mathcal{O}_{(\Phi, p)}$ *separates points. Then the following conditions are equivalent:*

(i) (Φ, p) *is an* $\langle E, F\rangle$*-domain of holomorphy.*

(ii) There exists α *such that* $\delta_\alpha(K) = \delta_\alpha(\hat{K})$ *for each compact set* $K \subseteq \Phi$.

(iii) There exists α *such that* $\delta_\alpha(\hat{K}) > 0$ *for each compact set* $K \subseteq \Phi$.

(iv) There exists α *such that the function* $-\log d_\alpha(\varphi)$ *is* $\mathcal{P}_{\langle E, F\rangle}$*-subharmonic on* Φ.

Proof. Assume that (Φ, p) is a domain of holomorphy. Then by the lemma there exists α such that $\Phi^{M_\alpha} = \Phi$. Set $M = M_\alpha$ and choose $L \subseteq E$ such that $\langle L, M\rangle \subseteq \langle E, F\rangle$. Then, as already noted, $d_\alpha(\varphi) = d_\alpha(\varphi_L)$ for each $\varphi \in \Phi$. Also, by Theorem 64.2 (Φ_L, p) is an $\langle L, M\rangle$-domain of holomorphy. Next let K be an arbitrary compact subset of Φ and denote by K_L, Z_K its projections into Φ_L, $M^\perp$ respectively. Then, as in the proof of Theorem 65.1, $\hat{K} \subseteq \hat{K}_L \oplus \hat{Z}_K$, where $\hat{K}_L$ and $\hat{Z}_K$ are the hulls of K_L and Z_K with respect to the holomorphic functions in Φ_L and $M^\perp$ respectively. Therefore

$$\delta_\alpha(K) \geq \delta_\alpha(\hat{K}) \geq \delta_\alpha(\hat{K}_L \oplus \hat{Z}_K) = \delta_\alpha(\hat{K}_L).$$

Furthermore by the classical Cartan-Thüllen results [N2, p. 109] $\delta_\alpha(\hat{K}_L) = \delta_\alpha(K_L)$. Since $\delta_\alpha(K_L) = \delta_\alpha(K)$ it follows that $\delta_\alpha(K) = \delta_\alpha(\hat{K})$, proving that (i) implies (ii). Also, since $\delta_\alpha(K) > 0$ for compact sets, (ii) implies (iii). Now observe that if (iii) is satisfied then $d_\alpha(\varphi) > 0$ for each $\varphi \in \Phi$, so $\Phi^{M_\alpha} = \Phi$. Moreover if (Φ, p) satisfies condition (iii) then so does the <L, M>-domain (Φ_L, p) and therefore, again by Cartan-Thüllen [N2, p. 114], (Φ_L, p) is a domain of holomorphy. Hence by Theorem 65.1, (Φ, p) is a domain of holomorphy, proving that (iii) implies (i). It only remains to prove the equivalence of (i) and (iv).

Observe first that if the function $-\log d_\alpha(\varphi)$ is $\mathcal{P}_{<E,\ F>}$-sh in Φ then by definition it does not assume the value $+\infty$. Therefore $d_\alpha(\varphi)$ is never zero, so $\Phi^{M_\alpha} = \Phi$. Thus we may assume that $\Phi^{M_\alpha} = \Phi$ under either (i) or (iv). Now let $<L, M> \subseteq <E, F>$, where $M = M_\alpha$. Then, as before, $d_\alpha(\varphi) = d_\alpha(\varphi_L)$, so $-\log d_\alpha(\varphi) = -\log d_\alpha(\varphi_L)$ for each $\varphi \in \Phi$. Hence, as in the proof of Theorem 65.5, the function $-\log d_\alpha(\varphi)$ will be $\mathcal{P}_{<E,\ F>}$-sh in Φ iff its restriction to Φ_L is $\mathcal{P}_{<L,\ M>}$-sh. Since (Φ_L, p) is a Riemann domain $\mathcal{P}_{<L,\ M>}$-sh is equivalent to psh in Φ_L. Therefore (Φ_L, p) will be an <L, M>-domain of holomorphy iff $-\log d_\alpha(\varphi)$ is $\mathcal{P}_{<E,\ F>}$-sh [G7, p. 283]. An application of Theorem 64.2 now gives the equivalence of (i) and (iv). ♦

CHAPTER XIII
DUAL PAIR THEORY APPLIED TO $[\Sigma, \mathfrak{G}]$-DOMAINS

§67. THE DUAL PAIR EXTENSION OF $[\Sigma, \mathfrak{G}]$. Δ-DOMAINS

In this chapter we return to the study of general $[\Sigma, \mathfrak{G}]$-domains using the results for dual pairs $\langle E, F\rangle$ obtained in the preceding two chapters. It is desirable to review briefly the setting in which the theory of dual pairs may be brought to bear on the study of an arbitrary system $[\Sigma, \mathfrak{G}]$. In the first place, the algebra $\mathfrak{G}$, as a linear space with the compact-open topology, is a locally convex topological space (CLTS). Therefore its dual space $\mathfrak{G}'$, consisting of all continuous linear functionals,

$$a' : \mathfrak{G} \to \mathbb{C},\ a \mapsto \langle a', a\rangle$$

together with $\mathfrak{G}$ is a dual pair $\langle \mathfrak{G}', \mathfrak{G}\rangle$. Since point evaluations are continuous linear functionals on $\mathfrak{G}$ and $[\Sigma, \mathfrak{G}]$ is assumed to be a system the mapping

$$\tau : \Sigma \to \mathfrak{G}',\ \sigma \mapsto \tau(\sigma)$$

where $\langle \tau(\sigma), a\rangle = a(\sigma)$ for all $a \in \mathfrak{G}$, is a homeomorphism of Σ into $\mathfrak{G}'$, provided $\mathfrak{G}'$ is given the $\sigma(\mathfrak{G}', \mathfrak{G})$-topology. Moreover we have the following proposition.

67.1 PROPOSITION. *The homeomorphism* $\tau : \Sigma \to \mathfrak{G}'$ *of* Σ *into* $\mathfrak{G}'$ *defines an extension*

$$\tau : [\Sigma, \mathfrak{G}] \Rightarrow [\mathfrak{G}', \mathfrak{P}_{\langle \mathfrak{G}', \mathfrak{G}\rangle}]$$

of the system $[\Sigma, \mathfrak{G}]$.

Proof. For a fixed element $a \in \mathfrak{G}$ denote by $\tilde{\mathbf{a}}$ the linear functional

$$\tilde{a} : \mathfrak{G}' \to \mathbb{C},\ a' \mapsto \langle a', a\rangle = \tilde{a}(a')$$

defined on $\mathfrak{G}'$ by a. Then, by Theorem 53.2 each element $P \in \mathfrak{P}_{\langle \mathfrak{G}', \mathfrak{G}\rangle}$ is of the form $P = \sum_i \prod_j \tilde{a}_{ij}$, where i and j range over finite sets. Observe also that for each $\sigma \in \Sigma$

$$(P\circ\tau)(\sigma) = \sum_i \prod_j \tilde{a}_{ij}(\tau(\sigma)) = \sum_i \prod_j \langle\tau(\sigma), a_{ij}\rangle$$
$$= \sum_i \prod_j a_{ij}(\sigma) = (\sum_i \prod_j a_{ij})(\sigma).$$

Therefore $P\circ\tau \in \mathfrak{G}$, so $\mathcal{P}_{\langle\mathfrak{G}', \mathfrak{G}\rangle}\circ\tau \subseteq \mathfrak{G}$. Furthermore since $\tilde{a}\circ\tau = a$ it follows that $\mathfrak{G} \subseteq \mathcal{P}_{\langle\mathfrak{G}', \mathfrak{G}\rangle}\circ\tau$, proving that $\mathcal{P}_{\langle\mathfrak{G}', \mathfrak{G}\rangle}\circ\tau = \mathfrak{G}$ and that τ does indeed define a pair extension. ♦

Now let us identify Σ with its homeomorphic image $\tau(\Sigma)$ in the space $\mathfrak{G}'$. Then $\mathfrak{G} = \mathcal{P}_{\langle\mathfrak{G}', \mathfrak{G}\rangle}|\Sigma$, so the system $[\Sigma, \mathfrak{G}]$ is realized as a subsystem of $[\mathfrak{G}', \mathcal{P}_{\langle\mathfrak{G}', \mathfrak{G}\rangle}]$. Therefore the study of general systems may be identified with the study of subsystems of special systems of type $[E, \mathcal{P}_{\langle E, F\rangle}]$, where $\langle E, F\rangle$ is a dual pair. With this fact in mind, we shift our attention to the dual pair set-up. We accordingly fix $\langle E, F\rangle$ along with a connected locally connected subspace Δ of the space E, under the $\sigma(E, F)$-topology, and study the subsystem $[\Delta, \mathcal{P}_{\langle E, F\rangle}]$. By Theorem 7.1 (ii), if the system $[\Delta, \mathcal{P}_{\langle E, F\rangle}]$ is natural then Δ must be a $\mathcal{P}_{\langle E, F\rangle}$-convex subset of E. Also by Theorem 7.1 (i), if $[E, \mathcal{P}_{\langle E, F\rangle}]$ is natural then $\mathcal{P}_{\langle E, F\rangle}$-convexity of Δ implies naturality of $[\Delta, \mathcal{P}_{\langle E, F\rangle}]$. However, if $[E, \mathcal{P}_{\langle E, F\rangle}]$ is not natural then Δ may be $\mathcal{P}_{\langle E, F\rangle}$-convex without $[\Delta, \mathcal{P}_{\langle E, F\rangle}]$ being natural.

Since the dual pair $\langle E, F\rangle$ and the set Δ are fixed in our discussion, we shall simplify terminology by calling a $[\Delta, \mathcal{P}_{\langle E, F\rangle}]$-domain (Φ, p) simply a *Δ-domain*. Now, by definition, a function in Φ will be $\mathcal{P}_{\langle E, F\rangle}$-holomorphic iff its local projections in Δ are $\mathcal{P}_{\langle E, F\rangle}$-holomorphic. Moreover, a function in Δ will be $\mathcal{P}_{\langle E, F\rangle}$-holomorphic, according to our general usage, iff it is $(\mathcal{P}_{\langle E, F\rangle}|\Delta)$-holomorphic. On the other hand, in the case of an arbitrary system $[\Sigma, \mathfrak{G}]$ with Σ regarded as a subset of $\mathfrak{G}'$, the algebra $\mathcal{P}_{\langle\mathfrak{G}', \mathfrak{G}\rangle}|\Sigma$ coincides with $\mathfrak{G}$. Therefore $\mathcal{P}_{\langle\mathfrak{G}', \mathfrak{G}\rangle}$-holomorphic in Σ is the same as $\mathfrak{G}$-holomorphic, so the result is *all* $\mathfrak{G}$-holomorphic functions. Hence, unless Δ is an open subset of E, so that a Δ-domain is actually an $\langle E, F\rangle$-domain, the special properties of $\mathcal{P}_{\langle E, F\rangle}$-holomorphic functions generally will not carry over to arbitrary Δ-domains. In order to exploit these properties, the ambient space E must be involved with the Δ-domain in a more essential way. The following theorem illustrates this point. First let us introduce, in terms of the $\langle E, F\rangle$-domain $(E \# (\mathcal{O}), \pi)$ of §61, a "universal" Δ-domain that does involve the special properties of $\mathcal{P}_{\langle E, F\rangle}$-holomorphic functions. Define the space

$$\Delta \# (\check{\mathcal{O}}) = \{(\delta, (\check{f})_\delta) \in E \# (\check{\mathcal{O}}) : \delta \in \Delta\}$$

where the topology in $\Delta \# (\check{\mathcal{O}})$ is that induced by the sheaf topology in $E \# (\check{\mathcal{O}})$. Then $(\Delta \# (\check{\mathcal{O}}), \pi)$ is a Δ-domain in which π is the projection

$$\pi : \Delta \# (\check{\mathcal{O}}) \to \Delta, \ (\delta, (\check{f})_\delta) \mapsto \delta.$$

Recall that $\check{\mathcal{O}}$ denotes the presheaf of all $\mathcal{P}_{\langle E, F\rangle}$-holomorphic functions defined on open subsets of E with values in $\mathbb{C}^\Lambda$ and $(\check{\mathcal{O}})$ denotes the corresponding sheaf of germs of these functions at points of E. Note that $\Delta \# (\check{\mathcal{O}})$ will generally fail to be a subsheaf of $E \# (\check{\mathcal{O}})$ since Δ is not required to be an open set in E.

67.2 THEOREM. *Assume that* $[\Delta, \mathcal{P}_{\langle E, F\rangle}]$ *is natural and let* Γ_0 *be an arbitrary component of* $\Delta \# (\check{\mathcal{O}})$. *Denote by* Γ *the component of* $E \# (\check{\mathcal{O}})$ *that contains* Γ_0 *and let* $\mathfrak{H}$ *be any algebra of complex-valued functions on* Γ_0 *such that*

$$\mathcal{O}_\Gamma | \Gamma_0 \subseteq \mathfrak{H} \subseteq \mathcal{O}_{\Gamma_0}.$$

Then $[\Gamma_0, \mathfrak{H}]$ *is a natural system.*

Proof. Observe that elements of $\mathfrak{H}$ are $\mathcal{P}_{\langle E, F\rangle}$-holomorphic and so, *a fortiori*, are $\mathcal{O}_\Gamma$-holomorphic on Γ_0. Therefore by Theorem 40.3 it will be sufficient to prove that $[\Gamma_0, \mathcal{O}_\Gamma]$ is natural. Observe that by Proposition 61.1 and Theorem 63.1 the pair $[\Gamma, \mathcal{O}_\Gamma]$, and hence $[\Gamma_0, \mathcal{O}_\Gamma]$, is a system. (Note, however, that $[\Gamma, \mathcal{O}_\Gamma]$ need not be natural.)

Now let

$$\chi_\Delta : \mathcal{O}_\Gamma | \Gamma_0 \to C, \ h|\Gamma_0 \mapsto \widehat{h|\Gamma_0}(\chi_\Delta)$$

be a homomorphism dominated by a compact set $K \subseteq \Gamma_0$. Defining

$$\chi : \mathcal{O}_\Gamma \to \mathbb{C}, \ h \mapsto \widehat{h|\Gamma_0}(\chi_\Delta)$$

we obtain a homomorphism of $\mathcal{O}_\Gamma$ also dominated by K. Furthermore, if

$$\chi_0 : \mathcal{P}_{\langle E, F\rangle} \to \mathbb{C}, \ P \mapsto \widehat{P \circ \pi}(\chi)$$

then χ_0 is a homomorphism of $\mathcal{P}_{\langle E, F\rangle}$ dominated by the compact set $\pi(K)$ in Δ. Since $[\Delta, \mathcal{P}_{\langle E, F\rangle}]$ is assumed to be natural, there exists a point $\delta_0 \in \Delta$ such that

$$\widehat{P\circ\pi}(\chi) = P(\delta_0),\ P \in \mathcal{P}_{\langle E, F\rangle}.$$

But (Γ, π) is an $\langle E, F\rangle$-domain of holomorphy (Theorem 49.4) and hence, by Theorem 65.4 (i) there must exist $\gamma_0 \in \Gamma$ such that

$$\hat{h}(\chi) = \widehat{h|\Gamma_0}(\chi_\Delta) = h(\gamma_0),\ h \in \mathcal{O}_\Gamma.$$

Moreover

$$P(\delta_0) = \widehat{P\circ\pi}(\chi) = (P\circ\pi)(\gamma_0) = P(\pi(\gamma_0))$$

for all $P \in \mathcal{P}_{\langle E, F\rangle}$, so $\pi(\gamma_0) = \delta_0$. Thus $\gamma_0 \in \Delta\ \#\ (\check{\mathcal{O}})$ and it only remains to prove that $\gamma_0 \in \Gamma_0$.

Denote by $\hat{K}$ the $\mathcal{O}_\Gamma$-convex hull of the compact set K in the space Γ. Then each point $\gamma \in \hat{K}$ determines, by point evaluation, a homomorphism of $\mathcal{O}_\Gamma$ dominated by K. As in the case of γ_0 above, we must have $\gamma \in \Delta\ \#\ (\check{\mathcal{O}})$, so $\hat{K} \subseteq \Delta\ \#\ (\check{\mathcal{O}})$. Observe that the above arguments show that the set $\hat{K}$ may be identified with the spectrum of the algebra $\mathcal{O}_\Gamma|K$ and is accordingly compact in the $\mathcal{O}_\Gamma$-topology. Since $[\Gamma, \mathcal{O}_\Gamma]$ is a system the $\mathcal{O}_\Gamma$-topology on $\hat{K}$ is equivalent to the topology induced on $\hat{K}$ by Γ. Therefore, setting $\hat{K} = (\hat{K} \cap \Gamma_0) \cup (\hat{K} \setminus \Gamma_0)$, we obtain a decomposition of $\hat{K}$ into disjoint compact sets. However, since $K \subseteq \Gamma_0$ it follows by the Šilov decomposition theorem (see [R1, Corollary (3.6.4)]) that $\hat{K}\setminus\Gamma_0 = \emptyset$, so $\hat{K} \subseteq \Gamma_0$. In particular $\gamma_0 \in \Gamma_0$. ♦

§68. GERM-VALUED FUNCTIONS

Another way of looking at the above situation is to observe that the set Δ is in general a relatively "thin" subset of E, so that a $\mathcal{P}_{\langle E, F\rangle}$-holomorphic function defined on a neighborhood of a point of Δ in E need not be determined by its restriction to Δ. For example, in the case of the space Σ embedded in $\mathfrak{a}'$ any non-trivial polynomial relation among the elements of the algebra $\mathfrak{a}$ defines a non-zero element of $\mathcal{P}_{\langle \mathfrak{a}', \mathfrak{a}\rangle}$ whose restriction to Σ is zero. This is precisely the situation that calls for the study of germs of functions on Δ. We shall adopt this approach. However, it needs to be "localized" before it can be applied to Δ-domains. To this end, let (Φ, p) be an arbitrary Δ-domain and, in analogy with previous notations, define

$$\Phi \# (\mathcal{O}) = \{(\varphi, (h)_{p(\varphi)}) : \varphi \in \Phi, (h)_{p(\varphi)} \in (\mathcal{O})\}$$

where $(\mathcal{O})$ denotes the sheaf of germs at points of E of $\mathcal{P}_{<E, F>}$-holomorphic complex-valued functions defined on open subsets of E. As usual, we topologize $\Phi \# (\mathcal{O})$ by giving it the product space topology induced by $\Phi \times (\mathcal{O})$, in which Φ carries its given topology and $(\mathcal{O})$ the sheaf topology. Then the topology in $\Phi \# (\mathcal{O})$ is generated by neighborhoods of the form

$$W_{(\varphi, (h)_{p(\varphi)})} = \{(\varphi', (h)_{p(\varphi')}) : \varphi' \in p^{-1}(N_{p(\varphi)} \cap \Delta)\}$$

where h is $\mathcal{P}_{<E, F>}$-holomorphic on the neighborhood $N_{p(\varphi)}$ in E and p_φ^{-1} is defined on $N_{p(\varphi)} \cap \Delta$, so $p_\varphi^{-1}(N_{p(\varphi)} \cap \Delta)$ is a p-neighborhood of φ. Define

$$\bar{\pi} : \Phi \# (\mathcal{O}) \to \Phi, (\varphi, (h)_{p(\varphi)}) \mapsto \varphi.$$

Then $\bar{\pi}$ maps $W_{(\varphi, (h)_{p(\varphi)})}$ one-to-one onto the p-neighborhood $p_\varphi^{-1}(N_{p(\varphi)} \cap \Delta)$. Therefore $(\Phi \# (\mathcal{O}), \bar{\pi})$ is a Φ-domain. It will be convenient also to have a notation for the projection of $\Phi \# (\mathcal{O})$ onto $(\mathcal{O})$, *viz*

$$\bar{\pi}' : \Phi \# (\mathcal{O}) \to (\mathcal{O}), (\varphi, (h)_{p(\varphi)}) \mapsto (h)_{p(\varphi)}.$$

Now let H denote a section of $(\Phi \# (\mathcal{O}), \bar{\pi})$ over an arbitrary set X in Φ. In other words, H is a $\bar{\pi}$-set and $\bar{\pi}_H^{-1}$ is continuous, so $\bar{\pi} : H \to X$ is a surjective homeomorphism. For each such section H we define the function

$$[h](\varphi) = \bar{\pi}'(\bar{\pi}_H^{-1}(\varphi)), \ \varphi \in X.$$

Denote by ${}_\Phi[\mathcal{O}]$ the collection of all such germ-valued functions associated in this way with sections of $(\Phi \# (\mathcal{O}), \bar{\pi})$ and by $[\mathcal{O}]_X$ those elements of ${}_\Phi[\mathcal{O}]$ defined on the set X. Elements of ${}_\Phi[\mathcal{O}]$ will be called $\mathcal{P}_{<E, F>}$-*holomorphic germ-valued functions*.

68.1 PROPOSITION. *A germ-valued function* $[h]$ *defined on a set* X *in* Φ *will belong to* ${}_\Phi[\mathcal{O}]$ *iff there exist functions* $\{h^\varphi : \varphi \in X\}$ *with the following properties: There corresponds to each* $\varphi \in X$ *a neighborhood* $N_{p(\varphi)}$ *in* E *and a p-neighborhood* U_φ *in* Φ, *with* $p(U_\varphi) \subseteq N_{p(\varphi)} \cap \Delta$, *such that (i)* h^φ *is* $\mathcal{P}_{<E, F>}$*-holomorphic on* $N_{p(\varphi)}$ *and* $[h](\varphi) = (h^\varphi)_{p(\varphi)}$.

(ii) $(h^{\varphi'})_{p(\varphi')} = (h^\varphi)_{p(\varphi')}$ *for all* $\varphi' \in U_\varphi \cap X$.

Proof. Assume that $[h]$ is associated with a section H of $(\Phi \# (\mathfrak{G}), \bar{\pi})$ over the set X and let $\bar{\pi}_H^{-1}(\varphi) = (\varphi, (h^\varphi)_{p(\varphi)})$, $\varphi \in X$. Then $[h](\varphi) = (h^\varphi)_{p(\varphi)}$ for $\varphi \in X$. Now choose the germ representative h^φ and a neighborhood $N_{p(\varphi)}$ on which h^φ is $\mathcal{P}_{<E, F>}$-holomorphic, so property (i) is automatically satisfied. Now choose a p-neighborhood U'_φ such that $p(U'_\varphi) \subseteq N_{p(\varphi)}$ and consider the neighborhood

$$W_{(\varphi, (h^\varphi)_{p(\varphi)})} = \{(\varphi', (h^\varphi)_{p(\varphi')}) : \varphi' \in U'_\varphi\}$$

of the point $\bar{\pi}_H^{-1}(\varphi) = (\varphi, (h^\varphi)_{p(\varphi)})$ in $\Phi \# (\mathfrak{G})$. Then since $\bar{\pi}_H^{-1}$ is continuous on X there exists a p-neighborhood $U_\varphi \subseteq U'_\varphi$ such that

$$\bar{\pi}_H^{-1}(U_\varphi \cap X) \subseteq W_{(\varphi, (h^\varphi)_{p(\varphi)})}.$$

In other words, $(\varphi', (h^{\varphi'})_{p(\varphi')}) = (\varphi', (h^\varphi)_{p(\varphi')})$ and hence $(h^{\varphi'})_{p(\varphi')} = (h^\varphi)_{p(\varphi')}$ for all $\varphi' \in U_\varphi \cap X$, proving (ii).

Next assume that $\{h^\varphi, \varphi \in X\}$ satisfies properties (i) and (ii) and define $H = \{(\varphi, (h^\varphi)_{p(\varphi)}) : \varphi \in X\}$. Then H is contained in $\Phi \# (\mathfrak{G})$ by property (i), and it is obvious that H is a $\bar{\pi}$-set. We must prove that $\bar{\pi}_H^{-1}$ is continuous. Let φ be an arbitrary point of X, so $\bar{\pi}_H^{-1}(\varphi) = (\varphi, (h^\varphi)_{p(\varphi)})$. Consider an arbitrary basic neighborhood

$$W_{(\varphi, (h^\varphi)_{p(\varphi)})} = \{(\varphi', (h^\varphi)_{p(\varphi')}) : \varphi' \in U'_\varphi\}$$

of the point $\bar{\pi}_H^{-1}(\varphi)$ in $\Phi \# (\mathfrak{G})$ with $U'_\varphi \subseteq U_\varphi$. By property (ii),

$$(h^{\varphi'})_{p(\varphi')} = (h^\varphi)_{p(\varphi')}, \quad \varphi' \in U'_\varphi \cap X.$$

Therefore

$$\bar{\pi}_H^{-1}(\varphi') = (\varphi', (h^\varphi)_{p(\varphi')}), \quad \varphi' \in U'_\varphi \cap X.$$

Hence

$$\bar{\pi}_H^{-1}(U'_\varphi \cap X) \subseteq W_{(\varphi, (h^\varphi)_{p(\varphi)})}$$

and this implies that $\bar{\pi}_H^{-1}$ is continuous. ♦

68.2 DEFINITION. *The expression* $[h](\varphi) = (h^{\varphi})_{p(\varphi)}$, $\varphi \in X$, *where* $\{h^{\varphi} : \varphi \in X\}$ *satisfies the properties in Proposition 68.1, is called a representation of the function* $[h] \in [\mathcal{O}]_X$.

Let U be an open subset of E, so $G = p^{-1}(U \cap \Delta)$ is an open set in Φ. Let h be any $\mathcal{P}_{<E,\ F>}$-holomorphic function defined on U and set $[h](\varphi) = (h)_{p(\varphi)}$, $\varphi \in G$. Then by Proposition 68.1 the function $[h]$ belongs to $[\mathcal{O}]_G$. Writing $[h] = (h)_p$, we have the result that $(\mathfrak{G}_U)_p \subseteq [\mathcal{O}]_G$ and, in particular, $(\mathcal{P}_{<E,\ F>})_p \subseteq [\mathcal{O}]_{\Phi}$.

Observe that the stalk $(\mathfrak{G})_x$ at each point $x \in E$ is an algebra under the usual operations induced on germs by the ordinary algebra operations for functions. Hence, for an arbitrary set $X \subseteq \Phi$ the set $[\mathcal{O}]_X$ is an *algebra* of germ-valued functions on X. In other words, ${}_{\Phi}[\mathcal{O}]$ is a presheaf of *algebras* of germ-valued functions over Φ. We may also extend to elements of ${}_{\Phi}[\mathcal{O}]$ the derivative operators $\mathfrak{D}_{\check{V}}^{\varkappa}$ introduced in Definition 57.1 for $\mathcal{P}_{<E,\ F>}$-holomorphic functions in E.

68.3 PROPOSITION. *Let* $[h]$ *be an element of* $[\mathcal{O}]_X$ *with representation*

$$[h](\varphi) = (h^{\varphi})_{p(\varphi)},\ \varphi \in X.$$

Define

$$\mathfrak{D}_{\check{V}}^{\varkappa}[h](\varphi) = (\mathfrak{D}_{\check{V}}^{\varkappa}h)_{p(\varphi)},\ \varphi \in X.$$

Then $\mathfrak{D}_{\check{V}}^{\varkappa}[h]$ *is an element of* $[\mathcal{O}]_X$.

Proof. By Theorem 57.2 the function $\mathfrak{D}_{\check{V}}^{\varkappa}h^{\varphi}$ is defined and $\mathcal{P}_{<E,\ F>}$-holomorphic on the domain of definition of the function h^{φ}. Therefore $\mathfrak{D}_{\check{V}}^{\varkappa}[h]$ is well-defined in X. In order to prove that $\mathfrak{D}_{\check{V}}^{\varkappa}[h] \in [\mathcal{O}]_X$, we must prove that the functions $\{\mathfrak{D}_{\check{V}}^{\varkappa}h^{\varphi} : \varphi \in X\}$ satisfy properties (i) and (ii) of Proposition 68.1, the first of which is already verified. Since the functions $\{h^{\varphi} : \varphi \in X\}$ satisfy property (ii) $(h^{\varphi'})_{p(\varphi')} = (h^{\varphi})_{p(\varphi')}$ for $\varphi' \in U_{\varphi} \cap X$. Therefore $h^{\varphi'}$ and h^{φ} coincide on a neighborhood of $p(\varphi')$ in E. Hence $\mathfrak{D}_{\check{V}}^{\varkappa}h^{\varphi'}$ and $\mathfrak{D}_{\check{V}}^{\varkappa}h^{\varphi}$ coincide on a neighborhood of $p(\varphi')$, so $(\mathfrak{D}_{\check{V}}^{\varkappa}h^{\varphi'})_{p(\varphi')} = (\mathfrak{D}_{\check{V}}^{\varkappa}h^{\varphi})_{p(\varphi')}$ for $\varphi' \in U_{\varphi} \cap X$. In other words, $\{\mathfrak{D}_{\check{V}}^{\varkappa}h^{\varphi} : \varphi \in X\}$ also satisfies property (ii). ◆

The next result is a strong uniqueness principle for elements of $_{\Phi}[\mathcal{O}]$.

68.4 PROPOSITION. *Let* $[h]_1$ *and* $[h]_2$ *be elements of* $_{\Phi}[\mathcal{O}]$ *defined on a connected open set* G *in* Φ. *If* $[h]_1$ *and* $[h]_2$ *are equal at a single point of* G *then they must be equal on all of* G.

Proof. Let H_1 and H_2 be the sections of $(\Phi \# (\mathfrak{G}), \bar{\pi})$ associated with $[h]_1$ and $[h]_2$. Then since G is connected H_1 and H_2 are connected open $\bar{\pi}$-sets and the condition that $[h]_1$ equal $[h]_2$ at some point of G implies that $H_1 \cap H_2 \neq \emptyset$. Therefore by Lemma 44.2, $H_1 \cup H_2$ is a $\bar{\pi}$-set. But this is possible only if $H_1 = H_2$ since $\bar{\pi}(H_1) = \bar{\pi}(H_2) = G$. In other words, $[h]_1 = [h]_2$ as claimed. ◆

We consider next certain complex-valued functions naturally associated with elements of $_{\Phi}[\mathcal{O}]$.

68.5 PROPOSITION. *Let* $[h]$ *be an arbitrary element of* $[\mathcal{O}]_X$, *with representation* $[h](\varphi) = (h^{\varphi})_{p(\varphi)}$, $\varphi \in X$, *and set* $h(\varphi) = h^{\varphi}(p(\varphi))$, $\varphi \in X$. *Then* h *is* $\mathcal{P}_{<E, F>}$*-holomorphic on* X *and the mapping*

$$r_X : [\mathcal{O}]_X \to \mathfrak{G}_X, \ [h] \to h$$

is an algebra homomorphism of $[\mathcal{O}]_X$ *into* $\mathfrak{G}_X$. *If* Δ *is an open subset of* E *(i.e.* (Φ, p) *is actually an* $<E, F>$*-domain* *and the set* X *is open then the homomorphism is an isomorphism.*

Proof. By Proposition 68.1 (ii) we have $(h^{\varphi'})_{p(\varphi')} = (h^{\varphi})_{p(\varphi')}$ for $\varphi' \in U_{\varphi} \cap X$. Therefore $h = h^{\varphi} \circ p$ on the set $U'_{\varphi} \cap X$, which means that h is locally $\mathcal{P}_{<E, F>}$-holomorphic on X. Hence $h \in \mathfrak{G}_X$. That $r_X : [\mathcal{O}]_X \to \mathfrak{G}_X$ is a homomorphism is obvious. Finally, let Δ be open in E and $X = G$ open in Φ. Suppose that $[h]_1$, $[h]_2$ are elements of $[\mathcal{O}]_G$ whose associated complex function h_1, h_2 are equal on G. Since G is open we may choose the neighborhoods U_{φ} in Proposition 68.1 to be contained in G. Also, the neighborhoods U_{φ} and $N_{p(\varphi)}$ may be chosen simultaneously for $[h]_1$ and $[h]_2$. Then by property (ii)

$$(h_1^{\varphi'})_{p(\varphi')} = (h_1^{\varphi})_{p(\varphi')}, \ (h_2^{\varphi'})_{p(\varphi')} = (h_2^{\varphi})_{p(\varphi')}, \ \varphi' \in U_{\varphi}.$$

Hence

$$h_1^\varphi(p(\varphi')) = h_1(\varphi') = h_2(\varphi') = h_2^\varphi(p(\varphi')), \quad \varphi' \in U_\varphi.$$

It follows that $h_1^\varphi = h_2^\varphi$ on a neighborhood of $p(\varphi)$ in E. This implies that $[h]_1(\varphi) = [h]_2(\varphi)$ and hence $[h]_1 = [h]_2$. Therefore $r_G : [\mathcal{O}]_G \to \mathfrak{G}_G$ is one-to-one.◆

68.6 DEFINITION. *The image of the presheaf* ${}_\Phi[\mathcal{O}]$ *in the presheaf* ${}_\Phi\mathfrak{G}$ *under the mappings* r_G *is called the* ***reduction*** *of* ${}_\Phi[\mathcal{O}]$ *and denoted by* ${}_\Phi\mathcal{O}$. *Elements of* ${}_\Phi\mathcal{O}$ *are called* ***strictly*** $\mathcal{P}_{\langle E, F\rangle}$***-holomorphic functions***.

Note that ${}_\Phi\mathcal{O}$ is also a presheaf of function algebras over Φ and is generally a proper subset of the full presheaf ${}_\Phi\mathfrak{G}$ of $\mathcal{P}_{\langle E, F\rangle}$-holomorphic functions in Φ. Elements of ${}_\Phi\mathcal{O}$ are in part distinguished in ${}_\Phi\mathfrak{G}$ by the fact that local projections admit $\mathcal{P}_{\langle E, F\rangle}$-holomorphic extensions to open neighborhoods in E. These local extensions must, of course, also be interrelated according to Proposition 68.1 (ii).

As suggested earlier, we regard ${}_\Phi[\mathcal{O}]$ as a generalization of the notion of germs of holomorphic functions from sets to spread domains. Theorem 68.7 below indicates more precisely the connections between the algebras of *germ-valued* functions and algebras of *germs* of functions. Consider first the product space $E \times \Phi$ and the homeomorphism

$$\mu : \Phi \to E \times \Phi, \quad \varphi \mapsto (p(\varphi), \varphi)$$

of Φ into $E \times \Phi$. Set $\tilde{\varphi} = \mu(\varphi)$, so $\tilde{X}$ is the homeomorphic image of a set $X \subseteq \Phi$ in $E \times \Phi$. Next let W be an arbitrary open subset of $E \times \Phi$ and denote by $\mathcal{H}_W$ the collection of all complex-valued functions $\tilde{h}$ defined on W and satisfying the following condition:

For each $(x_0, \varphi_0) \in W$, there exists a p-neighborhood U_{φ_0} in Φ and a neighborhood N_{x_0} in E, with $N_{x_0} \times U_{\varphi_0} \subseteq W$, such that

(1) $\tilde{h}(x, \varphi_0)$ is $\mathcal{P}_{\langle E, F\rangle}$-holomorphic as a function of $x \in N_{x_0}$.

(2) $\tilde{h}(x, \varphi) = \tilde{h}(x, \varphi_0)$, for $(x, \varphi) \in N_{x_0} \times U_{\varphi_0}$.

It is obvious that $\mathcal{H}_W$ is an algebra of continuous functions on W.

Let X be an arbitrary subset of Φ and W an arbitrary open set in $E \times \Phi$ that contains $\tilde{X}$. Also let $\tilde{h} \in \mathcal{H}_W$ and set

$$h^{\varphi}(x) = \tilde{h}(x, \varphi), \ (x, \varphi) \in W.$$

By definition of $\mathcal{H}_W$, there exists for each $\varphi \in X$ a p-neighborhood V_φ in Φ and a neighborhood $N_{p(\varphi)}$ in E with $N_{p(\varphi)} \times U_\varphi \subseteq W$, such that h^φ is $\mathcal{P}_{<E, F>}$-holomorphic in N_φ, and $h^{\varphi'} = h^\varphi$ on N_φ for $\varphi' \in U_\varphi$. Since $p(U_\varphi) \cap N_{p(\varphi)} \neq \emptyset$ we may assume that $p(U_\varphi) = N_{p(\varphi)} \cap \Delta$. It follows that, with this choice of the neighborhoods $N_{p(\varphi)}$ and U_φ, the properties in Proposition 68.1 are satisfied by the functions $\{h^\varphi : \varphi \in X\}$. Therefore if $[h](\varphi) = (h^\varphi)_{p(\varphi)}$, $\varphi \in X$, then $[h] \in [\mathcal{O}]_X$. Observe also that $[h]$ is independent of W, depending only on the germ $(\tilde{h})_{\tilde{X}}$ of the function $\tilde{h}$ on the set $\tilde{X}$. Denote by $(\mathcal{H})_{\tilde{X}}$ the algebra of all germs on $\tilde{X}$ of functions in the algebras $\mathcal{H}_W$ with $\tilde{X} \subseteq W$. Then

$$s_X : (\mathcal{H})_{\tilde{X}} \to [\mathcal{O}]_X, \ (\tilde{h})_{\tilde{X}} \to [h]$$

is an algebra isomorphism of $(\mathcal{H})_{\tilde{X}}$ with a subalgebra of $[\mathcal{O}]_X$. However, this isomorphism is not in general surjective. On the other hand, if X is compact then we have the following result, the proof of which is rather tedious.

68.7 THEOREM. *If* K *is a compact subset of* Φ *then the isomorphism* $s_K : (\mathcal{H})_{\tilde{K}} \to [\mathcal{O}]_K$ *is surjective.*

Proof. Let $[h]$ be an arbitrary element of $[\mathcal{O}]_K$ with representation $[h](\varphi) = (h^\varphi)_{p(\varphi)}$, $\varphi \in K$, given by Proposition 68.1. Then there exists a neighborhood $N_{p(\varphi)}$ in E and a p-neighborhood U_φ in Φ, with $p(U_\varphi) = N_{p(\varphi)} \cap \Delta$, such that h^φ is $\mathcal{P}_{<E, F>}$-holomorphic on $N_{p(\varphi)}$ and $(h^{\varphi'})_{p(\varphi')} = (h^\varphi)_{p(\varphi)}$, $\varphi' \in U_\varphi \cap K$. We may assume that $N_{p(\varphi)}$ is a basic neighborhood $N_{p(\varphi)}(\check{v}^\varphi, \delta_\varphi)$. For $0 < \delta \leq \delta_\varphi$ and $\check{v} \supseteq \check{v}^\varphi$ set

$$U_\varphi(\check{v}, \delta) = p_\varphi^{-1}(N_{p(\varphi)}(\check{v}, \delta) \cap \Delta)$$

and

$$W_{\tilde{\varphi}}(\check{v}, \delta) = N_{p(\varphi)}(\check{v}, \delta) \times U_\varphi(\check{v}, \delta).$$

Let φ', φ'' be arbitrary elements of K. If $\varphi' \neq \varphi''$ choose $\check{v}(\varphi', \varphi'') \supseteq \check{v}^{\varphi'} \cup \check{v}^{\varphi''}$ and $0 < \delta(\varphi', \varphi'') \leq \min(\delta_{\varphi'}, \delta_{\varphi''})$ such that

$$W_{\tilde{\varphi}'}(\check{v}(\varphi', \varphi''), \delta(\varphi', \varphi'')) \cap W_{\tilde{\varphi}''}(\check{v}(\varphi', \varphi''), \delta(\varphi', \varphi'')) = \emptyset$$

If $\varphi' = \varphi'' = \varphi$ define

$$\check{v}(\varphi, \varphi) = \check{v}^{\varphi}, \quad \delta(\varphi, \varphi) = \delta_{\varphi}.$$

Then since K is compact the set $\tilde{K} \times \tilde{K}$ may be covered by a finite number of neighborhoods of the form

$$W_{\tilde{\varphi}_k'}(\check{v}^k, \delta_k) \times W_{\tilde{\varphi}_k''}(\check{v}^k, \delta_k), \; k = 1,\ldots,n$$

where $\check{v}^k = \check{v}(\varphi_k', \varphi_k'')$ and $2\delta_k = \delta(\varphi_k', \varphi_k'')$. Now set

$$\check{v}^0 = \bigcup_{k=1} \check{v}^k, \quad \delta_0 = \min(\delta_1,\ldots,\delta_n)$$

and define

$$W = \bigcup_{\varphi \in K} W_{\tilde{\varphi}}(\check{v}^0, \delta_0).$$

Then W is an open set in $E \times \Phi$ that obviously contains $\tilde{K}$.

Next let φ', φ'' be an arbitrary pair of points in K. Then there exists k such that $(\varphi', \tilde{\varphi}'') \in W_{\tilde{\varphi}_k'}(\check{v}^k, \delta_k) \times W_{\tilde{\varphi}_k''}(\check{v}^k, \delta_k)$. It follows that

$$W_{\tilde{\varphi}'}(\check{v}^0, \delta_0) \subseteq W_{\tilde{\varphi}_k'}(\check{v}^k, 2\delta_k), \; W_{\tilde{\varphi}''}(\check{v}^0, \delta_0) \subseteq W_{\tilde{\varphi}_k''}(\check{v}^k, 2\delta_k).$$

Hence if $W_{\tilde{\varphi}'}(\check{v}^0, \delta_0) \cap W_{\tilde{\varphi}''}(\check{v}^0, \delta_0) \neq \emptyset$ then the points φ_k' and φ_k'' must be equal, say $\varphi_0 = \varphi_k' = \varphi_k''$. Then $\check{v}^k = \check{v}^{\varphi_0}$ and $\delta_k = \delta_{\varphi_0}$. Also $(h^{\varphi'})_{p(\varphi')} = (h^{\varphi_0})_{p(\varphi')}$ and $(h^{\varphi''})_{p(\varphi'')} = (h^{\varphi_0})_{p(\varphi'')}$. Since $N_{p(\varphi')}(\check{v}^0, \delta_0)$ and $N_{p(\varphi'')}(\check{v}^0, \delta_0)$ are contained in $N_{p(\varphi_0)}(\check{v}^{\varphi_0}, \delta_{\varphi_0})$ it follows by the uniqueness principle that $h^{\varphi'} = h^{\varphi''}$ on the set

$$N_{p(\varphi')}(\check{v}^0, \delta_0) \cap N_{p(\varphi'')}(\check{v}^0, \delta_0).$$

Now, for arbitrary $\varphi' \in K$ define

$$\tilde{h}(x, \varphi) = h^{\varphi'}(x), \; (x, \varphi) \in W_{\varphi'}(\check{v}^0, \delta_0).$$

Then by the preceding observations $\tilde{h}$ is a well-defined function on W. It is obvious that $\tilde{h} \in \mathcal{H}_W$ and that r_K maps $(\tilde{h})_{\tilde{K}}$ to the given element $[h] \in [\mathcal{O}]_K$. Therefore the mapping $s_K : (\mathcal{H})_{\tilde{K}} \to [\mathcal{O}]_K$ is surjective. ◆

§69. TOPOLOGIES FOR $[\mathcal{O}]_\Phi$

We consider next two topologies for the algebra $[\mathcal{O}]_\Phi$ of germ-valued functions on Φ. First we have a topology determined in $[\mathcal{O}]_\Phi$ by the homomorphism $r_\Phi : [\mathcal{O}]_\Phi \to \mathcal{O}_\Phi$. This is the coarsest topology with respect to which r_Φ is continuous when $\mathcal{O}_\Phi$ is given the compact open topology. Thus a basic neighborhood of an element $[h]_0 \in [\mathcal{O}]_\Phi$ is of the form

$$\mathfrak{J}_{[h]_0}(K, \delta) = \{[h] \in [\mathcal{O}]_\Phi : |r_\Phi([h]) - r_\Phi([h]_0)|_K < \delta\}$$

where $\delta > 0$ and K is an arbitrary compact subset of Φ. This topology is obviously generated by semi-norms,

$$|[h]|_K = |r_\Phi([h])|_K, \quad [h] \in [\mathcal{O}]_\Phi$$

associated with compact sets $K \subseteq \Phi$. We shall refer to this topology as the *(compact-open) semi-norm* topology. Observe that it need not be Hausdorff.

The second topology is more complicated and depends on the result in Theorem 68.7. Observe first that if K is any compact subset of Φ then the algebras $\{\mathcal{H}_W : \tilde{K} \subset W\}$, along with the restriction maps $\rho_{WW'} : \mathcal{H}_W \to \mathcal{H}_{W'}$ for $W' \subseteq W$, constitute an inductive system $\{\mathcal{H}_W, \rho_{WW'}\}$. In the usual way, the inductive limit may be identified with the algebra $(\mathcal{H})_{\tilde{K}}$ of germs on $\tilde{K}$. The associated mappings of the algebras $\mathcal{H}_W$ on the inductive limit are given by

$$\rho_W : \mathcal{H}_W \to (\mathcal{H})_{\tilde{K}}, \ \tilde{h} \mapsto (\tilde{h})_{\tilde{K}}.$$

Now give each of the algebras $\mathcal{H}_W$ the compact-open topology. Then the corresponding inductive limit topology on $(\mathcal{H})_{\tilde{K}}$ is the finest locally convex topology with respect to which each ρ_W is continuous. A basic neighborhood of zero in this topology is the union of all sets of the form

$$\mathfrak{G}(W,L,\delta) = \{(\tilde{h})_{\tilde{K}} : \tilde{h} \in \mathcal{H}_W, \ |\tilde{h}|_L < \delta\}$$

where $\tilde{K} \subset W$, L is a compact subset of W, and $\delta > 0$. Since the compact-open topologies in the algebras $\mathcal{H}_W$ are Hausdorff and the set K is compact it follows that the inductive topology in $(\mathcal{H})_{\tilde{K}}$ is also Hausdorff.

Next observe that the restriction map $[h] \to [h]|K$ defines an isomorphism of $[\mathcal{O}]_\Phi$ with a subalgebra of $[\mathcal{O}]_K$. Also, since the algebra $[\mathcal{O}]_K$ is isomorphic with

$(\mathcal{H})^{\sim}_{K}$ by Theorem 68.7, we obtain an isomorphism

$$s'_K : [\mathcal{O}]_\Phi \to (\mathcal{H})^{\sim}_{K}, \ [h] \mapsto s_K^{-1}([h]|K)$$

of the algebra $[\mathcal{O}]_\Phi$ with a subalgebra of $(\mathcal{H})^{\sim}_{K}$. We may therefore give to $[\mathcal{O}]_\Phi$ the projective topology associated *via* the mappings s'_K with the algebras $(\mathcal{H})^{\sim}_{K}$ under their inductive topologies. A basic neighborhood of zero for the topology is of the form

$$\mathfrak{n}_0(K, \mathcal{G}_0) = \{[h] \in [\mathcal{O}]_\Phi : s'_K(h) \in \mathcal{G}_0\}$$

where K is an arbitrary compact subset of Φ and $\mathcal{G}_0$ is a neighborhood of zero in $(\mathcal{H})^{\sim}_{K}$. This topology, which is also Hausdorff, we shall call the IP-*topology* for $[\mathcal{O}]_\Phi$.

69.1 PROPOSITION. *The* IP-*topology is finer than the* (c-o) *semi-norm topology in* $[\mathcal{O}]_\Phi$.

Proof. Let $\mathfrak{I}_0(K, \delta) = \{[h] \in [\mathcal{O}]_\Phi : |[h]|_K < \delta\}$, where $K \subset\subset \Phi$ and $\delta > 0$, be a basic neighborhood of zero in the semi-norm topology. Set

$$\mathcal{G}_0 = \{(\tilde{h})^{\sim}_{K} \in (\mathcal{H})^{\sim}_{K} : |\check{h}|^{\sim}_{K} < \delta\}.$$

Then $\mathcal{G}_0$ is a neighborhood of zero in $(\mathcal{H})^{\sim}_{K}$, so

$$\mathfrak{n}_0(K, \mathcal{G}_0) = \{[h] \in [\mathcal{O}]_\Phi : s'_K([h]) \in \mathcal{G}_0\}$$

is a neighborhood of zero in the IP-topology. Furthermore, $s'_K([h]) \in \mathcal{G}_0$ implies that $|r_K([h])|_K < \delta$, i.e. $|[h]|_K < \delta$ so $[h] \in \mathfrak{I}_0(K, \delta)$. Therefore $\mathfrak{n}_0(K, \mathcal{G}_0) \subseteq \mathfrak{I}_0(K, \delta)$, which gives the desired result. ♦

§70. NATURALITY OF ALGEBRAS OF GERM-VALUED FUNCTIONS

The next theorem asserts, in effect, that a homomorphism of $[\mathcal{O}]_\Phi$ onto $\mathbb{C}$ will be continuous in the IP-topology iff it is continuous in the (c-o) semi-norm topology. This continuity criterion generalizes a similar result for germs of holomorphic functions on a compact set [Z1, p. 277]. We make the proof for a subalgebra of $[\mathcal{O}]_\Phi$ with the topologies imposed on it by $[\mathcal{O}]_\Phi$.

70.1 THEOREM. *Let* $[\mathcal{Q}]$ *be a subalgebra of* $[\mathcal{O}]_\Phi$ *and let*

$$\chi : [\mathcal{Q}] \to \mathbb{C}, \ [h] \to \widehat{[h]}(\chi)$$

denote a homomorphism of $[\mathcal{Q}]$ *onto* $\mathbb{C}$. *Then* χ *will be continuous with respect to the* IP-*topology iff there exists a compact set* K *in* Φ *such that*

$$|\widehat{[h]}(\chi)| \le |[h]|_K, \ [h] \in [\mathcal{Q}].$$

Proof. The condition is obviously equivalent to continuity in the semi-norm topology. Therefore, in view of Proposition 69.1, it only remains to prove that continuity with respect to the IP-topology implies the condition. If χ is continuous in the IP-topology then the set

$$\mathfrak{n}_\chi = \{[h] \in [\mathcal{Q}] : |\widehat{[h]}(\chi)| < 1\}$$

must be open in $[\mathcal{Q}]$ with respect to the IP-topology. Therefore there must exist a compact set $K \subset \Phi$ and a neighborhood $\mathfrak{G}_0$ of zero in $(\mathcal{H})_{\tilde{K}}$ such that the set

$$\mathfrak{n}_0(K, \mathfrak{G}_0) = \{[h] \in [\mathcal{Q}] : s_K'([h]) \in \mathfrak{G}_0\}$$

is contained in $\mathfrak{n}_\chi$. Now let $[h]$ be an arbitrary element of $[\mathcal{Q}]$ and $\tilde{h}$ any representative of the germ $s_K'([h])$ on $\tilde{K}$. Assume that $\tilde{h}$ is defined on the open set $W_0 \supset \tilde{K}$. For arbitrary $\varepsilon > 0$ set

$$W_\varepsilon = \{(x, \varphi) \in W_0 : |\tilde{h}(x, \varphi)| < |\tilde{h}|_{\tilde{K}} + \varepsilon\}.$$

Since $\mathfrak{G}_0$ is a neighborhood of zero in $(\mathcal{H})_{\tilde{K}}$ there exists a compact set $L \subset W_\varepsilon$ and $0 < \delta_L \le 1$ such that $\tilde{h}' \in \mathcal{H}_{W_\varepsilon}$ and $|\tilde{h}'|_L < \delta_L$ imply that $(\tilde{h}')_{\tilde{K}} \in \mathfrak{G}_0$. We may obviously assume that $\tilde{K} \subseteq L$. Next, for arbitrary $\varepsilon' > 0$ and positive integer n define

$$[h]' = \frac{\delta_L [h]^n}{|\tilde{h}|_L^n + \varepsilon'}, \quad \tilde{h}' = \frac{\delta_L \tilde{h}^n}{|\tilde{h}|_L^n + \varepsilon'}.$$

Then $\tilde{h}' \in \mathcal{H}_{W_\varepsilon}$ and $s_K'([h']) = (\tilde{h}')_{\tilde{K}}$. Moreover, since

$$|\tilde{h}'|_L = \frac{\delta_L |\tilde{h}|_L^n}{|\tilde{h}|_L^n + \varepsilon'} < \delta_L$$

it follows that $(\tilde{h}')_{\tilde{K}} \in \mathfrak{G}_0$. In other words, $s_K'([h]') \in \mathfrak{G}_0$, so $[h]' \in \mathfrak{n}_\chi$. Thus

$$\frac{\delta_L |\widehat{[h]}(\chi)|^n}{|\tilde{h}|_L^n + \varepsilon'} = |\widehat{[h]}'(\chi)| < 1$$

and hence $\delta_L|\widehat{[h]}(\chi)|^n < |h|_L^n + \varepsilon'$. Since this holds for all $\varepsilon' > 0$ it follows that $\delta_L|\widehat{[h]}(\chi)|^n \le |\tilde{h}|_L^n$, or $\delta_L^{1/n}|\widehat{[h]}(\chi)| \le |\tilde{h}|_L$. Letting $n \to \infty$, we obtain $|\widehat{[h]}(\chi)| \le |h|_L$. But $L \subset W_\varepsilon$, so $|\tilde{h}|_L \le |\tilde{h}|_{\tilde{K}} + \varepsilon$ and hence $|\widehat{[h]}(\chi)| \le |\tilde{h}|_{\tilde{K}} + \varepsilon = \|[h]\|_K + \varepsilon$. Again, this holds for all $\varepsilon > 0$, so we conclude that $|\widehat{[h]}(\chi)| \le \|[h]\|_K$, $[h] \in [\mathcal{Q}]$. ♦

Consider the reduction homomorphism $r_\Phi : [\mathcal{O}]_\Phi \to \mathcal{O}_\Phi$, $[h] \mapsto h$, introduced in Definition 68.6 and let $\theta : \mathcal{O}_\Phi \to \mathbb{C}$, $h \mapsto \hat{h}(\theta)$, be a continuous homomorphism of the strictly $\mathcal{P}_{<E,\ F>}$-holomorphic functions $\mathcal{O}_\Phi$ onto $\mathbb{C}$, where $\mathcal{O}_\Phi$ is given the compact-open topology. Then there exists a compact set $K \subseteq \Phi$ such that $|\hat{h}(\theta)| \le |h|_K$, $h \in \mathcal{O}_\Phi$. If we define $\chi_\theta = \theta \circ r_\Phi$ then, since $|r_\Phi([h])|_K = |[h]|_K$, $\chi_\theta : [\mathcal{O}]_\Phi \to \mathbb{C}$, $[h] \mapsto \widehat{[h]}(\chi_\theta)$, is a continuous homomorphism of $[\mathcal{O}]_\Phi$ onto $\mathbb{C}$. On the other hand, let $\chi : [\mathcal{O}]_\Phi \to \mathbb{C}$ be a continuous homomorphism of $[\mathcal{O}]_\Phi$ with $|\widehat{[h]}(\chi)| \le |[h]|_K$ for a compact set $K \subseteq \Phi$. Then $r_\Phi([h]) = 0$ implies $\widehat{[h]}(\chi) = 0$. Therefore if we define $\theta_\chi : \mathcal{O}_\Phi \to \mathbb{C}$, $h \mapsto \widehat{[h]}(\chi)$, where $[h]$ is any element of $[\mathcal{O}]_\Phi$ such that $r_\Phi([h]) = h$, then θ_χ is a well-defined homomorphism of $\mathcal{O}_\Phi$ into $\mathbb{C}$. Furthermore $|\hat{h}(\theta_\chi)| \le |[h]|_K = |h|_K$, $h \in \mathcal{O}_\Phi$, and $\theta_\chi \circ r_\Phi = \chi$. Now denote by $\hat{\Phi}$ the space of all continuous homomorphisms $\theta : \mathcal{O}_\Phi \to \mathbb{C}$ and by $\hat{\hat{\Phi}}$ the space of all continuous homomorphisms $\chi : [\mathcal{O}]_\Phi \to \mathbb{C}$, where $\hat{\Phi}$ and $\hat{\hat{\Phi}}$ are given the topologies determined by $\hat{\mathcal{O}}_\Phi$ and $\widehat{[\mathcal{O}]}_\Phi$ respectively. Then we have the following result.

70.2 COROLLARY. *The mapping,* $\hat{\Phi} \to \hat{\hat{\Phi}}$, $\theta \mapsto \theta \circ r_\Phi$, *is a surjective homeomorphism.*

In the preceding discussion of germ-valued functions on a Δ-domain (Φ, p), the values of such a function at a point $\varphi \in \Phi$ lie in the algebra $(\mathfrak{G})_{p(\varphi)}$ of germs of $\mathcal{P}_{<E,\ F>}$-holomorphic functions at the point $p(\varphi) \in E$. Consider an arbitrary one of these algebras $(\mathfrak{G})_u$ for some $u \in E$. Note that $(\mathfrak{G})_u$ is a local algebra (with unique maximal ideal $\{(f)_u : f(u) = 0\}$) and is closed under differentiation, *i.e.* (Proposition 68.3) $\mathcal{D}_{\tilde{v}}^{\chi}(f)_u \in (\mathfrak{G})_u$ for each $(f)_u \in (\mathfrak{G})_u$ and arbitrary derivative $\mathcal{D}_{\tilde{v}}^{\chi}$. Furthermore, if $(f)_u$ is an arbitrary element of $(\mathfrak{G})_u$, where f is $\mathcal{P}_{<E,\ F>}$-holomorphic on some neighborhood of u in E, and if $f_u(x) = f(u+x)$ then f_u is $\mathcal{P}_{<E,\ F>}$-holomorphic on a neighborhood of 0 in E. It is obvious that the "translation" mapping $t_0 : (\mathfrak{G})_u \to (\mathfrak{G})_0$, $(f)_u \mapsto (f_u)_0$, is an algebra isomorphism of $(\mathfrak{G})_u$ onto $(\mathfrak{G})_0$. It follows that all of these algebras are isomorphic. Note that the above isomorphism t_0 also preserves derivatives. It is not difficult to prove, using Proposition 57.4, that any isomorphism of $(\mathfrak{G})_u$ with $(\mathfrak{G})_0$ that preserves derivatives must coincide with t_0. Thanks to these translation isomorphisms, we may regard the germ-valued functions considered above as $(\mathfrak{G})_0$-valued. We are, in fact, forced to take this point of view in order to extend certain standard constructions for complex-valued functions to the germ-valued case. This is so, for example, in forming the product of a family of algebras which was discussed for complex-valued functions in §4 and which will be needed for germ-valued functions in the next chapter (§72). In the preceding discussion it was convenient to treat the algebras $(\mathfrak{G})_u$ as being distinct since all operations on the functions ultimately involved only operations within the individual value algebras.

We consider briefly a notion of naturality for algebras of $(\mathfrak{G})_0$-valued functions. To begin with, let $[C]$ be an algebra of $(\mathfrak{G})_0$-valued functions defined on a Hausdorff space Σ. This situation obviously suggests looking at homomorphisms

$$\chi : [C] \to (\mathfrak{G})_0, \ [f] \mapsto \widehat{[f]}(\chi)$$

of $[C]$ onto $(\mathfrak{G})_0$. The homomorphism is said to be continuous if it is continuous with respect to the (c-o) semi-norm topologies (§69) in $[C]$ and $(\mathfrak{G})_0$. Observe that the (c-o) semi-norm topology for $(\mathfrak{G})_0$ is determined by the unique semi-norm

$$|(f)_0| = |f(0)|, \ (f)_0 \in (\mathfrak{G})_0.$$

We also write $(f)_0(0) = f(0)$ and call $(f)_0(0)$ the *evaluation* of $(f)_0$. Obviously $(f)_0(0)$ is independent of the choice of the germ representative f. Thus, χ will be continuous iff the associated complex-valued homomorphism

$$\chi_0 : [C] \to \mathbb{C}, \ [f] \mapsto \widehat{[f]}(\chi)(0)$$

is continuous. Therefore continuity of χ is equivalent to the existence of a compact set $K \subseteq \Sigma$ such that $|\widehat{[f]}(\chi)| \le |[f]|_K$, $[f] \in [C]$, where $|\widehat{[f]}(\chi)| = |\widehat{[f]}(\chi)(0)|$ and $|[f]|_K = \sup_{\sigma \in K} |[f](\sigma)(0)|$. The homomorphism is called a *point evaluation* if there exists $\sigma_\chi \in \Sigma$ such that

$$\widehat{[f]}(\chi) = [f](\sigma_\chi), \ [f] \in [C]$$

and is said to *preserve derivatives* if $[C]$ is closed under differentiation and

$$\widehat{\mathcal{D}_{\check{V}}^{\varkappa}[f]}(\chi) = \mathcal{D}_{\check{V}}^{\varkappa}\widehat{[f]}(\chi), \; [f] \in [C]$$

for all derivatives $\mathcal{D}_{\check{V}}^{\varkappa}$. Recall that for an arbitrary element $(f)_0$ of $(\mathfrak{G})_0$ the derivatives of $(f)_0$ are defined by the equation $\mathcal{D}_{\check{V}}^{\varkappa}(f)_0 = (\mathcal{D}_{\check{V}}^{\varkappa}f)_0$. It is obvious that point evaluations are always continuous and, if $[C]$ is closed under differentiation, also preserve derivatives. This suggests the following definition.

70.3 DEFINITION. $[\Sigma, [C]]$ *is said to be natural iff every continuous derivative preserving homomorphism of* $[C]$ *onto* $(\mathfrak{G})_0$ *is a point evaluation.*

Denote by C the algebra of complex-valued functions obtained by reducing elements of $[C]$ to Σ, *i.e.* C consists of all functions

$$f : \Sigma \to \mathbb{C}, \; \sigma \to f(\sigma) = [f](\sigma)(0)$$

for $[f] \in [C]$. We have the following result.

70.4 THEOREM. *If* $[C]$ *is closed under differentiation and* $[\Sigma, C]$ *is natural then* $[\Sigma, [C]]$ *is also natural.*

Proof. Let $\chi : [C] \to (\mathfrak{G})_0$ be a homomorphism of $[C]$ dominated by a compact set $K \subseteq \Sigma$, so $|\widehat{[f]}(\chi)| \leq |[f](\sigma)(0)|_K$, $[f] \in [C]$. Thus, if $[f]_1(\sigma)(0) = [f]_2(\sigma)(0)$ for all $\sigma \in \Sigma$ then $[f]_1(\sigma) = [f]_2(\sigma)$. Therefore if we set

$$\chi_0 : [C] \to \mathbb{C}, \; f \mapsto \widehat{[f]}(\chi)(0)$$

where $[f]$ is any element of $[C]$ such that $[f](\sigma)(0) = f(\sigma)$ for $\sigma \in \Sigma$, then χ_0 is a well-defined homomorphism of C into $\mathbb{C}$ dominated by K. Since $[\Sigma, C]$ is assumed to be natural there exists a point $\sigma_0 \in \Sigma$ such that $\widehat{[f]}(\chi)(0) = f(\sigma_0)$, $[f] \in [C]$. For arbitrary $[f] \in [C]$ choose a function $h \in \mathfrak{G}$ such that $\widehat{[f]}(\chi) = (h)_0$. Also, for each $\sigma \in \Sigma$ choose $f^\sigma \in \mathfrak{G}$ such that $[f](\sigma) = (f^\sigma)_0$. Then $f(\sigma) = f^\sigma(0)$ for each $\sigma \in \Sigma$, so $h(0) = \widehat{[f]}(\chi)(0) = f^{\sigma_0}(0)$, $[f] \in [C]$. Moreover, since χ preserves derivatives it follows that

$$(\mathcal{D}_{\check{V}}^{\varkappa}h)(0) = (\mathcal{D}_{\check{V}}^{\varkappa}f^{\sigma_0})(0)$$

for all derivatives $\mathcal{D}_{\check{V}}^{\varkappa}$. By Proposition 57.4, this implies that $(h)_0 = (f^{\sigma_0})_0$, so $\widehat{[f]}(\chi) = [f](\sigma_0)$, $[f] \in [C]$. ♦

CHAPTER XIV
HOLOMORPHIC EXTENSIONS OF Δ-DOMAINS

§71. EXTENSION RELATIVE TO GERM-VALUED FUNCTIONS

In this chapter we take up the study of extensions of a Δ-domain relative to algebras of germ-valued rather than complex-valued functions. If (Φ, p) and (Ψ, q) are two Δ-domains and $\rho : (\Phi, p) \to (\Psi, q)$ is a morphism of these domains then since ρ is an open local homeomorphism (Lemma 48.1) we always have $[\mathcal{O}]_\Psi \circ \rho \subseteq [\mathcal{O}]_\Phi$ just as for complex-valued functions. Moreover if Ψ is connected then by Proposition 68.3 the mapping of $[\mathcal{O}]_\Psi$ into $[\mathcal{O}]_\Phi$ is an isomorphism. Also, since $p = q \circ \rho$ the isomorphism obviously preserves derivatives. As in the case of complex-valued functions, the morphism is called an *extension of* (Φ, p) *relative to* $[\mathcal{O}]_\Psi$ provided $[\mathcal{O}]_\Psi \circ \rho = [\mathcal{O}]_\Phi$, and we write

$$\rho : (\Phi, p) \twoheadrightarrow (\Psi, q)$$

using the "closed" arrow "$\twoheadrightarrow$" in place of the "open" arrow "$\Rightarrow$" in order to distinguish the germ-valued from the complex-valued case. Maximal extensions and maximal Δ-domains for germ-valued functions may be defined exactly as in the case of complex-valued functions.

A critical result for the case of complex-valued functions was that maximal connected extensions always exist provided only that the functions involved satisfy the uniqueness principle. Such extension were constructed using a standard sheaf of germs approach (Theorem 49.5). A problem arises here, however, since this construction technique fails in general for germ-valued functions. The failure occurs because extensions must take into account the involvement of germ-valued functions with the ambient space E. This problem appears to be unavoidable. In order to deal with it to the extent possible, we resolve the algebra $[\mathcal{O}]_\Phi$ into certain well-behaved subalgebras with respect to which extensions are more manageable. For this purpose we introduce the following definition.

71.1 DEFINITION. *Any collection*

$$n = \{N^{\varphi} : \varphi \in \Phi\}$$

is called an extended cover for Φ *if* N^{φ}, *for each* $\varphi \in \Phi$, *is a connected neighborhood of the point* $p(\varphi)$ *in the space* E.

If $m = \{M^{\varphi}\}$ and $n = \{N^{\varphi}\}$ are extended covers for Φ then n is said to be *finer than* m, written $m \leq n$, if $N^{\varphi} \subseteq M^{\varphi}$ for each $\varphi \in \Phi$. Observe that for arbitrary m and n there obviously exists a cover ℓ such that $m \leq \ell$ and $n \leq \ell$. Therefore the family $\mathcal{E}_{\Phi}$ of all extended covers for Φ is a directed system under the partial ordering "$\leq$".

The special subalgebras of $[O]_{\Phi}$ which we require are defined in terms of extended covers.

71.2 DEFINITION. *A germ-valued function* $[h] \in [O]_{\Phi}$ *is said to be carried by an extended cover* $n = \{N^{\varphi}\}$ *if it admits a representation* $[h](\varphi) = (h^{\varphi})_{p(\varphi)}$ *such that* h^{φ} *is* $\mathcal{P}_{<E, F>}$*-holomorphic on* N^{φ} *for each* $\varphi \in \Phi$. *This is also called an* n*-representation of* $[h]$. *A subset of* $[O]_{\Phi}$ *is said to be uniformly carried by* n *if each of its elements is carried by* n. *The subalgebra of* $[O]_{\Phi}$ *consisting of all elements of* $[O]_{\Phi}$ *carried by* n *is denoted by* $[O]_{\Phi}^{n}$.

It follows from Proposition 68.1 that each element of $[O]_{\Phi}$ is carried by an extended cover for Φ. Therefore

$$[O]_{\Phi} = \bigcup_{n \in \mathcal{E}_{\Phi}} [O]_{\Phi}^{n}.$$

It is obvious that $m \leq n$ implies $[O]_{\Phi}^{m} \subseteq [O]_{\Phi}^{n}$. Moreover, each of the algebras $[O]_{\Phi}^{n}$ contains $(\mathcal{P}_{<E, F>})_{p}$ and is closed under differentiation.

We consider next the category consisting of all triples $(\Phi, p, [H])$, in which (Φ, p) is connected Δ-domain and $[H]$ is a uniformly carried subalgebra of $[O]_{\Phi}$. A *morphism*

$$\rho : (\Phi, p, [H]) \to (\Psi, q, [K])$$

within the category consists of a morphism $\rho : (\Phi, p) \to (\Psi, q)$ of Δ-domains such that $[K] \circ \rho \subseteq [H]$. Thus, in particular if $[H] = [K] \circ \rho$ then ρ defines an *extension of* (Φ, p) *relative to the algebra* $[H]$ denoted by

$$\rho\colon (\Phi,\ p,\ [H]) \twoheadrightarrow (\Psi,\ q,\ [K]).$$

If $[H] = [\mathcal{O}]^n_\Phi$ then this extension of (Φ, p) is called an *n-extension*.

A morphism $\rho : (\Phi, p) \to (\Psi, q)$ of Δ-domains is said to define a *uniform extension* if for each uniformly carried subalgebra $[H]$ of $[\mathcal{O}]_\Phi$ there exists a uniformly carried subalgebra $[K] \subseteq [\mathcal{O}]_\Psi$ such that

$$\rho : (\Phi,\ p,\ [H]) \twoheadrightarrow (\Psi,\ q,\ [K]).$$

An extension

$$\rho : (\Phi,\ p,\ [H]) \twoheadrightarrow (\Psi,\ q,\ [K])$$

is said to be *maximal* if

$$\rho' : (\Phi,\ p,\ [H]) \twoheadrightarrow (\Psi',\ q',\ [K]')$$

implies the existence of $\mu : \Psi' \to \Psi$ such that $\rho = \mu\circ\rho'$ and

$$\mu : (\Psi',\ q',\ [K]') \twoheadrightarrow (\Psi,\ q,\ [K]).$$

An element $(\Phi, p, [H])$ of the category is said to be *maximal* if

$$\rho : (\Phi,\ p,\ [H]) \twoheadrightarrow (\Psi,\ q,\ [K])$$

where Ψ is connected, implies that $\rho : \Phi \to \Psi$ is a surjective homeomorphism. In this case, we also say that (Φ, p) is *maximal relative to* $[H]$.

It follows from Lemma 48.2 that if $\rho : (\Phi, p, [H]) \twoheadrightarrow (\Psi, q, [K])$ and $\rho' : (\Phi, p, [H]) \twoheadrightarrow (\Psi', q', [K]')$ are both maximal then $(\Psi, q, [K])$ and $(\Psi', q', [K]')$ are isomorphic, *i.e.* there exists a surjective homeomorphism $\mu : \Psi \to \Psi'$ such that $q = q'\circ\mu$ and $[K] = [K]'\circ\mu$.

In constructing extensions of Δ-domains relative to algebras of germ-valued functions we use, as in the case of complex-valued functions treated in §49 and §61, the "universal" Δ-domain $(\Delta \# (\check{\mathcal{O}}), \pi)$ which was introduced in Chapter XIII for Theorem 67.2.

Now consider the collection of germ-valued functions

$$[F] = \{[F]_\lambda : \lambda \in \Delta\}$$

where

$$[F]_\lambda(\gamma) = (f_\lambda)_{\pi(\gamma)},\quad \gamma = (x,\ (\check{f})_x) \in E \# (\check{\mathcal{O}}).$$

Since $\check{f} = \{f_\lambda\} \in \check{\mathfrak{G}}$ there exists an open connected neighborhood N^γ of the point x in E such that f_λ is $\mathcal{P}_{<E,\ F>}$-holomorphic on N^γ for each λ. Note that N^γ is independent of λ. Hence $W_\gamma = \{(x', (\check{f})_{x'}) : x' \in N^\gamma\}$ is a basic neighborhood of the point γ in $E \# (\check{\mathfrak{G}})$. Therefore if $\gamma' \in W_\gamma$ then $[F]_\lambda(\gamma') = (f_\lambda)_{\pi(\gamma')}$ and it follows by Proposition 68.1 that $[F] \subseteq [\mathcal{O}]_{E \# (\check{\mathfrak{G}})}$. If $\gamma' = (x', (\check{f}')_{x'})$ and $\gamma'' = (x'', (\check{f}'')_{x''})$ are points of $E \# (\check{\mathfrak{G}})$ such that $[F]_\lambda(\gamma') = [F]_\lambda(\gamma'')$ for each $\lambda \in \Lambda$ then $(f'_\lambda)_{x'} = (f''_\lambda)_{x''}$ and in particular $x' = x'' = x$. Let N_x denote the connected component of $N^{\gamma'} \cap N^{\gamma''}$ that contains the point x. Then f'_λ and f''_λ may be chosen so that $f'_\lambda = f''_\lambda$ on N_x. Since N_x is independent of λ it follows that $\check{f}'$ and $\check{f}''$ belong to $\check{\mathfrak{G}}$ and $(\check{f}')_x = (\check{f}'')_x$. Therefore $\gamma' = \gamma''$, which proves that $[F]$ separates the points of $E \# (\check{\mathfrak{G}})$.

71.3 LEMMA. *Let* $\{W_\gamma : \gamma \in E \# (\check{\mathfrak{G}})\}$ *be an open covering of* $E \# (\check{\mathfrak{G}})$ *by connected basic* π*-neighborhoods in* $E \# (\check{\mathfrak{G}})$*. Then* $\mathcal{n}_0 = \{\pi(W_\gamma) : \gamma \in \Delta \# (\check{\mathfrak{G}})\}$ *is an extended cover for* $\Delta \# (\check{\mathfrak{G}})$ *and* $[F] | (\Delta \# (\check{\mathfrak{G}})) \subseteq [\mathcal{O}]^{\mathcal{n}_0}_{\Delta \# (\check{\mathfrak{G}})}$.

Proof. Since W_γ is a connected π-neighborhood, the projection $\pi(W_\gamma)$ is a connected neighborhood of $\pi(\gamma)$, so $\mathcal{n}_0$ is an extended cover for $\Delta \# (\check{\mathfrak{G}})$. Also, since W_γ is a basic neighborhood there exists a function f which is $\mathcal{P}_{<E,\ F>}$-holomorphic on $\pi(W_\gamma)$ such that $W_\gamma = \{(x, (\check{f})_x) : x \in \pi(W_\gamma)\}$. Therefore $[F]_\lambda(\gamma') = (f_\lambda)_{\pi(\gamma')}$, $\gamma' \in W_\gamma$. Thus $[F]_\lambda$ is carried by $\mathcal{n}_0$ for each $\lambda \in \Lambda$, which is the desired result.♦

The next theorem is an analogue of Theorem 49.4 for (uniformly carried) algebras of germ-valued functions.

71.4 THEOREM. *Let* Γ_0 *be an arbitrary connected component of the space* $\Delta \# (\check{\mathfrak{G}})$ *and* $[H]$ *any uniformly carried subalgebra of* $[\mathcal{O}]_{\Gamma_0}$ *that contains* $[F] | \Gamma_0$*. Then* (Γ_0, π) *is maximal relative to* $[H]$.

Proof. Let $\rho : (\Gamma_0, \pi, [H]) \to (\Psi, q, [K])$, where Ψ is connected. Since $[F] | \Gamma_0 \subseteq [H]$. Elements of $[H]$ separate the points of Γ_0. Therefore $\rho : \Gamma_0 \to \Psi$ is injective. The desired result will follow if we show that this map is also surjective, since it is always a local homeomorphism.

Let ψ_0 be a limit point of the open set $\rho(\Gamma_0)$ in Ψ and choose elements $\{[G]_\lambda\} \subseteq [K]$ such that

$$[F]_\lambda = [G]_\lambda \circ \rho, \ \lambda \in \Lambda.$$

Since $[K]$ is uniformly carried we may assume that it is carried by a cover $\mathcal{L} = \{L^\psi\}$ for Ψ such that q_ψ^{-1} is defined on $L^\psi \cap \Delta$, where $L^\psi \cap \Delta$, and hence $q_\psi^{-1}(L^\psi \cap \Delta)$, is connected. Choose an $\mathcal{L}$-representation

$$[G]_\lambda(\psi) = (g_\lambda^\psi)_{q(\psi)}, \ \psi \in \Psi$$

for each $[G_\lambda]$. Then $\check{g}^\psi \in \check{\mathfrak{G}}$, so

$$\gamma_0 = (q(\psi_0), (\check{g}^{\psi_0})_{q(\psi_0)})$$

defines a point in $\Delta \# (\check{\mathfrak{G}})$. Also, the set

$$W_0 = \{(x, (\check{g}^{\psi_0})_x) : x \in L^{\psi_0} \cap \Delta\}$$

is a connected π-neighborhood of γ_0 in $\Delta \# (\check{\mathfrak{G}})$. Consider the set

$$W = \rho^{-1}(q_{\psi_0}^{-1}(L^{\psi_0} \cap \Delta) \cap \rho(\Gamma_0)).$$

Since ψ_0 is a limit point of the set $\rho(\Gamma_0)$ in Ψ it follows that W is a non-empty open subset of Γ_0. Consider any point $\gamma_1 = (\delta_1, (\check{h})_{\delta_1}) \in W$. Then $\delta_1 = \pi(\gamma_1) = q(\rho(\gamma_1))$ and

$$(h_\lambda)_{\delta_1} = [F]_\lambda(\gamma_1) = ([G]_\lambda\ \rho)(\gamma_1) = [G]_\lambda(\rho(\gamma_1)) = (g_\lambda^{\rho(\gamma_1)})_{\delta_1}, \ \lambda \in \Lambda.$$

Also, since $\rho(\gamma_1) \in q_{\psi_0}^{-1}(L^{\psi_0} \cap \Delta)$ it follows that $\delta_1 \in L^{\psi_0} \cap \Delta$. Moreover

$$(g_\lambda^{\rho(\gamma_1)})_{\delta_1} = (g_\lambda^{\psi_0})_{\delta_1}, \ \lambda \in \Lambda.$$

Therefore $\gamma_1 \in W_0$, so $W \subseteq W_0$. In particular $W_0 \cap \Gamma_0 \neq \emptyset$. Since Γ_0 is a connected component of $\Delta \# (\check{\mathfrak{G}})$ and W_0 is connected it follows that $W_0 \subseteq \Gamma_0$. Finally observe that q maps both of the connected sets $\rho(W_0)$ and $q_{\psi_0}^{-1}(L^{\psi_0} \cap \Delta)$ onto $L^{\psi_0} \cap \Delta$. Therefore $\rho(W_0) = q_{\psi_0}^{-1}(L^{\psi_0} \cap \Delta)$, by Lemma 44.2. In particular $\psi_0 \in \rho(\Gamma_0)$. Thus $\rho(\Gamma_0)$ is both open and closed in Ψ so must equal Ψ. ♦

We are ready now to construct a maximal extension of a connected Δ-domain (Φ, p) relative to an arbitrary uniformly carried algebra $[H] \subseteq [O]_\Phi$. The construction parallels that for complex-valued functions given in §49. Denote by

$\{[h_\lambda] : \lambda \in \Lambda\} = [H]$ a Λ-indexing of the elements of $[H]$. Assume that $[H]$ is carried by the extended cover $\mathcal{n} = \{N^\varphi\}$ for Φ and, for each $\lambda \in \Phi$, let $[h]_\lambda(\varphi) = (h_\lambda^\varphi)_{p(\varphi)}$, $\varphi \in \Phi$, be an $\mathcal{n}$-representation of $[h]_\lambda$. Then since $\mathcal{n}$ is independent of λ it follows that $\check{h}^\varphi \in \check{\mathfrak{G}}$. Next define

$$\tau_{[H]} : \Phi \to \Delta \# (\check{\mathfrak{G}}),\ \varphi \mapsto (p(\varphi),\ (\check{h}^\varphi)_{p(\varphi)}).$$

Then $\pi \circ \tau_{[H]} = p$, so we have a Δ-domain morphism

$$\tau_{[H]} : (\Phi, p) \to (\Delta \# (\check{\mathfrak{G}}), \pi).$$

In particular $\tau_{[H]}(\Phi)$ is an open connected subset of $\Delta \# (\check{\mathfrak{G}})$ and hence is contained in a uniquely determined connected component $\Gamma_{[H]}$ of the space $\Delta \# (\check{\mathfrak{G}})$. Thus we have a morphism of connected Δ-domains $\tau_{[H]} : (\Phi, p) \to (\Gamma_{[H]}, \pi)$. Now observe that $[F]_\lambda \circ \tau_{[H]} = [h]_\lambda$, $\lambda \in \Lambda$, so $[F] \circ \tau_{[H]} = [H]$. By Lemma 71.3, $[F] | \Gamma_{[H]}$ is a uniformly carried algebra. Therefore we have an extension of (Φ, p) relative to $[H]$,

$$\tau_{[H]} : (\Phi, p, [H]) \Rightarrow (\Gamma_{[H]}, \pi, [F]).$$

This extension will be called a *canonical extension of* (Φ, p) *relative to* $[H]$. For the special case in which $[H] = [O]_\Phi^{\mathcal{n}}$ we shall call the extension a *canonical* $\mathcal{n}$*-extension* and denote it by

$$\tau_{\mathcal{n}} : (\Phi, p, [O]_\Phi^{\mathcal{n}}) \Rightarrow (\Gamma_{\mathcal{n}}, \pi, [F]).$$

It will also be convenient later to denote the algebra $[F] | \Gamma_{\mathcal{n}}$ by $[F]_{\mathcal{n}}$.

71.5 THEOREM. *A canonical extension of* (Φ, p) *relative to an arbitrary uniformly carried algebra* $[H] \subseteq [O]_\Phi$ *is maximal.*

Proof. Let $\rho : (\Phi, p, [H]) \Rightarrow (\Psi, q, [K])$ be an arbitrary connected extension of (Φ, p) relative to $[H]$. Then $[K] \circ \rho = [H]$, so there exist elements $\{[k]_\lambda\} \subseteq [K]$ such that $[k]_\lambda \circ \rho = [h]_\lambda$ for each $\lambda \in \Lambda$. Obviously $\{[k]_\lambda : \lambda \in \Lambda\}$ is a Λ-indexing of $[K]$. Since $[K]$ is required to be uniformly carried, say by an extended cover $\ell = \{L^\psi\}$ for Ψ, there exists an ℓ-representation $[k]_\lambda(\psi) = (k_\lambda^\psi)_{q(\psi)}$, $\psi \in \Psi$, for each $\lambda \in \Lambda$, so $\check{k}^\psi \in (\check{\mathfrak{G}})$. Now consider the canonical extension of (Ψ, q) relative to $[K]$ associated with $\{[k]_\lambda\}$.

$$\tau_{[K]} : (\Psi, q, [K]) \twoheadrightarrow (\Gamma_{[K]}, \pi, [F]).$$

Since $[h]_\lambda = [k]_\lambda \circ \rho$ and $p = q \circ \rho$, we have

$$(\check{h}^\varphi)_{p(\varphi)} = (\check{k}^{\rho(\varphi)})_{p(\varphi)} = (\check{k}^{\rho(\varphi)})_{q(\rho(\varphi))}$$

for each $\varphi \in \Phi$ and $\lambda \in \Lambda$. Therefore

$$(p(\varphi), (\check{h}^\varphi)_{p(\varphi)}) = (q(p(\varphi)), (\check{k}^{\rho(\varphi)})_{q(\rho(\varphi))}), \varphi \in \Phi.$$

In other words, $\tau_{[H]} = \tau_{[K]} \circ \rho$. This implies that the components $\Gamma_{[H]}$ and $\Gamma_{[K]}$ intersect so must be equal. Therefore $\tau_{[K]} : (\Psi, q, [K]) \twoheadrightarrow (\Gamma_{[H]}, \pi, [F])$, proving the desired maximality. ♦

Since maximal extensions are isomorphic, the canonical extension of (Φ, p) relative to $[H]$ is independent (up to isomorphism) of the particular Λ-indexing chosen for $[H]$.

We remark in passing that one can say a bit more about the canonical n-extension $\tau_n : (\Phi, p, [0]_\Phi^n) \twoheadrightarrow (\Gamma_n, \pi, [F])$, provided $\tau_n : \Phi \to \Gamma_n$ is injective. In this case we can define an extended cover $\ell = \{L^\gamma\}$ for Γ_n as follows:

$$L^\Phi = \begin{cases} N^{\tau_n^{-1}(\gamma)} & , \gamma \in \tau_n(\Phi) \\ \pi(W_\gamma) & , \gamma \in \Gamma_n \setminus \tau_n(\Phi) \end{cases}$$

where W_γ is any connected π-neighborhood of the point γ in $E \# (\check{\mathcal{O}})$. Then $n = \ell \circ \tau_n$, *i.e.* $N^\varphi = L^{\tau_n(\varphi)}$, $\varphi \in \Phi$. Moreover $[0]_{\Gamma_n}^\ell \circ \tau_n = [0]_\Phi^n$ and hence by the uniqueness principle $[F]|\Gamma_n = [0]_{\Gamma_n}^\ell$. Note that injectivity of $\tau_n : \Phi \to \Gamma_n$ is equivalent to the condition that $[0]_\Phi^n$ separate points in Φ.

The next result is an analogue of Proposition 50.1.

71.6 PROPOSITION. *Let* (Φ, p) *be a connected Δ-domain and* $[H]$ *a uniformly carried subalgebra of* $[0]_\Phi$. *Then an extension*

$$\rho : (\Phi, p, [H]) \twoheadrightarrow (\Psi, q, [K])$$

of (Φ, p) *relative to* $[H]$ *will be maximal iff* $(\Psi, q, [K])$ *is maximal.*

Proof. Assume first that the extension is maximal and let

$$\mu : (\Psi, q, [K]) \twoheadrightarrow (\Psi', q', [K]')$$

be an arbitrary connected extension of (Ψ, q) relative to $[K]$. Then

$$\mu\circ\rho : (\Psi, p, [H]) \twoheadrightarrow (\Psi', q', [K]')$$

is an extension of (Φ, p) relative to $[H]$. Therefore, by the assumed maximality there exists

$$\mu' : (\Psi', q', [K]') \twoheadrightarrow (\Psi, q, [K]'')$$

where $\rho = \mu'\circ\mu\circ\rho$. Since $\mu'\circ\mu$ leaves points of $\rho(\Phi)$ fixed, Lemma 48.2 implies that $\mu : \Psi \to \Psi'$ is a surjective homeomorphism with $\mu' = \mu^{-1}$. In other words, $(\Psi, q, [K])$ is maximal.

Now assume that $(\Psi, q, [K])$ is maximal and consider the canonical extension

$$\tau_{[H]} : (\Phi, p, [H]) \twoheadrightarrow (\Gamma_{[H]}, \pi, [F]).$$

Since this is maximal by Theorem 71.5, there exists

$$\mu : (\Psi, q, [K]) \twoheadrightarrow (\Gamma_{[H]}, \pi, [F])$$

with $\tau_{[F]} = \mu\circ\rho$. But $(\Psi, q, [K])$ is maximal, so $\mu : \Psi \to \Gamma_{[H]}$ must be a surjective homeomorphism. This means that the given extension is isomorphic with the canonical extension and is therefore maximal. ◆

§72. UNIFORM FAMILIES OF EXTENSIONS

We shall now use the results obtained above for uniformly carried subalgebras of $[O]_\Phi$ to study the full algebra $[O]_\Phi$. The main result, which will be proved in §73, is a maximal "pseudoextension" of (Φ, p) relative to $[O]_\Phi$. This is an extension involving a target space which is not quite a Δ-domain. Throughout the discussion, (Φ, p) will be a fixed Δ-domain.

Let $\{[H]_d : d \in \mathcal{D}\}$ be an arbitrary family of uniformly carried subalgebras of $[O]_\Phi$ directed by inclusion, *i.e.* $\mathcal{D}$ is a directed system under a partial ordering "$\le$" such that $d \le d'$ implies $[H]_d \subseteq [H]_{d'}$. Then for each $d \in \mathcal{D}$ we have the canonical extension

$$\tau_d : (\Phi, p, [H]_d) \twoheadrightarrow (\Gamma_d, \pi, [F]_d)$$

of (Φ, p) relative to $[H]_d$. Let $d \le d'$. Then, since $[H]_d \subseteq [H]_{d'}$ the canonical extension of (Φ, p) relative to $[H]_{d'}$ induces another extension of (Φ, p) relative to $[H]_d$, *viz.*

$$\tau_{d'} : (\Phi, p, [H]_d) \twoheadrightarrow (\Gamma_{d'}, \pi, [F]_{dd'})$$

where

$$[F]_{dd'} = \{[F] \in [F]_{d'} : [F]\circ\tau_{d'} \in [H]_d\}.$$

Since the canonical extensions are maximal there exists

$$\mu_{dd'} : (\Gamma_{d'}, \pi, [F]_{dd'}) \twoheadrightarrow (\Gamma_d, \pi, [F]_d)$$

such that $\mu_{dd'}\circ\tau_{d'} = \tau_d$. These remarks suggest another category of objects associated with the given Δ-domain (Φ, p).

Let $\mathcal{D} = \{d\}$ be a directed system under a partial ordering "$\leq$" and for each $d \in \mathcal{D}$ let

$$\rho_d : (\Phi, p, [H]_d) \twoheadrightarrow (\Psi_d, q_d, [K]_d)$$

be a connected extension of (Φ, p) relative to $[H]_d$, where $[H]_d$ and $[K]_d$ are uniformly carried subalgebras of $[O]_\Phi$ and $[O]_{\Psi_d}$ respectively such that $d \leq d'$ implies $[H]_d \subseteq [H]_{d'}$. Also assume for $d \leq d'$ the existence of Δ-domain morphisms

$$\nu_{dd'} : (\Psi_{d'}, q_{d'}) \to (\Psi_d, q_d)$$

such that $\nu_{dd'}\circ\rho_{d'} = \rho_d$. Then the collection

$$\{\rho_d : (\Phi, p, [H]_d) \twoheadrightarrow (\Psi_d, q_d, [K]_d); \nu_{dd'}\}$$

where d ranges over $\mathcal{D}$ and $d \leq d'$, is called a *uniform family of extensions of* (Φ, p). Note that $\nu_{dd} : \Psi_d \to \Psi_d$ must be the identity map, by Lemma 48.2.

A *morphism* from a given uniform family to a second

$$\{\rho'_d : (\Phi, p, [H]'_d) \twoheadrightarrow (\Psi'_d, q'_d, [K]'_d); \nu'_{dd'}\}$$

involving the same directed system $\mathcal{D}$, consists of a family of Δ-domain morphisms

$$\{\theta_d : (\Psi_d, q_d) \to (\Psi'_d, q'_d)\}$$

such that $\rho'_d = \theta_d\circ\rho_d$ for each $d \in \mathcal{D}$. Since the mappings involved are open local homeomorphisms it is easy to verify, using Lemma 48.2, that also $\nu'_{dd'}\circ\theta_{d'} = \theta_d\circ\nu_{dd'}$, for all $d \leq d'$.

Observe that the family of canonical extensions

$$\{\tau_d : (\Phi, p, [H]_d) \to (\Gamma_d, \pi, [F]_d); \mu_{dd'}\}$$

associated with an inclusion directed family $\{[H]_d : d \in \mathcal{D}\}$ of subalgebras of $[\mathcal{O}]_\Phi$, constitutes a uniform family of extensions of (Φ, p). Also, an arbitrary extension $\rho : (\Phi, p, [\mathcal{O}]_\Phi) \to (\Psi, q, [\mathcal{O}]_\Psi)$ of (Φ, p) relative to $[\mathcal{O}]_\Phi$ may be resolved into a uniform family of extensions. In this case, we take as the directed system the collection $\mathcal{E}_\Psi$ of all extended covers for the space Ψ and associate with each $\{N^\psi\} = n \in \mathcal{E}_\Psi$ the cover $n \circ \rho = \{N^{\rho(\varphi)}\}$ for Φ. Thus $\mathcal{E}_\Psi \circ \rho \subseteq \mathcal{E}_\Phi$. Moreover $[\mathcal{O}]_\Psi \circ \rho \subseteq [\mathcal{O}]_\Phi^{n \circ \rho}$, so $[\mathcal{O}]_\Psi^n \circ \rho$ is a uniformly carried subalgebra of $[\mathcal{O}]_\Phi$. Also, since $[\mathcal{O}]_\Phi = [\mathcal{O}]_\Psi \circ \rho$ the algebra $[\mathcal{O}]_\Phi$ is equal to the union of the increasing family $\{[\mathcal{O}]_\Psi^n \circ \rho : n \in \mathcal{E}_\Psi\}$ of uniformly carried subalgebras. Now, for each $n \in \mathcal{E}_\Psi$ define $(\Psi_n, q_n) = (\Psi, q)$ and $\rho_n = \rho$. Also, for $n \le n'$ define $\nu_{nn'} = \iota$, the identity map on Ψ. Thus we obtain trivially a uniform family of extensions for (Φ, p), which we shall denote simply by

$$\{\rho : (\Phi, p, [\mathcal{O}]_\Psi^n \circ \rho) \to (\Psi, q, [\mathcal{O}]_\Psi^n)\}.$$

This is called the *resolution* of the given extension $\rho : (\Phi, p, [\mathcal{O}]_\Phi) \to (\Psi, q, [\mathcal{O}]_\Psi)$ with respect to the extended covers for Ψ.

72.1 LEMMA. *Let*

$$\{\rho_d : (\Phi, p, [H]_d) \to (\Psi_d, q_d, [K]_d); \nu_{dd'}\}$$

be an arbitrary uniform family of extensions of (Φ, p). *Then the following are true:*

(i) $\nu_{dd'} \circ \nu_{d'd''} = \nu_{dd''}$ *for* $d \le d' \le d''$.

(ii) If $[K]_{dd'} = \{[k] \in [K]_{d'} : [k] \circ \rho_{d'} \in [H]_d\}$ *then* $[K]_{dd'}$ *is a subalgebra of* $[K]_{d'}$ *and* $\nu_{dd'} : (\Psi_{d'}, q_{d'}, [K]_{dd'}) \to (\Psi_d, q_d, [K]_d)$.

Proof. For the proof of (i) set

$$\Psi^0_{d''} = \{\psi \in \Psi_{d''} : (\nu_{dd'} \circ \nu_{d'd''})(\psi) = \nu_{dd''}(\psi)\}.$$

Since $\nu_{dd'} \circ \nu_{d'd''} \circ \rho_{d''} = \nu_{dd'} \circ \rho_{d'} = \rho_d = \nu_{dd''} \circ \rho_{d''}$, it follows that $\rho_{d''}(\Phi) \subseteq \Psi^0_{d''}$, so $\Psi^0_{d''}$ is nonempty. Also, since the maps $\nu_{dd'}$ are continuous $\Psi^0_{d''}$ is a closed subset of $\Psi_{d''}$. Now let ψ_0 be an arbitrary point of $\Psi^0_{d''}$ and V_0 a $q_{d''}$-neighborhood of ψ_0 in $\Psi_{d''}$. Set

$$V^{d'} = \nu_{d'd''}(V_0),\ V^d = \nu_{dd''}(V_0),\ W^d = \nu_{dd'}(V^{d'}).$$

Then $V^{d'}$ is a $q_{d'}$-neighborhood of $\nu_{d'd''}(\psi_0)$ in $\Psi_{d'}$ and V^d, W^d are q_d-neighborhoods of $\nu_{dd''}(\psi_0)$ in Ψ_d. Also $q_d \circ \nu_{dd''} = q_{d''}$ and $q_d \circ \nu_{dd'} \circ \nu_{d'd''} = q_{d'} \circ \nu_{d'd''} = q_{d''}$. Hence $q_d(V^d) = q_d(W^d)$. Since $\nu_{dd''}(\psi_0) \in V^d \cap W^d$ it follows by Lemma 44.2 that $V^d \cup W^d$ is a q_d-neighborhood. But this is possible only if $V^d = W^d$. Now let $\psi \in V_0$. Then $(\nu_{dd'} \circ \nu_{d'd''})(\psi)$ and $\nu_{dd''}(\psi)$ both belong to V^d and

$$q_d((\nu_{dd'} \circ \nu_{d'd''})(\psi)) = q_d(\nu_{dd''}(\psi))$$

so $(\nu_{dd'} \circ \nu_{d'd''})(\psi) = \nu_{dd''}(\psi)$. Therefore $V_0 \subseteq \Psi^0_{d''}$, proving that $\Psi^0_{d''}$ is both open and closed in $\Psi^0_{d''}$ so must equal $\Psi_{d''}$. In other words, (i) is true.

For property (ii), observe that $[H]_d \subseteq [H]_{d'} = [K]_{d'} \circ \rho_{d'}$ and hence

$$[K]_{dd'} \circ \rho_{d'} = [H]_d = [K]_d \circ \rho_d = [K]_d \circ \nu_{dd'} \circ \rho_{d'}.$$

Therefore $[K]_{dd'} = [K]_d \circ \nu_{dd'}$ by the uniqueness principle, so (ii) follows. ♦

§73. PSEUDOEXTENSIONS

Observe that $\{\nu_{dd'}, \Psi_{d'}\}$, in Lemma 72.1, is a projective (or inverse limit) system, so we have the projective limit

$$\Psi_\infty = \varprojlim \nu_{dd'}(\Psi_{d'})$$

of the spaces $\Psi_{d'}$. Recall that Ψ_∞ is that subspace of the product space $\Psi^{\mathcal{D}} = \prod_d \Psi_d$ consisting of all $\check{\psi} = \{\psi_d\}$ such that $\nu_{dd'}(\psi_{d'}) = \psi_d$ for arbitrary $d \le d'$. As we shall see in a moment, Ψ_∞ is nonempty. The canonical projection of $\Psi^{\mathcal{D}}$ onto Ψ_d will be denoted by ν_d, *i.e.*

$$\nu_d : \check{\psi} \to \psi_d,\ \check{\psi} = \{\psi_d\} \in \Psi^{\mathcal{D}}.$$

Restricted to Ψ_∞, this projection obviously satisfies the condition $\nu_{dd'} \circ \nu_{d'} = \nu_d$, $d \le d'$. If $\check{\psi}$ is any element of Ψ_∞ then, for $d \le d'$,

$$q_{d'}(\psi_{d'}) = (q_d \circ \nu_{dd'})(\psi_{d'}) = q_d(\psi_d).$$

Moreover, for arbitrary $d, d' \in \mathcal{D}$ we may choose $d'' \in \mathcal{D}$ such that $d \le d''$ and $d' \le d''$ and hence conclude that $q_d(\psi_d) = q_{d''}(\psi_{d''}) = q_{d'}(\psi_{d'})$. In other words,

$q_d(\psi_d)$ is independent of $d \in \mathcal{D}$. Therefore if

$$q_\infty : \Psi_\infty \to \Delta, \check{\psi} \mapsto q_d(\psi_d)$$

then q_∞ is a well-defined projection of Ψ_∞ into Δ. We also have $q_\infty = q_d \circ \nu_d$ for all $d \in \mathcal{D}$ and, in particular, q_∞ is continuous. It will be convenient to use the notation (Ψ_∞, q_∞), although this may not be a Δ-domain since we do not know that q_∞ is a local homeomorphism.

Now, as in the case of complex-valued functions, denote by ${}^{\mathcal{D}}[K]$ the product of the algebras $\{[K]_d : d \in \mathcal{D}\}$ whose elements are now regarded as $(\mathfrak{S})_0$-valued functions. (See comments following Corollary 70.2.) Then ${}^{\mathcal{D}}[K]$ may be regarded as an algebra of $(\mathfrak{S})_0$-valued functions on the product space $\Psi^{\mathcal{D}}$. Note that ${}^{\mathcal{D}}[K]$ is equal to the algebra generated on $\Psi^{\mathcal{D}}$ by the algebras $\{[K]_d \circ \nu_d : d \in \mathcal{D}\}$. For convenience of notation we set

$$[K]_\infty = {}^{\mathcal{D}}[K] | \Psi_\infty, \ [H]_\infty = \bigcup_d [H]_d.$$

Note that $[H]_\infty$ is a subalgebra of $[O]_\Phi$.

73.1 THEOREM. *There exists a continuous map* $\rho_\infty : \Phi \to \Psi_\infty$ *such that* $q_\infty \circ \rho_\infty = p$ *and* $[K]_\infty \circ \rho_\infty = [H]_\infty$. *Also,* $\nu_d \circ \rho_\infty = \rho_d$ *and* $([K]_d \circ \nu_d) \circ \rho_\infty = [H]_d$ *for each* $d \in \mathcal{D}$.

Proof. The following commutative diagrams record properties of the map ρ_∞ in the theorem along with other items already established in the preceding discussion.

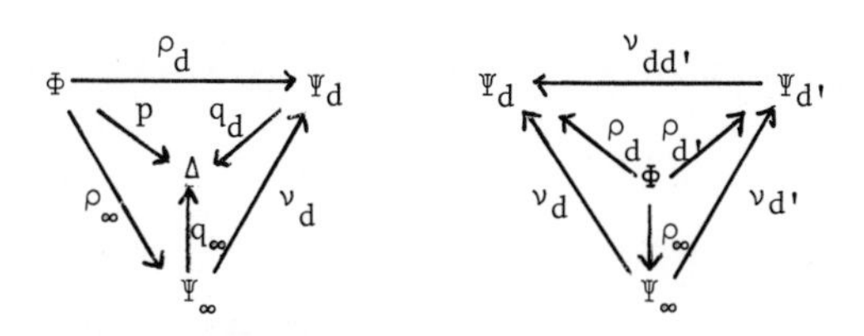

Now observe that $\nu_{dd'}(\rho_{d'}(\varphi)) = (\nu_{dd'} \circ \rho_{d'})(\varphi) = \rho_d(\varphi)$, for $d \le d'$ and $\varphi \in \Phi$. Therefore $\{\rho_d(\varphi)\} \in \Psi_\infty$ and we may define

$$\rho_\infty : \Phi \to \Psi_\infty, \varphi \mapsto \{\rho_d(\varphi)\}.$$

Furthermore, since each of the mappings $\rho_d : \Phi \to \Psi_d$ is continuous ρ_∞ is also continuous. By definition of ν_d

$$(\nu_d \circ \rho_\infty)(\varphi) = \nu_d(\{\rho_{d'}(\varphi)\}) = \rho_d(\varphi)$$

so $\nu_d \circ \rho_\infty = \rho_d$ for each $d \in \mathcal{D}$. Also, by definition of q_∞

$$(q_\infty \circ \rho_\infty)(\varphi) = q_\infty(\{\rho_d(\varphi)\}) = (q_d \circ \rho_d)(\varphi) = p(\varphi)$$

which proves $q_\infty \circ \rho_\infty = p$. Finally,

$$([K]_d \circ \nu_d) \circ \rho_\infty = [K]_d \circ \rho_d = [H]_d$$

and, since $[K]_\infty$ is generated by elements of the algebras $[K]_d \circ \nu_d$, we conclude that $[K]_\infty \circ \rho_\infty = [H]_\infty$. ♦

73.2 COROLLARY. *If* (Ψ_∞, q_∞) *is a* Δ*-domain then* ρ_∞ *defines an extension of* (Φ, p) *relative to* $[H]_\infty$.

In view of the above theorem, we shall say that $\rho_\infty : \Phi \to \Psi_\infty$ defines a *pseudo-extension of* (Φ, p) *relative to* $[H]_\infty$ and use the notation

$$\rho_\infty : (\Phi, p, [H]_\infty) \twoheadrightarrow (\Psi_\infty, q_\infty, [K]_\infty)$$

even when (Ψ_∞, q_∞) is not a Δ-domain. This will also be called the *limit* of the given uniform family of extensions of (Φ, p). We shall refer to (Ψ_∞, q_∞) as a *pseudo-*Δ*-domain*. The fact that it is in general not a Δ-domain is shown by the following example.

73.3 EXAMPLE. *A canonical pseudoextension in which the limiting pseudodomain is not a* Δ*-domain.*

Take the space E as well as the set Δ to be the entire complex plane $\mathbb{C}$. Thus, in particular, $[\Delta, \mathfrak{P}]$ is natural and Δ-domains are actually E-domains. Note also that the algebra of all germ-valued $\mathfrak{P}$-holomorphic functions on a Δ-domain will be isomorphic with the associated algebra of complex-valued functions obtained by reduction.

Consider in $\mathbb{C}$ the open discs $D_0 = \{\zeta : |\zeta| < 1\}$ and $D_n = \{\zeta : |\zeta| < 1 + \frac{1}{n}\}$ for $n = 1,2,\ldots$. Set $[H]_n = [O]_{D_n}|D_0$. Then $D_n \subseteq D_m$ and $[H]_m \subseteq [H]_n$ for $m \leq n$. Next denote by $\iota_n : D_n \to \mathbb{C}$ the identity map of D_n into $\mathbb{C}$ for $n = 0,1,2,\ldots$.

Then (D_n, ι_n) is a Δ-domain for each n. Also denote by $\tau_n : D_0 \to D_n$ the identity map of D_0 into D_n and by $\mu_{mn} : D_n \to D_m$ the identity map of D_n into D_m for $m \leq n$. Then

$$\{\tau_n : (D_0, \iota_0, [H]_n) \twoheadrightarrow (G_n, \iota_n, [O]_{D_n}); \mu_{mn}\}$$

is a uniform family of *canonical* extensions of (D_0, μ_0) relative to the algebras $\{[H]_n\}$. The projective limit

$$D_\infty = \varprojlim \mu_{mn}(D_n)$$

may be identified with the intersection of the discs D_n, *viz* the closed unit disc $\bar{D}_0$. Also, the limit projection ι_∞ is simply the identity map of $\bar{D}_0$ into $\mathbb{C}$. Since $\bar{D}_0$ is not an open subset of $\Delta = \mathbb{C}$ the pseudodomain (D_∞, ι_∞) is obviously not a Δ-domain.

Although (Ψ_∞, q_∞) need not be a Δ-domain, the fact that it is a limit of Δ-domains gives the following result.

73.4 PROPOSITION. *Each of the stalks* $q_\infty^{-1}(\delta)$ *for* $\delta \in q_\infty(\Psi_\infty)$ *is totally disconnected.*

Proof. Let G be a connected component of the stalk $q_\infty^{-1}(\delta)$ and let $\check{\psi}$ be an arbitrary point of G. For arbitrary $d \in \mathcal{D}$ choose a q_d-neighborhood V of the point $\nu_d(\check{\psi}) = \psi_d$ in Ψ_d. Then, since $q_d(\nu_d(G))$ contains only the single point δ, it follows that $V \cap \nu_d(G)$ contains only the point ψ_d. From the fact that $\nu_d(G)$ is connected in Ψ_d, this implies that $\nu_d(G)$ contains only the point ψ_d. Since d is arbitrary it follows that $G = \{\check{\psi}\}$, *i.e.* each connected component of $q_\infty^{-1}(\delta)$ is a singleton, so $q_\infty^{-1}(\delta)$ is totally disconnected. ◆

For a uniform family of canonical extensions of (Φ, p)

$$\{\tau_d : (\Phi, p, [H]_d) \twoheadrightarrow (\Gamma_d, \pi, [F]_d); \mu_{dd'}\}$$

the limit

$$\tau_\infty : (\Phi, p, [H]_\infty) \twoheadrightarrow (\Gamma_\infty, \pi_\infty, [F]_\infty)$$

is called a *canonical pseudoextension of* (Φ, p) *relative to* $[H]_\infty$.

It is trivial to verify that the limit of the resolution

$$\{\rho : (\Phi, p, [O]^n_\Psi \circ \rho) \twoheadrightarrow (\Psi, q, [O]^n_\Psi)\}$$

of an extension $\rho : (\Phi, p, [O]_\Phi) \twoheadrightarrow (\Psi, q, [O]_\Psi)$, with respect to extended covers for Ψ (see remarks preceding Lemma 72.1), is equal to the given extension.

Now let $\{[H_d] : d \in \mathcal{D}\}$ be a given family of uniformly carried subalgebras of $[O]_\Phi$ directed by inclusion. The next theorem contains a maximality property for the canonical pseudoextension $\tau_\infty : (\Phi, p, [H]_\infty) \twoheadrightarrow (\Gamma_\infty, \pi_\infty, [F]_\infty)$ of (Φ, p) relative to $[H]_\infty = \bigcup [H]_d$. We accordingly consider an arbitrary uniform family

$$\{\rho_d : (\Phi, p, [H]_d) \twoheadrightarrow (\Psi_d, q_d, [K]_d); \nu_{dd'}\}$$

of extensions of (Φ, p) relative to the algebras $[H]_d$ and the associated pseudo-extension limit

$$\rho_\infty : (\Phi, p, [H]_\infty) \twoheadrightarrow (\Psi_\infty, q_\infty, [K]_\infty).$$

73.5 THEOREM. *There exists a continuous mapping* $\theta_\infty : \Psi_\infty \to \Gamma_\infty$ *such that* $\theta_\infty \circ \rho_\infty = \tau_\infty$ *and*

$$\theta_\infty : (\Psi_\infty, q_\infty, [K]_\infty) \twoheadrightarrow (\Gamma_\infty, \pi_\infty, [F]_\infty)$$

i.e. $\pi_\infty \circ \theta_\infty = q_\infty$ *and* $[F]_\infty \circ \theta_\infty = [K]_\infty$.

Proof. Since each of the canonical extensions

$$\tau_d : (\Phi, p, [H]_d) \twoheadrightarrow (\Gamma_d, \pi, [F]_d),\ d \in \mathcal{D}$$

is maximal, there exists $\theta_d : \Psi_d \to \Gamma_d$ such that $\theta_d \circ \rho_d = \tau_d$ and

$$\theta_d : (\Psi_d, q_d, [K]_d) \twoheadrightarrow (\Gamma_d, \pi, [F]_d).$$

Thus $\{\theta_d\}$ defines a morphism of the given uniform families of extensions (§72). For $d \le d'$ we have

$$\theta_d \circ \nu_{dd'} \circ \rho_{d'} = \theta_d \circ \rho_d = \tau_d$$
$$= \mu_{dd'} \circ \tau_{d'} = \mu_{dd'} \circ \theta_{d'} \circ \rho_{d'}.$$

Therefore by Lemma 48.2

$$\theta_d \circ \nu_{dd'} = \mu_{dd'} \circ \theta_{d'}.$$

Now, for $\check{\psi} = \{\psi_d\} \in \Psi_\infty$ define

$$\theta_\infty(\check{\psi}) = \{\theta_d(\psi_d)\}.$$

Then since

$$\mu_{dd'}(\theta_{d'}(\psi_{d'})) = \theta_d(\nu_{dd'}(\psi_{d'})) = \theta_d(\psi_d), \; d \le d'$$

it follows that $\{\theta_d(\psi_d)\} \in \Gamma_\infty$, so θ_∞ maps Ψ_∞ into Γ_∞. Furthermore, since each of the maps $\theta_d : \Psi_d \to \Gamma_d$ is continuous $\theta_\infty : \Psi_\infty \to \Gamma_\infty$ is also continuous.

By definition of ρ_∞ and τ_∞, we have

$$\begin{aligned}(\theta_\infty \circ \rho_\infty)(\varphi) &= \theta_\infty\{\rho_d(\varphi)\}) = \{(\theta_d \circ \rho_d)(\varphi)\}\\ &= \{\tau_d(\varphi)\} = \tau_\infty(\varphi)\end{aligned}$$

for each $\varphi \in \Phi$. Hence $\theta_\infty \circ \rho_\infty = \tau_\infty$. Also, by definition of π_∞ and q_∞

$$\begin{aligned}(\pi_\infty \circ \theta_\infty)(\check{\psi}) &= \pi_\infty\{\theta_d(\psi_d)\}) = (\pi_d \circ \theta_d)(\psi_d)\\ &= q_d(\psi_d) = q_\infty(\check{\psi})\end{aligned}$$

for each $\check{\psi} \in \Psi_\infty$, so $\pi_\infty \circ \theta_\infty = q_\infty$.

Since

$$\begin{aligned}(\mu_d \circ \theta_\infty)(\check{\psi}) &= \mu_d(\{\theta_{d'}(\psi_{d'})\})\\ &= \theta_d(\psi_d) = (\theta_d \circ \nu_d)(\check{\psi})\end{aligned}$$

for each $\check{\psi} \in \Psi_\infty$ we have $\mu_d \circ \theta_\infty = \theta_d \circ \nu_d$, where μ_d denotes the canonical projection of $\Gamma^{\mathcal{D}}$ onto Γ_d restricted to Γ_∞. Therefore

$$\begin{aligned}([F]_d \circ \mu_d) \circ \theta_\infty &= [F]_d \circ (\mu_d \circ \theta_\infty)\\ &= [F]_d \circ \theta_d \circ \nu_d = [K]_d \circ \nu_d\end{aligned}$$

and hence

$$[F]_\infty = (\bigcup_d [F]_d \circ \mu_d) \circ \theta_\infty = \bigcup_d [K]_d \circ \nu_d = [K]_\infty$$

which completes the proof of the theorem. ♦

The most obvious uniformly carried algebras associated with Δ-domain (Φ, p) are the full subalgebras of $[O]_\Phi$ carried by an extended cover for Φ, *viz* the subalgebras $[O]_\Phi^n$, $n \in \mathcal{E}_\Phi$. Corresponding to these algebras we have the uniform family

$$\{\tau_m : (\Phi, p, [O]_\Phi^m) \twoheadrightarrow (\Gamma_m, \pi, [F]_m); \mu_{mn}\}$$

and its limiting pseudoextension

$$\tau_\infty : (\Phi, p, [O]_\Phi) \twoheadrightarrow (\Gamma_\infty, \pi_\infty, [F]_\infty).$$

Furthermore, if

$$\rho : (\Phi, p, [\mathcal{O}]_\Phi) \twoheadrightarrow (\Psi, q, [\mathcal{O}]_\Psi)$$

is an arbitrary *uniform* (§71) extension of (Φ, p) relative to $[\mathcal{O}]_\Phi$ then for each $m \in \mathcal{E}_\Phi$ there exists by definition a uniformly carried subalgebra $[K]_m$ of $[\mathcal{O}]_\Psi$ such that $[K]_m \circ \rho = [\mathcal{O}]^m_\Phi$. Hence, if we define $\rho_m = \rho$ and $(\Psi_m, q_m) = (\Psi, q)$ for each $m \in \mathcal{E}_\Phi$, and define ν_{mn} to be the identity map of Ψ_m onto Ψ_n for $m \le n$, the result is a uniform family

$$\{\rho_m : (\Phi, p, [\mathcal{O}]^m_\Phi) \twoheadrightarrow (\Psi_m, q_m, [K]_m); \nu_{mn}\}$$

whose limit is simply the given extension. Therefore an application of Theorem 73.5 gives the following result.

73.6 COROLLARY. *The canonical pseudoextension*

$$\tau_\infty : (\Phi, p, [\mathcal{O}]_\Phi) \twoheadrightarrow (\Gamma_\infty, \pi_\infty, [F]_\infty)$$

is maximal for uniform extensions of (Φ, p). *In other words, for each uniform extension*

$$\rho : (\Phi, p, [\mathcal{O}]_\Phi) \twoheadrightarrow (\Psi, q, [\mathcal{O}]_\Psi)$$

there exists a continuous map $\theta_\infty : \Psi \to \Gamma_\infty$ *such that* $\theta_\infty \circ \rho = \tau_\infty$ *and*

$$\theta_\infty : (\Psi, q, [\mathcal{O}]_\Psi) \twoheadrightarrow (\Gamma_\infty, \pi_\infty, [F]_\infty).$$

We are unable to prove a result analogous to that in the above corollary for extensions that are not uniform. On the other hand, for an arbitrary extension one can always construct a canonical pseudoextension which depends on the given extension but does majorize it in the sense of Corollary 73.6. In fact, let

$$\rho : (\Phi, p, [\mathcal{O}]_\Phi) \twoheadrightarrow (\Psi, q, [\mathcal{O}]_\Psi)$$

be an arbitrary extension of (Φ, p) relative to $[\mathcal{O}]_\Phi$ and let

$$\{\rho : (\Phi, p, [\mathcal{O}]^m_\Psi \circ \rho) \twoheadrightarrow (\Psi, q, [\mathcal{O}]^m_\Psi); m \in \mathcal{E}_\Psi\}$$

be its resolution (see remarks preceding Lemma 72.1). Then the limit of this uniform family is equal to the given extension. We also have the uniform family

$$\{\tau^\rho_m : (\Phi, p, [\mathcal{O}]^m_\Psi \circ \rho) \twoheadrightarrow (\Gamma^\rho_m, \pi, [F]^\rho_m); \mu'_{mn}\}$$

of canonical extensions, where $\Gamma_m^\rho = \Gamma_{[O]_\Psi^m \circ \rho}$ and $[F]_m^\rho = [F]|\Gamma_m^\rho$, with limit

$$\tau_\infty^\rho : (\Phi, p, [O]_\Phi) \twoheadrightarrow (\Gamma_\infty^\rho, \pi_\infty, [F]_\infty^\rho).$$

Therefore an application of Theorem 73.5 yields the following result.

73.7 COROLLARY. *There exists a continuous map* $\theta_\infty^\rho : \Psi \to \Gamma_\infty^\rho$ *such that* $\theta_\infty^\rho \circ \rho = \tau_\infty^\rho$ *and*

$$\theta_\infty^\rho : (\Psi, q, [O]_\Psi) \twoheadrightarrow (\Gamma_\infty^\rho, \pi_\infty, [F]_\infty^\rho).$$

§74. NATURALITY PROPERTIES

Consider again the uniform family

$$\{\tau_m : (\Phi, p, [O]_\Phi^m) \twoheadrightarrow (\Gamma_m, \pi, [F]_m); \mu_{mn}\}$$

of canonical extensions and the associated pseudoextension

$$\tau_\infty : (\Phi, p, [O]_\Phi) \twoheadrightarrow (\Gamma_\infty, \pi_\infty, [F]_\infty).$$

If the system $[\Delta, \wp_{\langle E, F\rangle}]$ is natural then the target domains in these extensions also exhibit naturality properties. In the case of the canonical extensions such properties are given by Theorems 67.2 and 70.4. In order to obtain analogous results for the pseudoextension we need a preliminary lemma.

74.1 LEMMA. *Let* Γ_0 *be any component of* $\Delta \# (\check{\mathfrak{G}})$, Γ *the component of* $E \# (\check{\mathfrak{G}})$ *that contains* Γ_0, *and* (Φ, p) *an arbitrary* Δ*-domain. Then for every morphism* $\rho : (\Phi, p) \to (\Gamma_0, \pi)$ *there exists an extended cover* $n \in \mathcal{E}_\Phi$ *such that* $[O]_\Gamma \circ \rho \subseteq [O]_\Phi^n$.

Proof. Note that the lemma asserts the existence of $n \in \mathcal{E}_\Phi$ such that ρ induces a morphism

$$\rho : (\Phi, p, [O]_\Phi^n) \to (\Gamma_0, \pi, [O]_\Gamma|\Gamma).$$

Let $\{W_\gamma : \gamma \in \Gamma_0\}$ be an arbitrary covering of Γ_0 by connected π-neighborhoods *contained in* Γ. Set $L^\gamma = \pi(W_\gamma)$ and define $\ell = \{L^\gamma : \gamma \in \Gamma_0\}$. Then $\ell \in \mathcal{E}_{\Gamma_0}$ and obviously $[O]_\Gamma|\Gamma_0 \subseteq [O]_{\Gamma_0}^\ell$. Furthermore $\ell \circ \rho \in \mathcal{E}_\Phi$ and $[O]_{\Gamma_0}^\ell \circ \rho \subseteq [O]_\Phi^{\ell \circ \rho}$, so $[O]_\Gamma \circ \rho \subseteq [O]_\Phi^n$, where $n = \ell \circ \rho$. ◆

Now apply the above lemma to the morphisms involved in the uniform family of canonical extensions. For notational convenience set $[O]_m = [O]_{\Gamma_m}$ and $[O]'_m = [O]_{\Gamma'_m}|\Gamma_m$, where Γ'_m denotes the component of $E \# (\check{\mathfrak{G}})$ that contains Γ_m. We obtain for each $m \in \mathcal{E}_\Phi$ an element $n \in \mathcal{E}_\Phi$ such that $[O]'_m \circ \tau_m \subseteq [O]^n_\Phi$. There is obviously no loss in assuming that $m \leq n$. Hence

$$[O]'_m \circ \tau_m \subseteq [O]^n_\Phi = [F]_n \circ \tau_n \subseteq [O]'_n \circ \tau_n.$$

Since $\mu_{mn} \circ \tau_n = \tau_m$ we have

$$[O]'_m \circ \mu_{mn} \circ \tau_n \subseteq [F]_n \circ \tau_n \subseteq [O]'_n \circ \tau_n.$$

Therefore by the uniqueness principle we obtain the following corollary.

74.2 COROLLARY. *For each* $m \in \mathcal{E}_\Phi$ *there exists* $n \in \mathcal{E}_\Phi$ *with* $n \geq m$ *such that*

$$[O]'_m \circ \mu_{mn} \subseteq [F]_n \subseteq [O]'_n$$

As before, we denote the limit of the projective system $\{\mu_{mn}, \Gamma_n\}$ by $\Gamma_\infty = \varprojlim \mu_{mn}(\Gamma_n)$, with canonical projections $\mu_n : \Gamma_\infty \to \Gamma_n$. Also, $[F]_\infty$ and $[O]'_\infty$ are the algebras generated on Γ_∞ by the sets of functions $\mathbf{U}([F]_n \circ \mu_n)$ and $\mathbf{U}([O]'_n \circ \mu_n)$ respectively. Observe that for each $n \in \mathcal{E}_\Phi$ the reduction (Definition 68.6) of the algebra $[O]'_n$ of germ-valued functions to the space Γ_n is simply the algebra $O'_n = O_{\Gamma'_n}|\Gamma_n$. Therefore the algebra O'_∞ of complex-valued functions generated on Γ_∞ by the set of functions $\mathbf{U}(O'_n \circ \mu_n)$ is the reduction of $[O]'_\infty$ to Γ_∞.

We may now prove the desired naturality results for the pseudoextension

$$\rho_\infty : (\Phi, p, [O]_\Phi) \twoheadrightarrow (\Gamma_\infty, \pi_\infty, [F]_\infty).$$

74.3 THEOREM. *If* $[\Delta, \rho_{<E,\, F>}]$ *is natural then* $[\Gamma_\infty, O'_\infty]$ *and* $[\Gamma_\infty, [F]_\infty]$ *are also natural.*

Proof. Observe first that each of the systems $[\Gamma_n, O'_n]$ is natural by Theorem 67.2. Therefore the system $[\Gamma_\infty, O'_\infty]$ is natural by Proposition 4.2. By Corollary 74.2 there exists for each $m \in \mathcal{E}_\Phi$ an $n \in \mathcal{E}_\Phi$ with $m \leq n$ such that

$$[O]'_m \circ \mu_{mn} \subseteq [F]_n \subseteq [O]'_n.$$

Since $\mu_{mn} \circ \mu_n = \mu_n = \mu_m$, it follows that

$$[O]_m'\circ\mu_m \subseteq [F]_n\circ\mu_n \subseteq [O]_n'\circ\mu_n.$$

Therefore

$$\bigcup_m [O]_m'\circ\mu_m \subseteq [F]_n\circ\mu_n\circ\mu_n \subseteq \bigcup_m [O]_n'\circ\mu_n$$

so the equality must hold. This implies that $[O]_\infty' = [F]_\infty$. Observe also that $[O]_\infty'$ is closed under differentiation. Since $[\Gamma_\infty, O_\infty']$ is natural, it follows that $[\Gamma_\infty, [O]_\infty']$, and hence $[\Gamma_\infty, [F]_\infty]$, is natural in the sense of Definition 70.3. ♦

BIBLIOGRAPHY

[A1] G. R. Allan. *An extension of Rossi's local maximum modulus principle.* J. London Math. Soc. (2) 3 (1971), 1-17.

[A2] R. Arens. *The problem of locally* A*-functions in a commutative Banach algebra* A. Trans. Amer. Math. Soc. 104 (1962), 24-36.

[A3] R. Arens and I. M. Singer. *Function values as boundary integrals.* Proc. Amer. Math. Soc. 5 (1954), 735-745.

[A4] V. Aurich. *The spectrum as envelope of holomorphy of a domain over an arbitrary product of complex lines.* Proc. on Inf. Dim. Holo., Univ. of Kentucky 1973 (Ed. T. L. Hayden and T. J. Suffridge). Lect. Notes in Math, No. 364 (109-122). Springer-Verlag, 1974.

[B1] R. F. Basener. *A generalized Šilov boundary and analytic structure.* Proc. Amer. Math. Soc. 47 (1975), 98-104.

[B2] L. Bers. *Introduction to several complex variables.* NYU Lecture Notes. Courant Institute of Mathematical Sciences, NYU, 1964.

[B3] F. Birtel. *Uniform algebras and unbounded functions.* Rice Studies, 1968, 1-13.

[B4] G. R. Blumenthal. *The geometric structure of the spectrum of a function algebra.* Dissertation, Yale University, 1968.

[B5] G. R. Blumenthal. *The spectrum of a function-algebra.* Proc. Amer. Soc. 25 (1970), 343-346.

[B6] S. Bochner and W. T. Martin. *Several complex variables.* Princeton Univ. Press, 1948.

[B7] F. Bonsall and J. Duncan. *Complete normed algebras.* Ergeb. der Math. B. 80. Springer-Verlag, 1973.

[B8] H. J. Bremerman. *On the conjecture of the equivalence of the plurisubharmonic functions and the Hartogs functions.* Math. Ann. 131 (1956), 76-86.

[B9] R. M. Brooks. *Boundaries for locally* m*-convex algebras.* Duke Math. J. 34 (1967), 103-116.

[B10] R. M. Brooks. *The structure space of a commutative locally* m*-convex algebra.* Pac. J. Math. 25 (1968), 443-454.

[B11] R. M. Brooks. *Boundaries for natural systems.* Indiana Univ. Math. J. 20 (1970/71), 865-875.

[B12] R. M. Brooks. *Boundaries for natural systems. II.* Math. Scand. 30 (1972), 281-289.

[B13] R. M. Brooks. *Analytic structure in the spectrum of a natural system.* Pac. J. Math. 49 (1973), 315-334.

[C1] G. Coeuré. *Analytic functions and manifolds in infinite dimensional spaces.* North-Holland Math. Studies, No. 11 (Notas de Matematica, No. 52). North-Holland/Amer. Elsevier, 1974.

[F1] B. A. Fuks. *Special chapters in the theory of analytic functions of several complex variables.* Fizmatziz, Moscow, 1963; Transl. Math. Monographs, vol. 14. Amer. Math. Soc., 1965.

[G1] T. W. Gamelin. *Uniform algebras*. Prentice-Hall, 1969.

[G2] T. W. Gamelin. *Hartogs series, Hartogs functions, and Jensen measures in spaces of analytic functions*. (Kristiansand, Norway, 1975). Lect. Notes in Math., No. 511, pp. 69-83. Springer-Verlag, 1976.

[G3] T. W. Gamelin. *Localization of subharmonicity with respect to a uniform algebra*. (Preprint.)

[G4] J. Garnett. *A topological characterization of Gleason parts*. Pac. J. Math. 20 (1967), 59-64.

[G5] I. Glicksberg. *Maximal algebras and a theorem of Rado*. Pac. J. Math. 14 (1964), 919-941. *Correction*. Ibid 19 (1966), 587.

[G6] S. L. Gulick. *The minimal boundary of* C(X). Trans. Amer. Math. Soc. 131 (1968), 303-314.

[G7] R. C. Gunning and H. Rossi. *Analytic functions of several complex variables*. Prentice-Hall, 1965.

[H1] F. Hausdorff. *Set theory*. (English Transl. of *Mengenlehre*, Berlin 1937). Chelsea, 1957.

[H2] M. Herve. *Analytic and plurisubharmonic functions in finite and infinite dimensions*. *Lecture Notes in Math.*, No. 198. Springer-Verlag, 1971.

[H3] A. Hirschowitz. *Remarques sur les ouverts d'holomorphic d'un produit denombrable de droite*. Ann. Inst. Fourier Grenoble. 19 (1969), 219-229.

[H4] L. Hörmander. *An introduction to complex analysis in several variables*. Van Nostrand, 1966. 2nd ed. North-Holland/Amer. Elsiver, 1973.

[K1] Eva Kallin. *A nonlocal function algebra*. Proc. Nat. Acad. Sci. 49 (1963), 821-824.

[K2] D. S. Kim. *A boundary for the algebra of bounded holomorphic functions*. Pac. J. Math. 45 (1973), 269-274.

[K3] B. Kramm. *Zur Analytischen Geometrie in den Spektren gewisser* Frechet-Schwartz-Algebren. (Preprint).

[K4] B. Kramm. *Analytische Struktur in Spektren - ein Zugang uber die ∞-dimensionale Holomorphie*. (Preprint).

[L1] P. Lelong. *Les fonctions plurisousharmoniques*. An. Sci. Ecole Norm Sup. (3) 62 (1945), 301-338.

[L2] P. Lelong. *Fonctions plurisousharmoniques et formes differentielles positives*. Gordon and Breach, 1968.

[M1] M. C. Matos. *Holomorphic mappings and domains of holomorphy*. Dissertation. Univ. of Rochester, 1970.

[M2] M. C. Matos. *The envelope of holomorphy of Riemann domains over a countable product of complex planes*. Trans. Amer. Math. Soc. 167 (1972), 379-387.

[M3] W. E. Meyers. *A uniform algebra with nonglobal peak points*. Studia Math. 33 (1969), 207-211.

[M4] W. E. Meyers. *Local peak sets of uniform algebras*. Duke Math. J. 37 (1970), 701-708.

[M5] E. A. Michael. *Locally multiplicatively-convex topological algebras*. Amer. Math. Soc. Memoir, No. 11, 1952.

[N1] L. Nachbin. *Uniformité d'holomorphie et type exponential*. Sem. Pierre Lelong (1970), 216-224. Lect. Notes in Math., No. 205, 1971.

[N2] R. Narasimhan. *Several complex variables*. Chi. Lect. Notes in Math. Univ. of Chi. Press, 1971.

[N3] Ph. Noverraz. *Pseudo-convexité, convexité polynomiale et domaines d'holomorphic en dimension infinite*. North-Holland Math. Studies, No. 3, 1973.

[O1] K. Oka. *Sur les fonctions analytiques de plusieurs variables.* Iwanami Shoten, Tokyo, 1961.

[Q1] F. Quigley. *Approximation by algebras of functions.* Math. Ann. 135 (1958), 81-92.

[Q2] F. Quigley. *Generalized Phragmen-Lindelof theorems. Function algebras.* (36-41). Proc. Int. Symp. on Funct. Alg., Tulane Univ., 1965 (Ed. F. T. Birtel), Scott-Foresman, 1966.

[R1] C. E. Rickart. *General theory of Banach algebras.* Van Nostrand, 1960. Reprint, Robert E. Krieger Pub. Co., 1974.

[R2] C. E. Rickart. *Analytic phenomena in general function algebras.* Pac. J. Math. 18 (1966), 361-377.

[R3] C. E. Rickart. *The maximal ideal space of functions locally approximable in a function algebra.* Proc. Amer. Math. Soc. 17 (1966), 1320-1326.

[R4] C. E. Rickart. *Holomorphic convexity for general function algebras.* Canad. J. Math. 20 (1968), 272-290.

[R5] C. E. Rickart. *Boundary properties of sets relative to function algebras.* Studia Math. 31 (1968), 253-261.

[R6] C. E. Rickart. *Analytic functions of an infinite number of complex variables.* Duke Math. J. 36 (1969), 581-598.

[R7] C. E. Rickart. *Plurisubharmonic functions and convexity properties for general function algebras.* Trans. Amer. Math. Soc. 169 (1972), 1-24.

[R8] C. E. Rickart. *A function algebra approach to infinite dimensional holomorphy. Analyse fonctional et applications,* (245-260). Proc. Colloq. of Anal., Rio, 1972 (Ed. L. Nachbin). Act. Sci. et Ind., Herman, 1974.

[R9] C. E. Rickart. *Holomorphic extensions and domains of holomorphy for general function algebras.* Proc. on Inf. Dim. Holo., Univ. of Kentucky 1973 (Ed. T. L. Hayden and T. J. Suffridge). Lect. Notes in Math., No. 364 (80-91). Springer-Verlag, 1974.

[R10] C. E. Rickart. *Some function algebras with common spectrum.* J. of Algebra (to appear).

[R11] H. Rossi. *The local maximum modulus principle.* Ann. of Math. 72 (1960), 1-11.

[R12] H. Royden. *Function algebras.* Bull. Amer. Math. Soc. 69 (1963), 281-298.

[R13] J. V. Ryff. *The support of representing measures for the disc algebra. Function algebras.* (112-117). Proc. Int. Symp. on Funct. Alg., Tulane Univ., 1965, (Ed. F. T. Birtel), Scott-Foresman, 1966.

[S1] H. H. Schaefer. *Topological vector spaces.* Grad. Texts in Math. Springer-Verlag, 1966 (3rd printing, 1971).

[S2] M. Schottenloher. *Analytische Fortsetzung in Banachraumen.* Dissertation, Munich, 1971. Math. Ann. 199 (1972), 313-336.

[S3] M. Schauk. *Algebras of holomorphic functions in ringed spaces.* Dissertation, Tulane Univ. 1966.

[S4] S. Sidney. *Properties of the sequence of closed powers of a maximal ideal in a sup-norm algebra.* Trans. Amer. Math. Soc. 131 (1968), 128-148.

[S5] S. Sidney. *High-order nonlocal function algebras.* Proc. London Math. Soc. (3) 23 (1971), 735-752.

[S6] S. Sidney. *More on high-order nonlocal function algebras.* Ill. J. Math. 18 (1974), 177-192.

[S7] G. Stolzenberg. *The maximal ideal space of functions locally in a function algebra.* Proc. Amer. Math. Soc. 14 (1963), 342-345.

[S8] G. Stolzenberg. *A hull with no analytic structure.* J. Math. Mech. 12 (1963), 103-111.

[S9] G. Stolzenberg. *Polynomial and rationally convex sets.* Acta Math. 109 (1963), 259-289.

[S10] E. L. Stout. *The theory of uniform algebras.* Bagden and Quigley, 1971.

[V1] A. D. Varsavskii. *A function algebra of second order degree of nonlocalness.* Uspehki Mat. Nauk. 24 (1969), No. 2 (146), 223-224. Mat. Sbornik (N.S.) 80 (122)(1969), 266-280.

[W1] J. Wermer. *On algebras of continuous functions.* Proc. Amer. Math. Soc. 4 (1953), 866-869.

[W2] J. Wermer. *Banach algebras and several complex variables.* Markham Pub. Co., 1971, 2nd Ed. Grad. Texts in Math., Springer-Verlag, 1976.

[W3] D. R. Wilken. *Maximal ideal spaces and A-convexity. Function algebras.* (120-121). Proc. Int. Symp. on Funct. Alg., Tulane Univ., 1965 (Ed. F. T. Birtel), Scott-Foresman, 1966.

[Z1] W. R. Zame. *Algebras of analytic germs.* Trans. Amer. Math. Soc. 174 (1972), 275-288.

INDEX OF SYMBOLS

The numbers indicate the pages on which the symbols are introduced.

GENERAL INDEX